An Introduction to
Three-Dimensional Climate Modeling

An Introduction to
Three-Dimensional Climate Modeling

Warren M. Washington
NATIONAL CENTER FOR
ATMOSPHERIC RESEARCH

Claire L. Parkinson
NASA GODDARD
SPACE FLIGHT CENTER

University Science Books
Mill Valley, California

Oxford University Press
Oxford   New York

University Science Books
20 Edgehill Road
Mill Valley, CA 94941

Copy Editor: *Charles B. Hibbard*
Designer: *Robert Ishi*
Composition: *Eileen Boettner, NCAR T_EX System*
Illustrators: *NCAR Graphics Group*
Production: *R. David Newcomer Associates*
Printing and Binding: *Maple-Vail Book Manufacturing Group*
Reprinted 1992

Library of Congress Catalog Card Number 86-050345

ISBN 0-935702-52-0

Printed in the United States of America

10 9 8 7 6 5 4 3 2

Dedicated to the memories of

Jo Washington
and
Thomas I. Parkinson

Contents

Preface

The purpose of this book is to provide an introductory guide to the development and use of three-dimensional computer models of the global climate system, including its four major components of atmosphere, oceans, land, and sea ice. Processes in each of these components, as well as interactions among them, are examined from the point of view of basic principles and in the context of recent developments. In dealing with each aspect we attempt to show how the theory grew historically from fundamental beginnings and how well it is able to account for known aspects of the climate system. For example, the current models describing the general circulation of the atmosphere evolved from early numerical weather prediction models, and they in turn from simple theoretical models. Today's simulations and predictions are far better than those of a few decades ago and succeed in simulating major features of the atmosphere, oceans, and sea ice.

Most review articles that have dealt with the development of climate models have presumed of the reader a great deal of prerequisite knowledge of atmospheric or ocean dynamics. However, as widespread use of these models is growing in new areas of atmospheric sciences, geography, geology, hydrology, and oceanography, there is a need to explain the models from a more elementary level. This book attempts to fill that gap.

The book is organized so that a reader initially only vaguely aware of climate models will be able to gain an understanding of what the models are attempting to simulate, how the models are constructed, what the models have succeeded in simulating, and how the models are being used for evaluative and predictive purposes. After a short introduction, provided in Chapter 1, Chapter 2 describes the climate system, largely from observational evidence. Chapters 3 and 4 are the two most theoretical chapters, describing the equations and numerics, respectively, behind global climate models, as well as the various approximations made, both physical and numerical. Chapter 5 provides numerous examples of how well the models have simulated various aspects of the climate system, including the basic global patterns of wind, temperature, precipitation, ocean currents, and sea ice distributions. In many instances, the chapter presents simulated and observational fields side by side for easy comparison. Chapter 6 illustrates what the models have suggested would be the response of the climate to various changes imposed on it, such as those due to changes in solar output or to increases in atmospheric carbon dioxide. Many of the responses are predictable in a qualitative sense, but due to the multiple feedbacks in the climate system many are not, so that the numerical modeling approach provides a method of obtaining answers not obtainable by other means. Finally, since this is a rapidly changing field, Chapter 7 provides some discussion of expected future developments. The book is supplemented by a set of appendices providing more technical details on specific aspects, and in particular providing computer coding to allow the reader with access to a desktop computer to obtain hands-on experience with simple numerical models of the climate system.

Although it is assumed that the reader has some knowledge of calculus and introductory physics, the orientation of the book is pedagogical. It can be used within a graduate-level course on general circulation modeling of the climate system, as a supplement to advanced courses in atmospheric and ocean sciences, or as a self-study text. It is hoped that the book will prove in large part self-contained for the nonexpert who desires an introduction to the basic principles upon which three-dimensional computer models of the climate system are based. It can thereby serve as the basis for understanding more detailed discussions in scientific journals and advanced texts. We hope that the pedagogical approach will help scientists who have not been involved in climate modeling appreciate both the strengths and the weaknesses of the models and enable them to consider potential applications of the models for their own research.

Acknowledgments

The authors wish to thank several colleagues for reading parts of the manuscript and making valuable suggestions, particularly Robert Chervin and John Walsh for reading most or all of the book and Byron Boville, Jeffrey Kiehl, Gerald Meehl, Albert Semtner, Leo Donner, Steve Hodge, James Hack, Larry Gates, and Åke Johansson for reading various sections within their individual fields of expertise. Akira Kasahara has contributed to the book both directly and indirectly by being an inspiration, by allowing the authors to use unpublished materials, by having written key review papers, and by providing valuable comments and suggestions. Eileen Boettner typed the manuscript and numerous revisions on two word processing systems, including expertly typesetting the final text using TₑX on the VAX computer system at the National Center for Atmospheric Research (NCAR). Her patience and willingness to learn the new NCAR system and to use this book as an example of what the system can do are gratefully appreciated by the authors. She also made many valuable suggestions for making the equations more presentable and adapted the book-style program developed by Joseph Klemp. Lynda VerPlank made many appreciated editorial suggestions, while Lisa Cirbus and Susan Marshall offered several suggestions on the level of presentation, and Scott Bringen, as well as many of the above-named people, helped in the final stages of production. Both authors also appreciate greatly the support and help provided throughout the publication process by Bruce Armbruster of University Science Books.

Justin Kitsutaka, Barbara Mericle, and other members of the NCAR Graphics Department along with Melanie Pappas professionally prepared many of the figures, making the book more attractive and easier to understand. The figures for the book's jacket were produced separately and came from the following sources. David Kennison created the figure for the front cover from an NCAR computer simulation. The satellite image of global surface temperatures was provided by Moustafa Chahine of the Jet Propulsion Laboratory, California Institute of Technology, and Joel Susskind of the National Aeronautics and Space Administration's (NASA's) Goddard Space Flight Center (GSFC). The satellite image of Arctic sea ice is from the volume *Arctic Sea Ice, 1973–1976* (Parkinson et al., 1987) under production at GSFC. The satellite image of North Atlantic sea surface temperatures was obtained from Otis Brown, Robert Evans, Jim

Brown, Angel Li, and Mark Carle of the University of Miami's Rosenstiel School of Marine and Atmospheric Science.

One of the authors (WMW) is supported by the National Science Foundation through a contract to NCAR, and the other (CLP) is employed at NASA's GSFC and supported by NASA's Oceanic Processes Branch. A portion of the research involved is supported by a contract to NCAR from the Department of Energy as part of its Carbon Dioxide Research Division, Office of Basic Energy Science. We especially appreciate Fred Koomanoff and Michael Riches of the Department of Energy and Stan Wilson of NASA Headquarters for being enthusiastic about our efforts to make three-dimensional climate modeling more understandable to a wider audience.

An Introduction to
Three-Dimensional Climate Modeling

CHAPTER 1

Introduction and Historical Development

For centuries individuals have dreamed of being able to understand how the climate system works and from that understanding being able to forecast and perhaps even modify the future climate. This distant goal has become more approachable as a result of the invention of the modern electronic computer, since it is now possible to solve numerically many of the equations encapsulating the physical laws that govern climate. This provides us with exciting new possibilities, although we must move with caution, being careful to understand the limitations as well as the potentials of our new capabilities.

Improved understanding of the climate system could have substantial impact on the economic well-being of the nations of the world. For instance, better predictions of future climate states could help in the determination of more opportune times for the planting and harvesting of crops, and could warn of possible harm to the environment from various human activities such as deforestation of the landscape and industrial insertion of carbon dioxide and other trace gases into the earth's atmosphere. On shorter climatic time scales (months to years), many uncertainties about climate change remain which could be expected to lessen as computer modeling studies continue. For instance, still unknown are the full climatic impacts of ocean temperature anomalies such as those that occur during El Niño episodes and the full impacts of volcanic eruptions on ozone amounts in the stratosphere and on atmospheric cooling, the latter arising as volcanoes insert into the atmosphere particles that reflect sunlight back out of the earth/atmosphere system. Without computer models it would be virtually impossible to give quantitative answers to the questions raised by such issues, since the interactions are so complex. Computer models already have become powerful investigative tools for climate research and will become more so in the future, especially as the observational data base improves both in quality and in spatial coverage, due to the expanded satellite and conventional data networks.

The realization that the physical laws governing the atmosphere, oceans, and sea ice could be used to determine future conditions was inherent in much of the philosophical reasoning in the late eighteenth and early nineteenth centuries, following the resounding early success of Newtonian mechanics. This was stated most vividly by Pierre Simon de Laplace in 1812 when he suggested that complete knowledge of the masses, positions, and velocities of all particles at any single instant would enable precise calculation of all past and future events. At the time, the classical laws of mechanics were firmly in hand, having been derived largely from Isaac Newton's monumental *Principia Mathematica* in 1687, but still undiscovered were the fundamental laws of thermodynamics, which would be crucial for accomplishing predictions for fluid fields such as the atmosphere and oceans. The 1840s witnessed the independent development of the concept of the conservation of energy by Robert Mayer, James Joule, and Hermann von Helmholtz. This led Rudolf Clausius in 1850 to identify the conservation of energy as the first law of thermodynamics and to formulate the second law of thermodynamics, that in the absence of external constraints the net flow of heat between two bodies is from the warmer to the cooler one. By the late 1800s the fundamental laws of classical physics were known and thus the goal of accurate prediction through numerical calculation was closer to becoming realizable.

In the context of the hydrodynamical equations that govern the flow of the atmosphere or the oceans, Vilhelm Bjerknes (Fig. 1.1) planned a course of action that in many ways remains as valid today as when he first stated it in 1904 (Bjerknes, 1904). His words were:

> If it is true, as every scientist believes, that subsequent atmospheric states develop from the preceding ones according to physical law, then it is apparent that the necessary and sufficient conditions for the rational solution of forecasting problems are the following:

Fig. 1.1 Vilhelm Bjerknes (1862–1951), who stated the basic principles for solving the mathematical equations governing the flow of the atmosphere and ocean. [From Platzman (1967).]

1. A sufficiently accurate knowledge of the state of the atmosphere at the initial time.
2. A sufficiently accurate knowledge of the laws according to which one state of the atmosphere develops from another.

Later in the book we will discuss limitations on our ability to determine the two broad conditions stated by Bjerknes.

During World War I, while resting between battles during his work as a volunteer ambulance driver for the Red Cross, English scientist Lewis Fry Richardson (Fig. 1.2) attempted to use the basic equations of atmospheric motions to develop a capability of forecasting weather, using only a mechanical calculator. By that time it was well known that the equations were highly complex and were not amenable to simple solution, and that the only recourse for solving the equations was by the use of numerical approximations. In essence, a set of continuous equations was handled approximately by obtaining a numerical solution to a corresponding set of discrete equations. In his book *Weather Prediction by Numerical Process*, Richardson (1922) describes in

Fig. 1.2 Lewis Fry Richardson (1882–1953), who first attempted a numerical solution of atmospheric flow equations. [From Platzman (1967).]

a step-by-step manner his predictive method for one small area in Europe using available observed data. Since Richardson did not know the relative importance of all the various factors in the atmosphere, he attempted to include a great deal of atmospheric physics. This served to encumber his calculations, and the resulting prediction was highly unrealistic. However, many of the problems he encountered in the process of trying to construct this unprecedented calculation are the same ones that climate and weather forecast modelers continue to face today. A fascinating historical account is provided by Platzman (1967). Richardson's book in many ways serves as a blueprint for constructing a numerical model. From time to time we will make reference to it in the context of present understanding, thereby illuminating both some of the progress that has been made in this field and possible worthwhile future directions of research as more of the problem areas become less severe. Climate models now are capable of simulating almost all the major observed features of the atmosphere, oceans, and cryosphere, albeit with considerable room for improvement, especially in view of the sometimes questionable adjustments or "tuning" factors needed to obtain the desired simulations.

Returning to the historical perspective, not much was done on the direct numerical solution to the equations after the work of Richardson until the late 1940s, when the first electronic computers were developed, starting with the ENIAC (Electronic Numerical Integrator and Computer) at Aberdeen Proving Grounds. Within months, the IAS (Institute for Advanced Studies) computer was constructed at Princeton University under the direction of mathematician John von Neumann. One of the first problems addressed by von Neumann with his new calculating tool was the forecasting of weather patterns, for which purpose he formed a team of scientists under the leadership of Jule Charney (see Charney et al., 1950, and Thompson, 1978, for interesting accounts of the history of that group).

The computer program, or model, that Charney and others tested for their first numerical weather prediction forecast did not contain the general set of equations used by Richardson but rather a simpler set. The general equations admit a wide spectrum of motions in the atmosphere, ranging from sound waves and gravity waves to the very slow-moving large-scale meteorological waves. In the late 1930s Carl Gustav Rossby found that by making suitable approximations to the more general equations he could find simpler equations to solve for the weather patterns that in effect filtered out the unwanted, numerically troublesome, and meteorologically less important waves. It was this simpler set of equations that was used by Charney and his colleagues. A few years later Norman Phillips (1956) added simple forcing terms to this same set of equations and carried out a long-term integration that became independent of the initial state and showed many of the general features of the atmospheric circulation. This single experiment was in some sense the beginning of the long development of general circulation models. More details on Phillips's effort will be noted in later chapters.

Recent development of general circulation modeling has included a shift back to a more general set of equations very close to that used by Richardson. This became possible as the reasons for Richardson's failure, tied both to erroneous input fields and to the specific numerical scheme, were diagnosed and remedies were found. In the mid-1950s several groups of scientists dedicated to the development of numerical weather prediction and general circulation modeling of the atmosphere were formed.

In particular, a major general circulation modeling effort in the United States Weather Bureau under the direction of Joseph Smagorinsky became the nucleus of what is now the Geophysical Fluid Dynamics Laboratory (GFDL) at Princeton University. Since the formation of GFDL, now part of the National Oceanic and Atmospheric Administration (NOAA), numerous other climate modeling centers have been established in the United States and elsewhere throughout the world.

Large-scale ocean modeling was developed somewhat later, in the early 1960s, and involves distinctive problems, many of which will be mentioned in the course of this book. Modeling of the important large-scale roles of sea ice, snow, land processes, and the biosphere has developed since the mid-1960s, and their incorporation in climate models is only beginning to take place. We will describe the relevant formulations, limitations, and coupling with the atmospheric and oceanic calculations.

Preparatory to the presentation of the details of the various models of the climate system, the next chapter presents a basic description of some of the observational aspects of the atmosphere, ocean, and cryosphere. This discussion provides a basis for the later discussions of the numerical models and comparisons between the modeling results and the observed atmosphere, ocean, and sea ice distributions.

Chapter 3 shows how the basic laws of physics can be made appropriate for the atmosphere, oceans, and sea ice, including the important processes of cloud radiation, convection, precipitation, boundary transfers, and small-scale mixing. Chapter 4 illustrates some of the numerical methods presently used for the solution of the model equations. The remainder of the book concentrates on simulation results, with Chapter 5 presenting results of simulations that attempt to reproduce aspects of today's climate and Chapter 6 presenting results of attempts to simulate climate under conditions altered from the present. In this way we show examples of what can be learned from present computer models, and in the concluding chapter, Chapter 7, we suggest likely advances for the future. Although references are included, no effort is made to provide a complete bibliography of the field. Where possible, reference is made to suitable review papers that are readily available. The appendices cover various technical details and provide explanations of, and the coding for, a few simple computer programs appropriate for desktop personal computers. Although these sample programs do not approach the complexity of most of the models discussed in this book, they provide those readers with access to a desktop computer the opportunity for hands-on experience in numerically simulating various atmospheric phenomena. Information on how to purchase the computer programs on floppy disks is provided in Appendix F.

CHAPTER 2

Physical Description
of the Climate System

The central purpose of this book is to explain the physical and mathematical principles behind three-dimensional general circulation climate models and describe some of the results of model simulations. In order to provide the reader with some insight into what is being attempted in modeling, this chapter contains a brief description of the climate system in terms of three major components—atmosphere, oceans, and sea ice—and the interactions among them. Since other books already descriptively cover details of each of the components, the discussion here will emphasize general features. Later in the book, in Chapter 5, examples will be given of simulations of each of the three mediums, with comparisons shown between the simulated results and the observed atmosphere, oceans, and sea ice.

Atmosphere

Atmospheric composition

The earth's atmosphere is composed of a mechanical mixture of several gases: nitrogen (N_2) constituting 78.08% of the volume of dry air, oxygen (O_2) 20.95%, argon (A) 0.93%, carbon dioxide (CO_2) 0.0332%, neon (Ne) 0.0018%, helium (He) 0.00052%, methane (CH_4) 0.00015%, krypton (Kr) 0.00011%, and hydrogen (H_2) 0.00005%. The amount of water vapor in the air varies widely, depending on location, evaporation, temperature, and other factors. It constitutes anywhere from 0 to 3% of the total atmospheric volume. There are several other gases with even smaller trace amounts in the atmosphere that also influence atmospheric radiation. Primary among these are carbon monoxide (CO), sulfur dioxide (SO_2), nitrogen dioxide (NO_2), and ozone (O_3). Certainly the two principal gases are nitrogen and oxygen, together constituting more than 99% of the dry atmosphere. Although the other gases contribute relatively small percentages to the total atmospheric volume, their importance in terms of atmospheric processes can be considerable. This is particularly true of carbon dioxide, ozone, and water vapor, but is also true of other trace gases (for instance, see Ramanathan et al., 1985).

Since the atmosphere is composed of material particles, it exerts a pressure due to the effects of gravity. Since the atmosphere is gaseous, the pressure decreases with height, so that the gas is most dense near the ground, with a sea level pressure of approximately 1.013×10^3 mb, and at the top its density approaches zero. This considerable variation of density with height distinguishes the atmosphere from the ocean, where the density at the bottom is very nearly the same as at the top. The difference derives from the ready compressibility of a gas and the near incompressibility of a liquid.

Temperature profiles

In the lower part of the atmosphere, to a height of about 17 km at the equator and 6–9 km at the poles, the general tendency is for the air temperature to decrease with height, doing so at a fairly uniform rate averaging 6.4°C per kilometer. This lower region, termed the *troposphere*, is the portion of the atmosphere in direct contact with the earth's surface and is where most atmospheric processes usually associated with day-to-day weather take place. Above the troposphere, in the atmospheric region called the *stratosphere*, the temperature generally stays constant or increases with height. The division between the two regions, where the temperature often is near its minimum for the given vertical air column, is the *tropopause*.

Figure 2.1 depicts in more detail typical climatological (long-term time-mean) temperature profiles with height and shows the contrast between summer and winter seasons and polar, temperate (mid-latitude), and tropical latitudes. In the tropics the profile is similar throughout the year, with a temperature near the ground of about 300 K, decreasing to about 190 K at a height

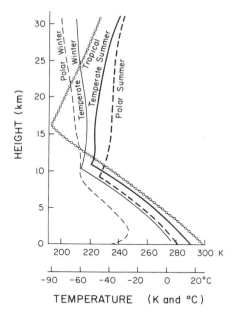

Fig. 2.1 Atmospheric temperature profiles as a function of height. [From Dobson (1968).]

of about 17 km, then increasing with height in the stratosphere. Due to the great height over which the temperature decrease occurs in the tropical troposphere, the 190 K tropopause temperature in the tropics is typically the coldest temperature anywhere in the atmosphere, in spite of the warm surface temperatures there.

The polar regions experience the largest summer/winter tropospheric temperature contrast (Fig. 2.1). In summer, polar temperatures are about 280 K near the ground and decrease to about 230 K at a height of 9 km, above which there is a general warming with height in the lower stratosphere. In winter, temperatures are about 235 K at the ground and increase rather than decrease with height for the first 1.5–2 km, above which the more typical tropospheric decrease with height occurs up to the tropopause at about 9 km. In this case, above the tropopause the temperature continues to decrease with height, although at a much lesser rate than in the troposphere. The anomalous temperature increase with height near the surface is termed an *inversion*, because of its being inverted from the usual tropospheric temperature decrease with height. Such a condition can exist whenever there is strong surface cooling of the air.

The temperate or mid-latitude temperature profiles are very similar in shape to that of the polar summer, especially in the troposphere, where the main contrasts are the higher surface temperatures in the temperate summer and the greater height of the temperate tropopause (Fig. 2.1). In the mid-latitude stratosphere above about 15 km the summer temperatures slowly increase with height, while the winter temperatures slowly decrease with height.

It is important to understand that the temperature profiles of Fig. 2.1 are climatological averages and do not depict exactly the actual temperature structure in the atmosphere at any specific time or place. The variability about the vertical means is greater for the polar regions than for the tropics and for winter than for summer. For an indication of the interannual variations in the height of the tropical tropopause, the reader is referred to Reid and Gage (1985).

Energy balances

In order to understand how the atmosphere maintains the various temperature structures depicted in Fig. 2.1, it is necessary to consider the processes by which the atmosphere gains and loses energy and how these processes vary with latitude and season. Essentially all the energy that enters the earth's climate system comes from the sun. Since a portion of this solar energy is absorbed in the atmosphere/ocean/sea ice system, this must be balanced with outgoing energy to space in order to maintain the overall climate system in its observed approximate equilibrium state. Such outgoing energy is emitted by the earth's surface, the oceans, and the atmosphere. Due to the tremendous contrast in solar and terrestrial temperatures (the sun has a surface temperature of about 6000 K), there is also a sharp contrast in the energy emitted by the two bodies. The warmer a body is, the shorter is the wavelength of its peak electromagnetic emission (Fig. 2.2), and thus the sun's emission peaks at much shorter wavelengths than does the earth's. In fact, most of the energy that arrives from the sun is in the wavelength range from 0.1 to 2 μm, centered on the visible portion of the electromagnetic spectrum, while most of the outgoing terrestrial radiation to space is in the range from 4 to 40 μm, which is entirely within the infrared portion of the spectrum.

The spatial and temporal distribution of the receipt of solar energy at the earth's surface is highly dependent upon the earth's annual revolution about the sun, its daily rotation about

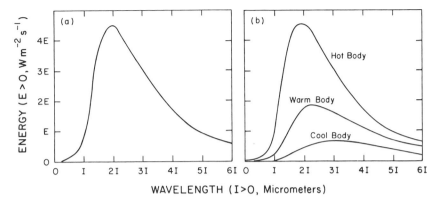

Fig. 2.2 (a) Generalized shape of the "blackbody" emission curve for any body, showing the maximum possible emission from the body (at constant temperature) for each wavelength. The curve is described by an equation formulated by Max Planck in 1900 and is termed Planck's radiation curve. The equation is called Planck's law. (b) Comparative blackbody radiation curves for bodies of different temperatures. The warmer a body is, the greater is its blackbody emission at each wavelength and the shorter is the wavelength at which its emission peaks.

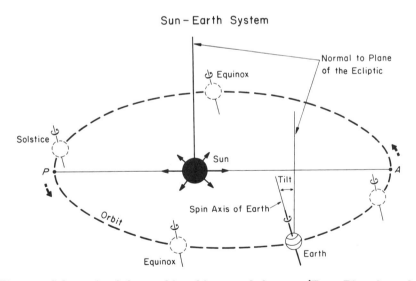

Fig. 2.3 Schematic of the earth's orbit around the sun. [From Pittock et al. (1978).]

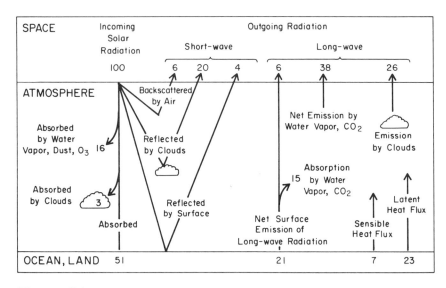

Fig. 2.4 Schematic of the relative amounts of the various energy inputs and outputs in the earth/atmosphere system. [From data in NRC (1975).]

its axis, and the tilt of its axis with respect to the plane of its orbit (Fig. 2.3). When the earth is in that portion of its orbit where the Northern Hemisphere is tilted toward the sun, the Northern Hemisphere receives the majority of the direct sunlight and experiences summer, while the Southern Hemisphere experiences winter. Naturally the reverse occurs when the earth is in the opposite portion of its orbit, with the Southern Hemisphere tilted toward the sun. This accounts for the majority of the local seasonal change in solar input. As expected, the largest seasonal difference in the amount of energy received from the sun takes place at the poles, where the sun is above the horizon for a full 6 months, then below the horizon for the following 6 months. The solar radiation input varies least from season to season in the tropics.

Figure 2.4 depicts a quantitative breakdown of the absorption, reflection, and emission of the globally averaged solar and terrestrial radiation. The numbers are scaled such that the incoming solar radiative energy at the top of the atmosphere is set at 100 units. Of this incoming energy, about 16% is absorbed by ozone (mostly in the stratosphere), water vapor, and dust, only about 3% is directly absorbed by clouds, and about 51% is absorbed at the earth's surface (ocean and land). The remaining 30% of the incoming solar radiation is reflected back to space and lost to the climate system. Most of the reflection is by clouds (accounting for 20% of the total incoming radiation), with the remainder being by air-molecule backscattering (6%) and by the earth's surface (4%). (These numbers are generalized global averages and would differ somewhat from one year to another. Even greater differences would appear if the figure were adjusted for any particular time or place, as large departures from the values in Fig. 2.4 occur spatially and temporally and even within the diurnal [daily] cycle.)

As mentioned earlier, for an equilibrium to exist, the outgoing radiation from the earth/atmosphere system must balance the incoming radiation. Since essentially all the incoming radiation is the 100 units of incoming solar radiation depicted in Fig. 2.4 and 30 units of that are reflected back to space, there must be 70 units of terrestrial radiation emitted to space also. There are three main sources of this terrestrial emission: water vapor and CO_2 in the atmosphere, accounting for 38 units scaled on the same annually, globally averaged basis as for the values in the rest of Fig. 2.4; clouds, accounting for 26 units; and the earth's surface, accounting for 6 units. At the earth's surface, the 51 units of absorbed solar radiation are balanced by a net upward emission of 21 units of infrared radiation, 7 units of sensible or thermal heat transfer from the surface to the atmosphere, and 23 units of evaporative or latent heat transfer from the surface to the atmosphere. The 45 total units absorbed by the atmosphere from the earth's surface, plus the 19 units absorbed by the atmosphere from the incoming solar radiation, are balanced by the 38 + 26 units emitted to space. More on this aspect of the climate system will be given in Chapter 3 when the radiation and other physical processes incorporated in climate models are discussed.

Average surface temperature patterns

Figure 2.5 shows the January and July long-term mean global distributions of surface air temperature. In January, with winter in the Northern Hemisphere and summer in the Southern Hemisphere, the maximum surface temperatures occur just south of the equator, with peak values exceeding 300 K for a broad strip between 40°E and 130°W and also for small regions over the South American continent. The tendency for the Southern Hemisphere temperature contours to run parallel to the latitude circles is perturbed mainly by the continents, as the summer temperature contours dip poleward particularly over South America and Australia. The reverse effect occurs in July, as the temperature

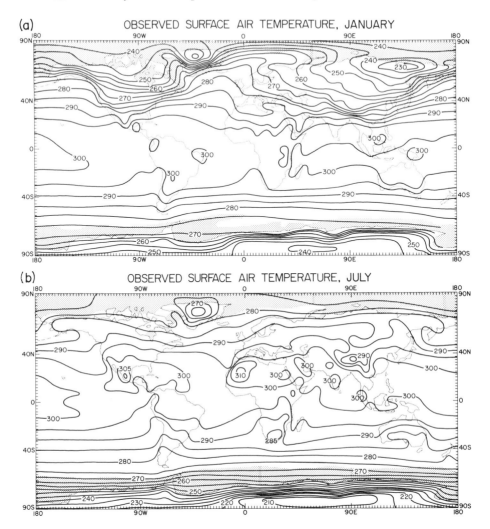

Fig. 2.5   (a) January and (b) July long-term mean geographical distributions of surface atmospheric temperature. [From Washington and Meehl (1984).]

contours curve equatorward over the Southern Hemisphere land areas. Both deflections reflect the modifying influence of the oceans, making continental summers warmer and winters cooler than the adjacent ocean areas. The oceans have this effect because of their large heat storage capacity and because part of the energy that they retain from solar input during the summer they return to the atmosphere in the forms of sensible and latent heat and longwave radiation the following winter. Conversely, one can think of the continents as having the reverse effect, strengthening the seasonal temperature contrasts. This is termed the *continentality effect*.

In the Northern Hemisphere similar oceanic and continentality effects can be seen, although the lesser expanse of oceans in the mid- and low latitudes results in a much lesser tendency for the temperature contours to follow latitude circles than in the Southern Hemisphere (Fig. 2.5). The great expanse of the European/Asian continent produces a very strong continentality effect in winter, with some of the coldest temperatures ever recorded occurring in northern Siberia. This continentality contributes, along with higher elevations, to the coldest average January temperatures occurring over Greenland and northern Siberia rather than over the North Pole (Fig. 2.5). These minimum average temperatures are about 230 K, which is roughly 20 K warmer than the winter minima in the Southern Hemisphere, where over the glacial ice of Antarctica the continentality effect combines with even higher latitudes and higher elevations than in Greenland and Siberia.

Another effect reflected in the monthly average surface air temperatures (Fig. 2.5) is the influence of ocean currents. As is discussed in the next section, ocean currents transport considerable heat from low to high latitudes, a phenomenon especially pronounced with the Gulf Stream of the North Atlantic. This heat transport further contributes to much higher winter temperatures over mid- to high latitude oceans than over adjacent land areas at the same latitude. The average Southern Hemisphere ocean circulation does not have as strong a north/south component as does the Northern Hemisphere circulation. This contrast, in combination with the higher ratio of ocean to land area, helps create a temperature structure in the Southern Hemisphere that shows less deviation along latitude circles than exists in the Northern Hemisphere.

Large-scale hemispheric circulation patterns:
three-cell structure

Since heated air tends to rise, the relative heating and cooling of different areas of the earth's surface in large part drives local winds and large-scale atmospheric circulation. For instance, the greater heating in the tropics versus that of the subtropics causes rising atmospheric motions over the tropics and compensatory sinking motions in the subtropics. Similarly, the heating of the Northern Hemisphere continents during summer causes rising motions over the land areas and compensatory sinking motions

over the adjacent oceans. This in turn causes relatively low sea
level pressures over land and high pressures over mid-latitude
ocean regions during summer.

The processes by which these vertical circulation patterns
arise can be understood conceptually with elementary physics:
If a fluid or gas is heated, it tends to become less dense than
its surroundings and hence to rise due to buoyancy forces. As it
rises, the cooler surrounding gas or fluid moves in to replace it,
necessitating a sinking motion at some distance from the original rising motion. This generates a simple vertical cellular circulation pattern often referred to as a *direct thermal circulation.*
Early researchers contemplating such processes envisioned a very
large-scale, hemispheric circulation cell in which heated air rose
in the tropics, traveled horizontally poleward in the upper atmosphere, sank in the polar regions, and finally circulated back
toward the equator before rising again and beginning the jour-
ney anew. This proved to be an oversimplification, but was an
important early step toward our current understanding.

Historically, two important individuals in the development
of the concept of the large-scale cellular circulation pattern were
Edmond Halley in the seventeenth century and George Hadley in
the eighteenth century. Halley was interested in explaining the
observed trade wind phenomenon, in which low latitude winds in
the Northern Hemisphere blow from the northeast and low latitude winds in the Southern Hemisphere blow from the southeast,
both with a persistence in marked contrast to the seemingly more
erratic behavior of the winds in higher latitudes. Halley in 1686
explained the equatorward flow of the trade winds as resulting
from solar heating, producing rising air near the equator, and the
consequent surface inward flow of air toward the updraft region.
His explanation of the trade winds' east-to-west flow was erro-
neously based on the east-to-west movement of the sun. A much
improved explanation was provided by Hadley in 1735, based on
the concept of the conservation of angular momentum.

Hadley's theory includes a full-fledged, large-scale, single-cell
pattern: solar heating causes air near the equator to rise, the
risen air moves poleward, cools, sinks, and returns equatorward,
producing a vertical circulation cell in each hemisphere. The
east-to-west motion of the trade winds is explained with the aid of
the principle of the conservation of angular momentum through
the following reasoning: Due to the earth's rotation about its
axis, each point on the earth's surface has a linear component
of west-to-east motion. The lower the latitude, the larger this
component because of the greater circumference of travel. (The
rate at each point is approximately the length of the latitude

circle divided by 24 hours.) Hence the low latitude surface air moving toward the equator has a smaller linear component of west-to-east motion due to the earth's rotation about its axis than does the air at the equator. From the perspective of the earthbound observer, this results in an easterly component to the low latitude trade winds, thus accounting for northeasterly trades in the Northern Hemisphere and southeasterly trades in the Southern Hemisphere. (Winds are customarily labeled by the direction from which they flow, so that a northeasterly wind flows from northeast to southwest.)

The single-cell pattern of circulation with rising air in the tropics and sinking air at the poles was revised to a three-cell pattern by William Ferrel in 1855, by which time it was clear observationally that a single-cell, equator-to-pole pattern did not exist and that the poleward-flowing air at upper altitudes in the tropics was descending far short of the polar regions. Ferrel suggested three major vertical circulation cells in each hemisphere: (1) air rising near the equator and descending near 30° latitude, (2) air descending near 30° latitude and rising near 60° latitude, and (3) air rising near 60° latitude and descending at the pole. The first cell has been termed the *Hadley cell* and the second the *Ferrel cell* in honor of the respective conceptual advances made by George Hadley and William Ferrel. Ferrel also assimilated theoretical studies done by Gaspard de Coriolis in 1835 on the motion of objects relative to a separate moving surface and applied them to the atmosphere, concluding that the rotation of the earth causes moving air to be deflected to the right in the Northern Hemisphere and to the left in the Southern Hemisphere. This *Coriolis deflection* is a primary reason why the poleward-flowing upper air in the tropics does not travel from equator to pole and create a single-cell circulation pattern.

Ferrel applied the Coriolis deflection to explain the easterly trade winds in low latitudes, as the near-surface air of the Hadley cell is deflected from its 30°-to-equator path, and to explain the westerly winds in mid-latitudes, as the surface air of the Ferrel cell is deflected from its low-to-high latitude path. He similarly explained the counterclockwise rotation of cyclones (major low pressure weather systems) in the Northern Hemisphere and the clockwise rotation of cyclones in the Southern Hemisphere by the respective right and left directions of the Coriolis deflection as air converges toward the low pressure at the center of the cyclonic system.

The three-cell circulation pattern envisioned by William Ferrel still forms an important part of our understanding of the overall pattern of atmospheric circulation (Fig. 2.6). The tropical

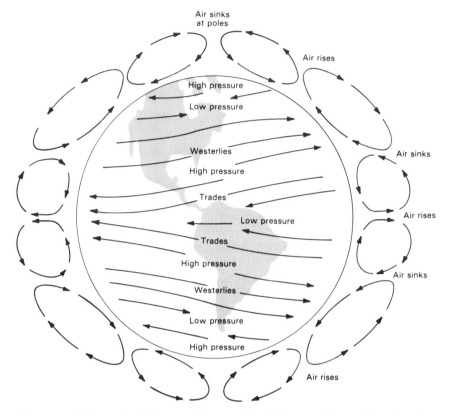

Fig. 2.6   Schematic of the earth's generalized, large-scale meridional and low-level circulation patterns. [From Battan (1979).]

Hadley cell is driven by solar heating, causing rising motion near the equator, then by the release of latent heat as the rising air leads to precipitation in the upward branch of the cell. This upward branch of the Hadley cell is not centered consistently on the equator but migrates north and south with season, following the seasonal shifts in the maximum intensity of solar heating. Moreover, the strength of the Hadley circulation varies with longitude, being strongly affected by such factors as whether the earth's surface is land or ocean. The seasonal temperature change over ocean is lessened by the ocean's significant heat storage capacity compared to that of land.

After rising near the equator, the air in the Hadley cell moves poleward in the upper region of the troposphere, sinking near 30°N and 30°S and thereby generating belts of high pressure near 30°N and 30°S. The subtropical jet streams arise near the upper, poleward portions of the north and south Hadley circulations (see

Fig. 2.9). Since the sea level pressure is low near the region of rising motion at the thermal equator, the high pressure produced by the sinking air at about 30°N and 30°S creates a surface pressure gradient leading to the movement of a portion of the sinking air back toward the low pressure near the equator. The Coriolis deflection (produced by the rotation of the earth) then deflects this air toward the west in both hemispheres, creating the low-level, low-latitude trade wind system mentioned previously.

Various indications of the Hadley cell circulation are visible on or can be inferred from satellite images and daily tropical weather maps. In particular, the low-level convergence is reflected in surface wind patterns, the upper-level divergence is reflected in cloud movements, and the central region of upward motions is generally conspicuous for its high humidity, cloudiness, and precipitation. The region of low-level convergence toward the upward branch of the Hadley cell is called the *intertropical convergence zone* (ITCZ). The heavy precipitation in the ITCZ leads over land to the world's major tropical rain forests, while the corresponding dry conditions in the downward branch of the Hadley cell lead to desert conditions, in particular creating the Mojave Desert in the southwestern United States, the Great Sandy, Gibson, and Great Victorian deserts in western Australia, the Sahara Desert in northern Africa, the Kalahari and Namib deserts in southern Africa, the Gobi Desert in China, and the Atacama Desert along the west coast of South America.

In the mid-latitude Ferrel cell, termed *indirect* because of having rising air in its cooler branch, the low-level flow is toward the poles, away from the relatively high pressure produced by the descending arms of the Hadley and Ferrel cells at about 30° latitude and toward the relatively low pressure at about 60° latitude. This low-level flow also is deflected by the rotation of the earth, giving rise, in an average sense, to surface westerlies as shown in Fig. 2.6. These westerlies, however, are not nearly as persistent as the trade wind easterlies but instead reflect the general west-to-east passage of high and low pressure systems within the mid-latitudes.

The third cell in the three-cell pattern is the polar cell. Strong radiational cooling near the poles causes polar air to become cold and dense, which in turn causes it to sink. Thus there is relatively high pressure at the pole, which, combined with the low pressure near 60°N and 60°S discussed in connection with the upward branch of the Ferrel cell, produces surface flow equatorward from the pole. This polar cell, however, in great contrast to the Hadley cell, is extremely weak, although it remains detectable in time averages of the air circulation.

The Hadley cell and polar cell are termed *direct* because they are driven basically by heating and cooling patterns. By contrast, the Ferrel cell is driven indirectly by the circulation system and particularly by the other two cells. More will be mentioned on these and other aspects of the circulation patterns when comparisons are made in Chapter 5 between model simulations and the observed atmosphere.

The three-cell pattern is basically a generalized schematic representing certain large-scale aspects of the atmospheric circulation system visible in zonal averages. The actual circulation entails significant variations from this schematic on all spatial scales, so that daily- and even monthly-averaged winds and pressures show notable differences from the simplified picture that the three-cell pattern would suggest (compare Fig. 2.7 with Fig. 2.6). In particular, there are marked seasonal variations and marked variations produced by the land/sea contrast. Seasonally, the whole three-cell pattern shifts northward during the Northern Hemisphere summer and southward during the Northern Hemisphere winter, as the ITCZ moves in response to the changing patterns of solar heating. The land/sea contrast results in many local and regional deviations from the three-cell pattern, particularly the sea breeze phenomenon on a local basis and the seasonal monsoons on a regional basis.

The north-south circulations of the Hadley, Ferrel, and polar cells are contrasted with another major vertical circulation cell, which has a mean east-west motion. This is the Walker circulation of the equatorial Pacific, involving rising air over the region of Indonesia and descending air over the eastern Pacific. The Walker circulation was originally examined by G. T. Walker and colleagues in the 1920s as a result of their discovery of strong negative correlations between surface pressure anomalies in the two regions. Variations in the strength of the Walker circulation produce a large-scale fluctuation with an irregular period known as the Southern Oscillation. This phenomenon is tied closely to atmosphere/ocean interactions and will be discussed further in the section entitled Atmosphere/Ocean/Ice Interconnections.

Land/sea breezes and monsoons

The term *monsoon*, derived from the Arabic word for *season*, is used generally to refer to any large-scale seasonal reversal of wind regime. Monsoons typically occur in coastal or

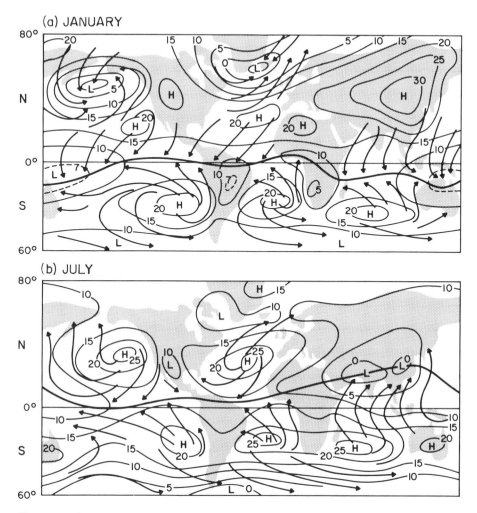

Fig. 2.7   Average surface winds and sea level pressures (in mb above 1000 mb) for (a) January and (b) July. [Redrawn from Riehl (1978).]

near-coastal regions, with the seasonal wind reversal arising due to the land/sea contrast. Although monsoons occur in many parts of the world, the Asian monsoon centered over India is particularly influential both for the atmosphere and for humanity and hence it will be used here to illustrate monsoon phenomena.

As monsoons have some resemblance to a giant land/sea breeze, the simpler land/sea breeze will be described first (Fig. 2.8). Land heats and cools more rapidly than water, so that at midday, with rapid solar heating of the ground, the air over the land tends to be heated and hence to rise. This contributes to

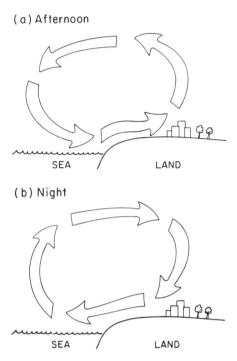

(a) Afternoon

SEA   LAND

(b) Night

SEA   LAND

Fig. 2.8   Land/sea breeze phenomenon. Differential heating of the land and sea generates relatively warm, rising air over the land in the afternoon and relatively cool, sinking air over the land at night, leading to a surface-level sea breeze in the day and land breeze at night.

an offshore horizontal motion at upper levels and to the creation of a surface pressure gradient, with relatively high pressures over the sea and low pressures over the land. The pressure gradient in turn leads to an onshore afternoon sea breeze at lower, near-surface levels, as cooler air flows landward from the adjacent water area. The cellular circulation pattern is completed by the descending air over the water (Fig. 2.8). At night the situation is reversed, with the land cooling more rapidly than the ocean, thereby leading to descending air and relatively high surface pressures over the land, rising air and relatively low surface pressures over the ocean, and a surface land breeze (Fig. 2.8). Because the stability of an air column tends to be greater at night than in the day, the land breeze tends to be much weaker than the sea breeze.

The Asian monsoon shows some similarities to the sea breeze concept, except transferred from a local to a near continental event and from a day/night contrast to a summer/winter contrast. In summer the large land mass of Asia heats more quickly than the surrounding oceans, thereby generating rising air and a large low pressure system centered over the land. This is reinforced by the summertime northward shift of the ITCZ (Fig. 2.9). The low pressure over land leads to the convergence of air toward the low, so that over much of southern Asia the air flows from the Indian or Pacific oceans, bringing warm moist air that leads to heavy precipitation as the air rises over the foothills of the Himalayas, Ghats, and Annamese Highlands.

In winter the situation is reversed. The rapidly cooling continent develops a strong high pressure center (the persistent Siberian High) from which cold, dry air flows outward. As a result, southern Asia typically experiences several months of dry, clear weather each winter, in sharp contrast to the heavy rainfall experienced in summer.

The analogy between the monsoon and land/sea breezes is only partial, however, because monsoon circulations typically encompass several complications not included in the above description. For instance, being much larger-scale phenomena than typical land/sea breezes, monsoon circulations often interact with upper atmospheric circulations, particularly the jet stream. The reader interested in monsoons is encouraged to read more detailed discussions, such as the description given by Barry and Chorley (1971; pp. 269–283).

Oceans

Approximately 71% of the earth's surface area is covered by seas and oceans, and 29% by land. The waters of the oceans transfer heat, salt, nutrients, and momentum from one location to another, making an understanding of the dynamics of the ocean, like the dynamics of the atmosphere, essential for understanding the global climate system. Furthermore, at the air/water interface the oceans and atmosphere interact in ways that significantly affect both media.

The three major oceans are the Pacific, Atlantic, and Indian oceans, occupying 46%, 23%, and 20%, respectively, of the total world ocean area. With an average ocean depth of approximately 4000 m, the oceans contain roughly 1350×10^6 km^3 of water. Surrounding most continents there is a relatively shallow *continental shelf* area extending outward anywhere from a few to several hundred kilometers. Worldwide, the shelf width averages about 65 km and has a gradient of about 1 in 500. The seaward edge of the shelf is generally marked by a dramatic increase in

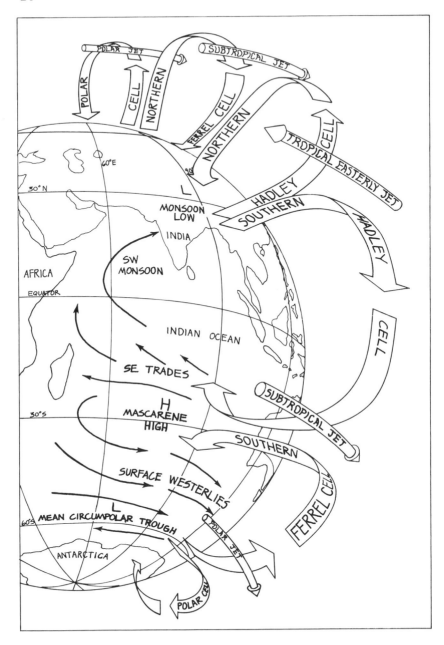

Fig. 2.9 Basic three-cell circulation pattern as shifted for Northern Hemisphere summer conditions, with the concomitant southwest Asian monsoon winds. The vertical cross section is meant as a schematic representation for all longitudes. [From Meehl (1987).]

gradient as the shelf gives way to a very steep *continental slope*, which extends to or near the deep-sea bottom. The ocean floor topography, like the land topography above sea level, is uneven, marked notably by underwater mountains or *seamounts*, ridges, trenches, and in some locations fairly smooth basins covering hundreds of square kilometers. The largest, most extensive feature of the world ocean is the mid-ocean ridge, extending north to south along the mid-Atlantic Ocean, where it is termed the mid-Atlantic Ridge, and into both the Indian and Pacific oceans.

Similarly to the situation in the atmosphere, the bulk of the external heating of the oceans is due to solar insolation in the region of the tropics. Ocean currents then redistribute part of this heat to higher latitudes, where large amounts of energy are released to the atmosphere in the form of sensible and latent heat and longwave radiation. A portion of this energy eventually is lost to space. The ocean also transfers sensible and latent heat to the atmosphere in the tropics and mid-latitudes, and at individual places and times these amounts can also be significant.

Seawater composition

Seawater contains, in addition to its liquid water content, a quantity of dissolved material collectively termed *salinity*. This material is a combination of chemicals such as chloride ion (55%), sodium ion (30.6%), sulfate ion (7.6%), magnesium ion (3.7%), calcium ion (1.2%), potassium ion (1.1%), bicarbonate ion (0.41%), bromide ion (0.19%), boric acid ion (0.075%), strontium ion (0.038%), and fluoride ion (0.003%), where the percentages indicated are the percentages of total salt content. The salinity of the ocean is usually expressed in units of grams (g) of solid material per kilogram (kg) of sea water. The average or mean salinity of the oceans is about 35 g/kg, which converts to 35 parts per thousand (35‰) since 1 kg = 1000 g. Details of salinity measurement can be found in standard oceanography textbooks such as Neumann and Pierson (1966).

When the annual mean salinities at the sea surface are mapped globally (Fig. 2.10), values are seen to range from about 29‰ in the central Arctic to about 37.5‰ in the subtropical North Atlantic. Low values in the Arctic result partly from fresh water runoff and from the melting of sea ice; high values in the subtropical North Atlantic are influenced by high-salinity inflows from the Mediterranean. Peak salinity values in the Pacific are somewhat lower than those in the Atlantic, although in the

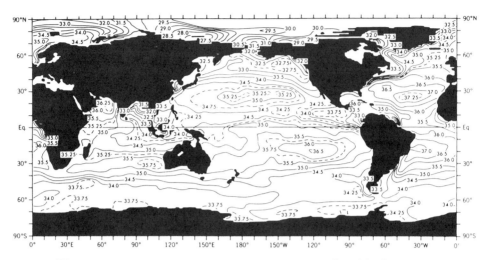

Fig. 2.10 Annually averaged sea surface salinities (in ‰). [From Levitus (1982), with relabeling.]

mid-subtropical South Pacific annually averaged values do reach 36.5‰ (Fig. 2.10).

Ocean temperatures

At the sea surface, water temperatures tend to range from about $-1°C$ in the polar regions to 20–30°C in the tropics (Fig. 2.11). On a seasonally averaged basis, the temperatures are approximately zonal, with the isotherms, or lines of equal temperature, running basically east-west. Notable divergences from the zonal structure appear in the tropical Pacific, where temperatures tend to be warmer in the west than in the east, and in regions where persistent currents have a strong nonzonal component. One of the most noted examples of the latter is the Gulf Stream in the North Atlantic, transporting warm waters to the northeast and hence moderating the climates of Great Britain and western Scandinavia (Fig. 2.11).

Ocean temperatures also vary significantly with depth, as indicated by plotting the zonally and annually averaged temperature structure as a function of depth (Fig. 2.12). The averaged temperatures at the lower levels of the ocean are nearly uniform, with values generally between 0.5 and 1.25°C. The coldest bottom temperatures of just below $-0.25°C$ occur at 60–70°S, near the Antarctic continent. Much of the cold bottom water

(a) MEAN TEMPERATURE (°C) FOR AUGUST, SEPTEMBER, AND OCTOBER

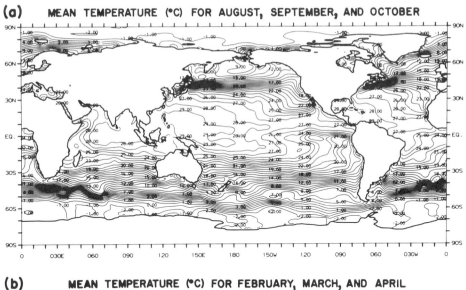

(b) MEAN TEMPERATURE (°C) FOR FEBRUARY, MARCH, AND APRIL

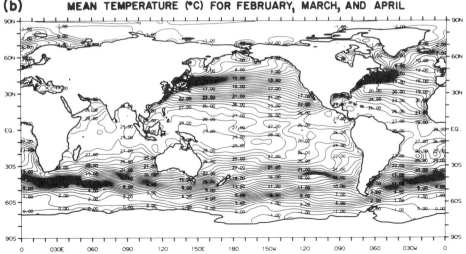

Fig. 2.11 Seasonally averaged sea surface temperatures for (a) August, September, and October and for (b) February, March, and April. [From Levitus (1982), with relabeling.]

throughout the oceans derives from surface cooling and sinking of water in the polar regions of both hemispheres, particularly near the open-ocean and shoreline edges of sea ice. This results because as freezing occurs, much of the salt content of the water does not freeze but is expelled as brine to mix with the water below. The resulting cold, saline water is more dense than the surrounding water, so that it sinks to a depth where the buoyancy

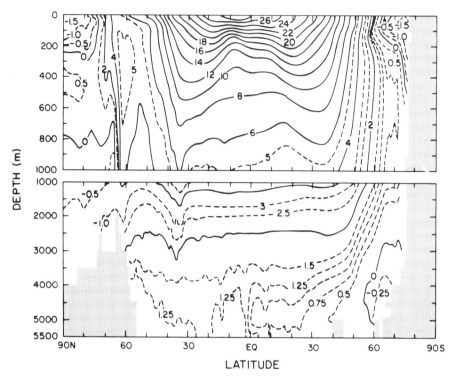

Fig. 2.12 Annually averaged ocean potential temperature (in °C) as a function of latitude and depth. [Redrawn from Levitus (1982).]

matches that of the water around it. On occasion this implies sinking to the ocean bottom. In the tropical regions, surface water temperatures often exceed 25°C, and the temperature profile with depth exhibits a much stronger gradient than in polar regions. Typical tropical ocean temperatures decrease to 8–10°C by a depth of 500 m.

Figure 2.13 depicts schematically the seasonal variation of the ocean's vertical temperature profile in Northern Hemisphere mid-latitudes. In summer, strong warming and vertical mixing caused by wind stirring result in a nearly isothermal or mixed layer, about 20–30 m thick, which is considerably warmer than the water underneath. This upper mixed layer is underlain by a *thermocline*, characterized by a sharp decrease in temperature with depth. In winter, the upper isothermal layer deepens to 60–100 m because of surface cooling and stronger wind stirring than in summer. Because the significant temperature gradient apparent at 20–80 m depth in summer ceases to exist in winter, the summertime temperature gradient is labeled a *seasonal*

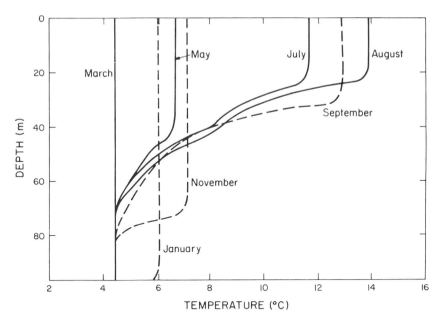

Fig. 2.13 Standard ocean temperature profiles in the Northern Hemisphere mid-latitudes in January, March, May, July, August, September, and November. [After Knauss (1978).]

thermocline. Much deeper in the ocean, extending from about 200 m to 800–1000 m, is a *permanent thermocline,* which does not have considerable seasonal variation. In the polar regions where sea ice is being formed each winter season, the temperature structure at times becomes isothermal throughout the entire depth of the water column, with the mixed layer extending to the bottom of the ocean.

The contrast between the summer and winter temperature profiles (Fig. 2.13) also reflects the net heat absorption in the surface waters during summer and its release in winter. The intervening oceanic heat storage and the basic thermal inertia of the ocean cause a significant lag in the ocean temperature cycle behind the cycle of solar insolation. Maximum mid-latitude surface water temperatures occur in August and minimum surface temperatures in March, about 2 months behind the maximum and minimum solar inputs and about 1/2 to 1 month behind the maximum and minimum atmospheric temperatures.

Overall, the world ocean consists primarily of a stable fluid in which the warmer, less dense portion is near the surface over the bulk of the mid-latitude to tropical oceans, especially in summertime (Fig. 2.13). This limits the thermal coupling between the

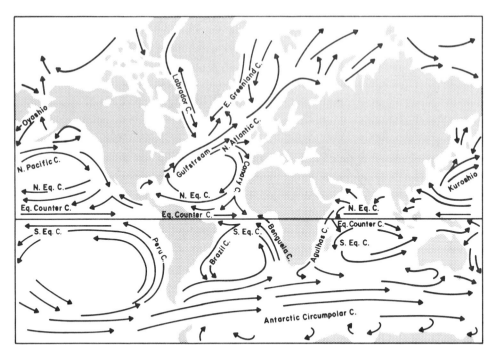

Fig. 2.14 Schematic of the earth's major surface ocean currents. [From Knauss (1978).]

surface waters and the lower levels in the low and mid-latitudes. A much greater amount of vertical convection occurs near the poles. This is quite different from the case in the atmosphere, where strong surface heating, especially in summer, causes significant vertical convection at all latitudes, and in particular causes the large-scale occurrence of rising air in the equatorward branch of the Hadley cell.

The density of ocean water, generally in the vicinity of 1.025 g cm^{-3}, is determined in large part by the temperature and salinity. Even though the density differences are small, it is these differences that in turn determine pressure differences and thereby drive the ocean circulation. Ocean models therefore require accurate density calculations, necessitating in turn realistic temperatures and salinities.

Ocean circulation

Figure 2.14 shows the primary surface currents of the world's oceans. On the very large scale, there tends to be a clockwise circulation in the North Pacific and a counterclockwise circulation in the South Pacific. The Pacific equatorial countercurrent flows from west to east near the equator, while the north equatorial and south equatorial currents, associated with the trade winds, flow from east to west. Major named currents along the continents are the north-flowing Kuroshio and south-flowing Oyashio currents along the east coast of Asia, the north-flowing Peru current along the west coast of South America, and the Antarctic Circumpolar current. The north and south equatorial currents are basically wind-driven currents driven by the atmospheric trade winds. The North Pacific, Kuroshio, and Antarctic Circumpolar currents are also largely wind-driven, resulting from the surface westerly wind patterns in mid-latitudes.

In the Atlantic the current systems are similar to those in the Pacific, with the westerly equatorial countercurrent, the easterly north and south equatorial currents, and a major clockwise pattern in the North Atlantic and counterclockwise pattern in the South Atlantic. Major named currents are the Gulf Stream moving northward along the east coast of North America and northeastward across the Atlantic, the North Atlantic current extending northeastward from the Gulf Stream, the Canary current moving southward off the southern west coast of Europe and northern west coast of Africa, and the Brazil and Benguela currents in the South Atlantic, moving southward along the east coast of South America and northward along the west coast of Africa, respectively. The North Atlantic has some water mass exchange with the Arctic Ocean to the north, as the North Atlantic current transports North Atlantic surface water into the Arctic, both to the west and east of Spitsbergen, and the southward-flowing East Greenland current transports surface water (and sea ice) from the Arctic to the North Atlantic through the passage between Spitsbergen and Greenland. Similarly, to the west of Greenland the West Greenland current brings relatively warm North Atlantic water into Davis Strait and Baffin Bay, while the Labrador current brings colder water and ice from those regions back to the North Atlantic (Fig. 2.14).

The basic current pattern in the Indian Ocean differs somewhat from those in the Pacific and Atlantic oceans because of the small ocean area north of the equator. Hence, while the Indian Ocean contains the equatorial countercurrent, the north and south equatorial currents, and a counterclockwise circulation to the south of the equator, it does not contain a major clockwise circulation to the north of the equator as do the Atlantic and Pacific oceans (Fig. 2.14). One of the interesting seasonal events in the Indian Ocean current system is the reversal of the currents off the east coast of Africa near Somalia. During winter the northeast trades from India drive the waters toward the south, whereas when the summer Indian monsoon sets in, the southwest trades in the Arabian Sea drive the waters toward the north. This total reversal dramatically illustrates the seasonal response of current systems to major wind patterns. The forces driving these currents and the mechanisms maintaining them will be discussed in Chapter 5 in conjunction with discussions of computer simulations of the ocean circulation.

During the past several decades it has been discovered that there are important local swirling circulation systems in the ocean with horizontal scales on the order of 100 to 300 km. These *ocean vortices* or *meso-eddies* are superimposed upon the large-scale general ocean circulation system. Figure 2.15 shows in schematic form a typical instantaneous surface flow pattern for the North Atlantic. The Gulf Stream is a constantly changing, meandering stream that often forms large loops that occasionally cut off from the main current. From the direction of flow in the Gulf Stream it is clear that the eddies or rings that form from cut-off meanders to the north of the Gulf Stream will rotate in a clockwise (anticyclonic) direction, whereas those that form from cut-off meanders to the south of the Gulf Stream will rotate in a counterclockwise (cyclonic) direction (Fig. 2.15). In addition to these cutoff rings there are other mesoscale eddies, also illustrated in Fig. 2.15, which sometimes survive for months to years and can transmit kinetic energy and other properties of the water over great distances. During the past decade studies have shown that mesoscale eddies cover much of the surface area of the oceans and contain perhaps 99% of the total kinetic energy in the world ocean current system. Thus the simplified picture of the ocean with rather smooth circulation patterns obscures some important complications in the circulation system. The full role of the mesoscale eddies in the circulations of the oceans is not yet fully understood, but it is clear that these eddies are capable of transporting momentum, heat, and salt throughout the upper levels of the oceans, both horizontally and vertically. Such mesoscale circulation systems appear to have close parallels with the much larger low and high pressure systems seen on daily weather maps. Both the pressure systems and the ocean eddies seem to play an important although not entirely understood role in transporting properties within the atmosphere and oceans, respectively.

Measurement of water velocities, especially low velocities, is difficult in the oceans and does not have the advantage of large observational networks such as have long been in place for

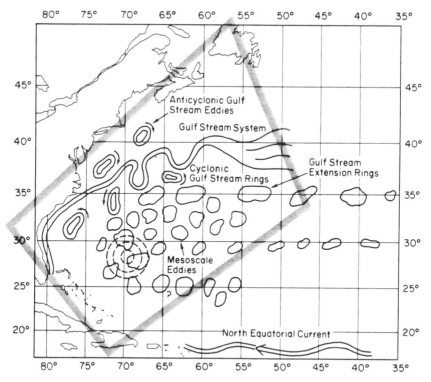

Fig. 2.15 Schematic of a typical instantaneous surface flow pattern in the North Atlantic. The corresponding typical variation of the zonal averaged wind stress with latitude would indicate a general westerly direction of winds in mid-latitudes and a general easterly direction of winds in the tropics and high latitudes. [From Semtner and Mintz (1977).]

atmospheric observations. Another complicating reality in analyzing ocean data is that most of the ocean measurements are not continuous in time, making it difficult to construct synoptic maps of the ocean circulation analogous to the routinely generated synoptic atmospheric weather maps. Perhaps such synoptic maps will be possible in the future, at least for the surface ocean features, when a more complete array of satellite-based observational systems is in place.

Because of the paucity of oceanic data, the amount of daily or seasonal variability in the ocean circulation is far more difficult to determine than the variability in the atmosphere. However, it is known that the subsurface ocean circulations generally have rising (or upwelling) motion near the equator and sinking motion in the subtropics, creating a Hadley-type circulation as in the atmosphere. In the mid-latitudes there is a Ferrel-type cell primarily driven by the mean and eddy motions. A second direct cell (second to the tropical Hadley circulation) occurs in high latitudes, driven in part by the sinking of the cool, dense waters in its poleward portion. The sinking motion is complemented by a rising motion near 60°N. The details of this circulation system are not known, but measurements have suggested the possibility of such a three-cell overall flow pattern (Bryan and Lewis, 1979; Peixóto and Oort, 1984).

Important source regions for the deep (abyssal) waters appear to be in the high latitudes of the North Atlantic and in the vicinity of Antarctica. Downwelling of upper layer waters in these high latitude regions is induced by surface cooling and salt rejection during the formation and growth of sea ice. The downwelled waters move equatorward largely in western boundary currents but eventually spread to cover much of the global ocean basin. The waters are believed to reemerge at the surface largely by slow upwelling through the main thermocline.

Tracers, both natural and man-made, provide one method of determining ocean circulation patterns observationally. For example, since waters of a given temperature and salinity structure often do not mix very rapidly with other waters, it often becomes possible to trace the trajectory of a water mass along isopycnic (constant density) surfaces. This type of water mass analysis allows a determination of ocean circulation without any direct measurement of the current structure. Further knowledge of the circulation can be obtained from the flow and distribution of certain nonnatural radioactive materials such as tritium, which entered the ocean in the early 1950s as a result of nuclear bomb explosions. Analysis of the flow patterns from the tritium distributions is aided because the only known source of tritium in the oceans is man-made and the times and locations of its injection into the oceans are approximately known. The circulation patterns surmised from the tritium distributions indeed confirm such predicted qualitative aspects as the formation of cold dense water in the polar regions, the rapid convective overturning it induces, the concomitant vertical mixing of the surface water with intermediate and deeper waters, and the slow horizontal spreading of the cold bottom waters toward the equator. The patterns of flow revealed by the tritium distributions also give some indication of the typical time scales for which individual particles of water will reside and circulate in the subsurface waters before upwelling to the surface levels. These time scales range from decades to hundreds of years. More will be mentioned on the ocean circulation when model simulations are discussed in Chapter 5.

Sea Ice

Sea ice covers roughly 7% of the earth's oceans and forms a partial but at times very effective barrier between those oceans and the atmosphere. It exists as a result of the cooling of polar waters below their freezing point and varies from fine ice particles suspended in water to expansive stretches of ice 3 m or more in thickness. Although often very compact, the ice cover is never complete, because local heat in the ocean and differential motion and cracking of the ice create open water areas. Frequent deformation of the ice pack tends to break uniform expanses of ice into irregular shapes called *floes* (Fig. 2.16).

In marked contrast to the distribution of the atmosphere and the oceans, the distribution of sea ice undergoes large variations both seasonally and interannually, thereby introducing additional complications to climate modeling studies. The considerable impacts that the ice and its variable distribution have on the climate system will be identified and discussed in the section on Atmosphere/Ocean/Ice Interconnections.

Global sea ice distributions

The areal extent of floating ice in the Southern Hemisphere typically varies within the year from a minimum of 4×10^6 km^2 to a maximum of 20×10^6 km^2, a variation from roughly 1.6% to roughly 8% of the Southern Hemisphere, or from 2.5% to 13% of the Southern Hemisphere oceans. Sea ice extent in the Northern Hemisphere varies far less, ranging from a summer minimum of 8×10^6 km^2 to a winter maximum of 15×10^6 km^2, the latter being roughly 10% of the ocean area in the Northern Hemisphere.

The Arctic Ocean itself, ignoring the peripheral seas, remains covered with ice during most of the year (Fig. 2.17). The central 70% of the ocean, covering roughly 5.2×10^6 km^2, consists of a fairly compact ice mass called the polar cap. At any particular time, this cap area tends to have regions of extensive ridging, an irregular placement of linear openings or cracks called *leads*, and occasionally large nonlinear areas of open water called *polynyas*. In addition to the cap ice, fast ice grounded to the shore horizontally or to the ocean floor vertically also exists in the Arctic and may extend outward from the coast for a distance of up to 500 km in the region of the Siberian continental shelf. Sea ice moves into the North Atlantic by way of the East Greenland and Labrador currents and, to a much lesser extent, into the North Pacific through the Bering Strait. It also forms in situ in the

38

(a)

(b)

Fig. 2.16 (a) Arctic pack ice in the Beaufort Sea, August 1975, from an altitude of 12,000 m. [Photo by C. Parkinson from the NASA CV-990 aircraft.] (b) Sparsely dispersed sea ice floes at the ice edge in the Bering Sea, March 1981. [Photo by C. Parkinson from the NOAA ship *Surveyor*.]

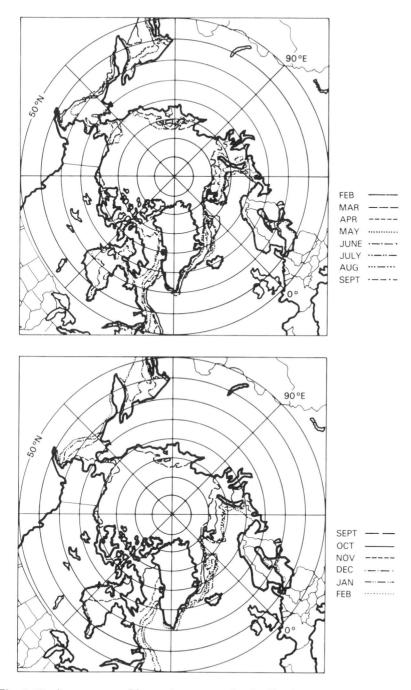

Fig. 2.17 Average monthly sea ice extents in the Northern Hemisphere. These are averaged, by month, for the four years 1973–1976 from passive microwave data from NASA's Nimbus 5 satellite. [From Parkinson et al. (1987).]

North Pacific and surrounding waters, although this is generally restricted to the Bering Sea, the Sea of Okhotsk, and the Sea of Japan. Ice rarely is sighted in the Pacific outside the period from November to June. Ice also forms in Baffin Bay, Davis Strait, and Hudson Bay, almost completely covering both bays for much of the winter season.

The distribution of ice in the Antarctic is more symmetric and less compact than that in the Arctic (Fig. 2.18 versus Fig. 2.17). Antarctic sea ice extends to about 60°S in winter and about 66°S in summer, with the ice generally extending farthest equatorward north of the Weddell Sea at longitudes approximately 30°E–40°W. The lessened compactness of Antarctic versus Arctic ice—i.e., the larger percentage of open water within the boundaries of the ice edge—is a consequence of at least three major factors: the lower latitudes of Antarctic ice, the contribution to divergent motion from predominant winds, and the nonenclosed nature of the ice cover. Interannually, the total areal extent of Antarctic ice at its winter maximum varies from about 17×10^6 km^2 to about 21×10^6 km^2, at least for the period of the 1970s and early 1980s.

Ice thicknesses in the central Arctic average about 3–4 m, while those in the surrounding seas and in the Antarctic are more commonly 0.5–1 m. The sea ice floes are distinct from and far smaller than the occasional icebergs found in the polar waters of both hemispheres. Icebergs calve from land glaciers, are composed of fresh rather than salt water, and can be over 600 m thick. Sea ice, by contrast, forms within the sea, and even with extensive ridging only very rarely has thicknesses exceeding 20 m.

Following is a description of the processes of sea ice formation and ablation and of various properties, including the general topography of the sea ice cover. Some of the complexities mentioned are commonly ignored in computer models, especially those constructed for large-scale studies, but the discussion is included in order to provide the reader with a basis for better understanding the sea ice component of the climate system.

Sea ice formation and growth

Basically, sea ice forms because water reaches its freezing temperature (or freezing point) and solidifies from its liquid state. Even this, however, is not entirely straightforward, because the freezing point is a function of salinity, and it is not uncommon to have supercooling, a condition in which water remains liquid

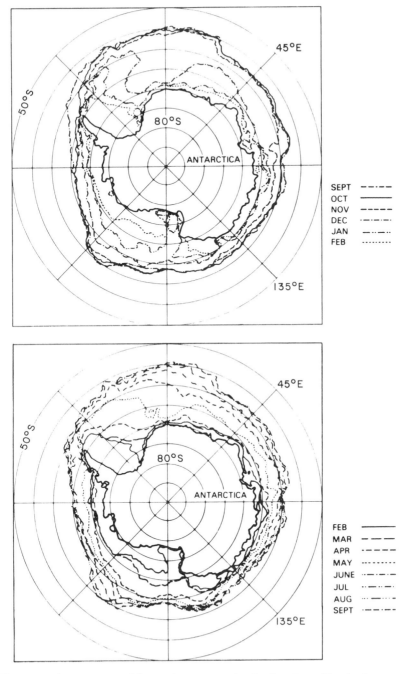

Fig. 2.18 Average monthly sea ice extents in the Southern Hemisphere. These are averaged, by month, for the four years 1973–1976 from passive microwave data from NASA's Nimbus 5 satellite. [Modified from Zwally et al. (1983).]

although at a temperature below normal freezing. The freezing point of seawater decreases with an increase in salinity, so that low surface salinities encourage and high surface salinities delay ice formation. Absence of nucleation centers about which ice can form also discourages ice formation, with the result that under calm conditions with few or no nucleation centers, a small amount of supercooling (on the order of 0.1 K) is common in both Arctic and Antarctic waters. However, once an ice crystal is formed or introduced into supercooled water, the freezing for the entire supercooled area tends to be rapid, as the water needs no further cooling and each new ice crystal serves as an additional nucleation center.

Salinity further complicates the situation beyond its immediate impact on the freezing point. As the salinity rises, the temperature of density maximum for the water decreases at a faster rate than the decrease in the freezing point, so that for water with a salinity greater than 24.7‰, the temperature of density maximum is below the freezing point, while for water with a salinity less than 24.7‰, the freezing point is below the temperature of density maximum. Hence if the salinity exceeds 24.7‰ in an initially uniform water column, then as the surface water is cooled by the atmosphere it continues to become denser until it freezes. This implies that the cooling water should convect down, for its lowered surface temperature yields densities greater than those of the waters beneath. As a consequence, in contrast to the situation with pure water or low-salinity water, in this case the entire depth of vertically uniform water tends to be cooled to or near the freezing point before freezing commences at the surface. This further reveals convective depth as an important factor for ice formation, since greater convective depths necessitate the cooling of a larger volume of water and therefore delay ice formation. If the ocean mixed layer extends to the ocean floor and the salinity exceeds 24.7‰, initial ice formation can be delayed until the entire water column cools. When this occurs, subsequent freezing proceeds rapidly since no warm water remains underneath to hinder it.

Ocean mixed layers generally do not extend to the ocean floor, however, but instead are more commonly underlain by a pycnocline with salinities and densities increasing with depth. Stable density structures in the Arctic and much of the Antarctic limit the actual amount of convective mixing required before freezing commences at the surface. The Antarctic shelf waters are somewhat anomalous in that they often are well mixed to the shelf bottom, so that in these regions cooling does tend to convect down to the bottom before freezing begins (Untersteiner, 1975).

As seawater begins to freeze, individual ice crystals, spicules, and platelets form first. As they coalesce, a portion of the initial brine escapes to the water beneath, with the remainder trapped in cavities formed during coalescence. Thus brine within the ice tends to be concentrated in entrapped brine pockets. The structure of the newly formed ice and the amount of the entrapped brine are notably affected by the state of the sea surface and atmospheric conditions.

As the ice grows, the rate of additional freezing tends to decrease with increasing ice thickness, all other factors being equal, because the thicker ice tends to have a lesser upward conductive heat flux. This will become apparent in the modeling equations of Chapter 3, as the heat flux is inversely dependent on ice thickness. This thickness dependence is crucial to the role of sea ice in climate (and in climate simulations involving atmosphere, oceans, and ice) because the conductive flux is the major mechanism by which heat passes between ocean and atmosphere in the presence of an ice cover.

Sea ice ablation

Langleben (1972) describes a typical summer scenario of melt at the edge of the snow-covered Arctic ice pack. As the summer solstice approaches and the days lengthen, the increased insolation causes rapid snow melt, which decreases the surface albedo and thereby allows absorption of a greater percentage of shortwave radiation. With the decreased or vanished snow cover the ice starts ablating, a typical rate of ablation being roughly 0.04 m per day. Drainage canals, vertical melt holes, and scattered melt ponds begin to develop. After melting reduces the ice thickness to approximately 1–1.5 m, erosion has become intense enough to crack the ice along many of the drainage-canal flaws, the cracking being aided by the action of tides and storms. The lessened continuity of the ice cover then accelerates the decay by the increased thermal interaction with the water. In most of the peripheral seas and bays and in many of the inlets and fiords along the Arctic coastline the disintegration is complete, with the ice cover vanishing at some point during the summer and then not reforming until the fall freeze.

During ice ablation the salt crystals and immediately surrounding ice melt first. This increases the volume of the brine cavities, making the ice more porous and frequently allowing some of the brine to drop to the water underneath. This partially explains the altered properties, in particular the lowered salinity, of older ice, parts of which often have melted and then refrozen several times.

Sea ice composition and properties

As suggested above, the structure and composition of sea ice depend on many factors, including the salinity of the water from which it formed, the speed of freezing, the amount of any melting subsequent to freezing, the temperature, and the age of the ice. The more quickly freezing occurs, the more brine is likely to be trapped and the larger the expected size of the brine cavities. Thus for a given water salinity, the sea ice salinity tends to increase with a decrease in the temperature of ice formation, the lower temperature encouraging faster freezing. A reasonable bulk value to use in numerical models for the salinity of sea ice is 4‰, although new, rapidly formed ice might have salinities as high as 20‰.

The salinity of an individual ice floe is neither uniform through the floe nor constant over time. It is common for salt to settle through the ice—doing so sometimes through gravitational sinking, sometimes through meltwater flushing, and sometimes through thermal expulsion—and hence for salinities to increase with depth and to decrease as the ice floe ages. Weeks (1976) estimates salinities of 12–15‰ for newly formed ice, salinities of 4–5‰ for ice after a year's growth, and salinities increasing from near 0‰ at the top of multiyear ice (ice which has survived a summer melt season, during which meltwater and its salt content percolate downward through the ice) to 2–3‰ at the bottom of the ice. However, ice in the Antarctic, particularly in the Weddell Sea area, does not appear to have surface melt to the extent that it occurs in the Arctic, so that much of the older ice in the Antarctic seems to retain higher salinities than the multiyear ice of the Arctic.

The most important fact regarding sea ice density is that it is always lower than the density of the surrounding water, allowing the ice to float rather than sink. Typical values range from 880 to 910 kg m^{-3}, with the differences depending primarily on the air and salt content. Ice density tends to increase with an increase in salinity and to decrease with an increase in porosity.

Other sea ice properties depend largely on salinity and temperature. For instance, specific heat increases significantly with both temperature and salinity, and latent heat of fusion decreases with them both. Dielectric properties, optical depths, and emissivities at various electromagnetic wavelengths also vary depending on the temperature and salt content of the ice, but these dependencies are not explicitly included in large-scale sea ice models since bulk coefficients and formulae are adequate for most modeling purposes.

Sea ice topography

A large ice pack reveals a rough topography both on the bottom surface and the top. Thin ice especially is capable of extensive ridging when acted upon by opposing horizontal forces. As large ice floes move toward each other, intervening thin ice and ice rubble mass together and, under sufficient pressure, pile to heights and depths significantly larger than the original floes. Although ridges greater than 4 m in height are rare, occasionally a ridge extends 10 m above sea level and the keel beneath it extends 50 m below sea level. From two years of aerial surveys, Wittmann and Schule (1966) estimate that pressure ice covers 13–18% of the ice area in the Canadian Basin of the Arctic Ocean. Spatially, the greatest amount of ridging in the Arctic is near the coasts, with the maximum just north of Greenland and Ellesmere Island. Observations are far fewer in the Antarctic, but Antarctic ice is believed to have less ridging than Arctic ice, because of the less confined nature of the ice cover.

Although ridges are the most spectacular relief form on the ice, there are also other irregularities throughout a normal ice pack, such as fissures, hills, depressions from melt ponds, and floes rafted one over another. Irregularities tend to lose their angularity with time, and former ridges often become chains of smoothed ice mounds after several years. Among the approximations assumed in many sea ice models, the rough ice topography of the actual world is often eliminated by forcing all the ice in a grid square to have a uniform ice thickness at any given time. This simplification is probably adequate for most large-scale studies examining first-order effects.

Sea ice concentration and velocity

Two of the most important aspects of the sea ice cover simulated in ice models are the ice concentrations (*concentration* being defined as the percentage of an ocean area or grid square which is overlain by ice) and the ice velocities. The ice concentrations have a major impact on the amount of heat transferred from ocean to atmosphere and thereby on the climatic influence of the sea ice cover. The ice velocities are important for their contributions to determining the changing distributions of the ice and for the fact that as the ice moves it transports cold, low-salinity water from one location to another. This latter essentially provides a "negative transport" of heat and salinity.

Sea ice concentrations in much of the central Arctic exceed 85% year-round, with lower concentrations in some of the peripheral seas. The unconfined nature of the Southern Ocean waters contributes to generally lower ice concentrations in the Southern Ocean than in the central Arctic, although half the area of winter ice coverage still generally contains concentrations exceeding 80%. In both hemispheres the ice edge tends to be fairly sharp, with the transition from open water to fairly heavily compacted ice (70% ice concentration or above) frequently occurring within a distance of 100 km (Zwally et al., 1983; Parkinson et al., 1987).

General patterns of ice drift are better known in the Arctic than in the Antarctic. The two major aspects of the Arctic drift pattern are a clockwise gyre in the Beaufort Sea, termed the Beaufort Gyre, and the Transpolar Drift Stream flowing from near the Siberian coast at about 100–140°E, across the pole, and out of the Arctic through the Fram Strait between Greenland and the Norwegian island of Spitsbergen (Fig. 2.19). As the ice and water of the Transpolar Drift Stream proceed out of the Arctic they become the East Greenland Drift, flowing southward along the east coast of Greenland. In the Antarctic, observational studies on sea ice have been predominantly in the Weddell Sea, where the ice drift tends to have a clockwise direction approximately following the coast and flowing equatorward along the Antarctic Peninsula, then curving toward the northeast. As this outflow eventually melts in warmer waters, it effectively transports cold, relatively fresh water equatorward, as does also the outflow from the Arctic.

Atmosphere/Ocean/Ice Interconnections

Although until recently the oceans, atmosphere, and ice generally were considered independently in climate studies, it has long been recognized that there are many interconnections among the three. For instance, important transfers of energy, mass, and momentum occur between the oceans and atmosphere and are greatly modified in the presence of an ice cover. Such phenomena as the monsoons that affect massive human populations in Asia

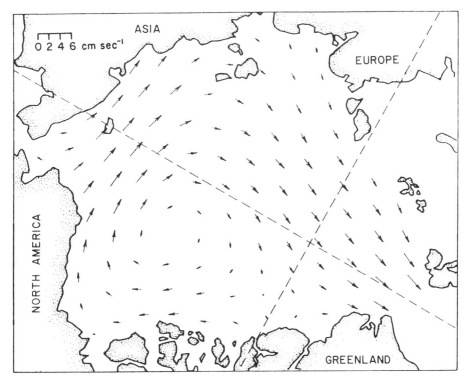

Fig. 2.19 Annual average ice drift vectors in the Arctic Basin. All available data from 1893 to 1983 were incorporated in the averaged field. [From Colony and Thorndike (1984), with labels added.]

and Africa derive in large part from ocean/atmosphere interactions; and the El Niño/Southern Oscillation is an even larger-scale phenomenon also of immediate practical consequence and also dependent upon a close coupling between the ocean and atmosphere. This section reviews some of the major interactions among the ocean, atmosphere, and ice.

Impacts of the atmosphere

The atmosphere significantly affects both the upper layers of the ocean and the sea ice covers. Wind forcing is a prime driver for upper ocean circulation and thereby impacts deeper ocean circulation as well. It is also a prime driver for sea ice, thereby impacting ice motions and ice edge location. In addition, air temperatures and moisture amounts contribute to determining the energy fluxes across the interface between the atmosphere and surface, and thereby contribute to ice maintenance, growth, and melt, and to the temperature distribution of the upper ocean.

One of the most direct impacts of the atmosphere on both the oceans and the ice is the impact of the wind on ice and ocean velocities. This impact has been recognized for centuries on a qualitative level and was put into a mathematical form in the early part of the twentieth century by Vagn Ekman. In examining data from the Norwegian North Polar Expedition of 1893–1896, Fridtjof Nansen noted that sea ice does not drift in the direction of the wind but at an angle of 20° to 40° to the right of the wind. He attributed the directional deviation to the earth's rotation and speculated that the motions in the water beneath the ice deviate even more from the wind direction, the motion of each water layer deviating to the right of the layer above (or in the Southern Hemisphere, to the left). Nansen then encouraged Ekman to examine the dynamics mathematically. Ekman published the results of his ensuing investigation in a paper entitled "On the Influence of the Earth's Rotation on Ocean Currents"(Ekman, 1905). Ekman not only shows a theoretical spiraling effect, with waters moving slower and farther to the right with depth, but also shows that the wind-driven current should extend only to a depth of about 200 m. In honor of Ekman's work, the theoretical spiral is generally termed the Ekman spiral. The spiraling is leftward rather than rightward in the Southern Hemisphere.

The impact of the winds on ocean circulation has been illustrated above in the discussion of the seasonal reversal of ocean currents off the east coast of Africa. Like many of the seasonal and interannual contrasts in the surface waters of the ocean, these contrasts are basically wind-driven.

The atmospheric impact on the sea ice edge of the Southern Ocean is illustrated in Fig. 2.20 for a 29-day period during the austral autumn. The figure displays the mean sea level atmospheric pressure field for the 29 days plus the ice edge position at the start, middle, and end of the period. Approximate wind directions can be inferred from the pressure field, with winds circulating in a clockwise direction around the low pressure centers due to the direction of the Coriolis deflection in the Southern Hemisphere. The most rapid ice advance occurs in the Weddell Sea immediately to the west of the most intense low pressure system, where there should be the strongest wind-driven equatorward ice advection and a temperature cooling through cold air advection from the south. This is not uncommon for the Weddell Sea. The average low pressure system centered at about 0°E, 70°S results in an equatorward flow of air from the continent

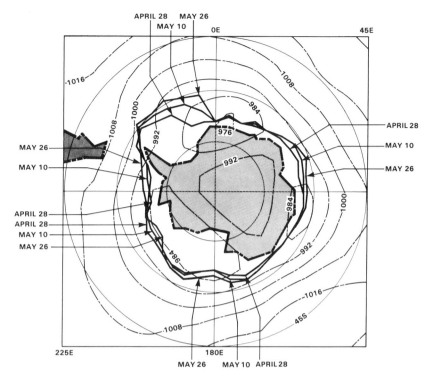

Fig. 2.20 Advance/retreat of the Southern Ocean sea ice edge from
28 April through 26 May 1974, along with the average sea level pressure
field over the 28 April–26 May period. The sea ice edge, outlined by heavy
solid lines, was determined from microwave imagery from NASA's Nimbus 5
satellite. [After Cavalieri and Parkinson (1981).]

over the western portion of the sea. The cold air limits ice melt
thermodynamically, and the winds dynamically drive ice equatorward, therefore advancing the ice edge while also opening leads
in which further ice production occurs. The ice edge depicted in
Fig. 2.20 also moves equatorward to the west of the two other low
pressure systems in the vicinity of the ice edge, while it is more
stationary to the east of the lows, where the pressure patterns
would suggest air flow from lower rather than higher southern
latitudes.

Impacts of the ocean

The ocean also has significant impacts on the other two components of the climate system. It is essentially a limitless moisture
source, and supplies the overwhelming majority of atmospheric

water vapor. The ocean/land contrast thus exerts a major control over the distribution of evaporation and thereby greatly affects the distribution of global precipitation. Furthermore, the large thermal inertia of the ocean results in a much lesser seasonal temperature range at the open ocean surface than at a land surface or a sea ice surface. This creates a lesser atmospheric temperature range over open ocean than over land or sea ice, as cold winter air is warmed by the underlying ocean and warm summer air is cooled. Since the ocean also absorbs a significant amount of the atmosphere's carbon dioxide (CO_2), it can be expected to slow the buildup of atmospheric CO_2 concentration, although the extent to which it will do so is uncertain since the ocean processes that regulate CO_2 concentration are still not completely known.

The oceans contribute to the poleward transport of heat by means of the major ocean currents, influencing the heat balances in both high and low latitudes and the average temperature gradient between equator and poles. The temperature distribution of the ocean surface is the prime determinant of the sea ice distribution, because basically the ice forms where the water temperature has reached the freezing point. Similarly, the ocean salinity distribution is important since the salt content of the water affects the freezing temperature. Once ice is formed, warm currents entering an ice-covered region will tend to melt the ice cover, and cold currents moving away from the ice region will tend to carry ice with them. The impact of ocean currents on the positioning of the sea ice edge is well illustrated in the North Atlantic region, where the cold south-flowing Greenland Current extends the ice edge to the south immediately to the east of Greenland and the warm north-flowing Gulf Stream and North Atlantic Current restrict ice formation in the Norwegian Sea, leaving the ice edge far poleward of the ice edges in other regions (see Fig. 2.17). Similarly, the one region where Antarctic sea ice frequently is seen decidedly north of its roughly symmetric pattern is in the region of the cold, north-flowing Falkland Current to the east of South America.

An example of atmosphere/ocean interconnections:
the El Niño/Southern Oscillation

The El Niño/Southern Oscillation (ENSO), one of the most remarkable large-scale patterns clearly identifiable in meteorological data, will be used to illustrate a closely interconnected situation in which the atmosphere and ocean strongly influence each other. The ENSO, characterized by consistent sea level pressure, temperature, and precipitation changes in the South Pacific, was first discussed by Walker (1924), who found an alternating pressure pattern between the normal southeast Pacific high pressure and the low pressure region near the Indian Ocean and western Pacific regions. Rasmusson and Wallace (1983) have outlined our present knowledge about the essential features of this phenomenon.

There is usually a strong sea surface temperature difference between warm water in the western Pacific and relatively cool water in the eastern Pacific. When the cool water in the eastern Pacific disappears with the cessation of upwelling colder water, El Niño, characterized by warm water, sets in. It occurs with differing intensity in episodes lasting from 2 years to a decade and is called El Niño (or Christ Child) because almost every year there is a minor seasonal warming of the coastal water near Peru at about Christmas time. In the present scientific literature the term El Niño usually refers to a moderate or large warming event.

Under non-El Niño conditions, precipitation normally occurs over the warmer waters of the western Pacific, where there is convection and strong heat input into the atmosphere. These in turn generate a large, directly driven circulation with air rising over the western Pacific and sinking over the eastern Pacific. In contrast to the north-south Hadley circulation pattern discussed above, El Niño/Southern Oscillation reflects fluctuations in a Walker-type circulation characterized by an east-west circulation pattern. ENSO has special significance since it can affect large regions of the tropics and, at times, of the mid-latitudes. Over the region of significant precipitation, the air in the Walker circulation rises, while there is strong subsidence in the downward branch. This produces low atmospheric pressure in the western Pacific and high atmospheric pressure in the eastern Pacific, which in turn drives a strong east-to-west trade wind pattern at the surface. During an ENSO event the sea surface temperatures rise in the eastern Pacific and cool in the central or western Pacific, so that the rising air and the heavy precipitation usually found in the western Pacific shift eastward and the usually strong trade wind pattern is disrupted. Depending on the extent of the eastward shift, various other dramatic effects can occur. For instance, if the shift is approximately to the International Date Line then there tend to be drought patterns in the Indonesia-northern Australia regions. The trades weaken or reverse in the western Pacific, which can cause the water that has been piled up in the west due to the persistent trade winds to start flowing to

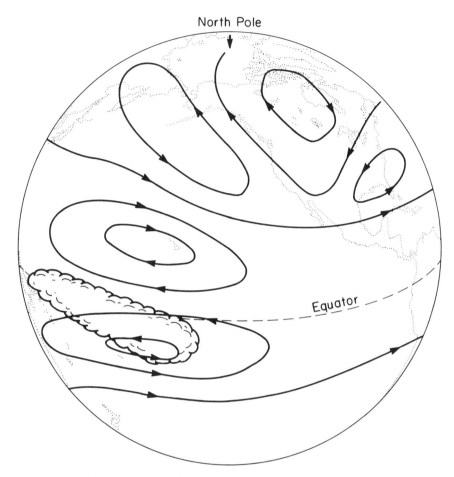

Fig. 2.21 Schematic anomaly pattern of the upper tropospheric geopotential height for a typical ENSO episode. Also indicated, by a cloud outline, is a region of enhanced precipitation in the central equatorial Pacific. [From Rasmusson and Wallace (1983).]

the east. This eastward-propagating movement or wave is called a Kelvin wave, named after Lord Kelvin in recognition of his early studies of wave dynamics. (Gill [1982] provides a thorough description of this phenomenon and a theoretical discussion of its mechanics.) A Kelvin wave can travel across the Pacific to the west coast of South America in several months and help cause a cessation of the upwelling, which in turn can cause a major manifestation of the El Niño event along the west coast of South America.

Figure 2.21 shows a schematic of an anomaly pattern considered typical of an ENSO event. Note the region of clouds in the central Pacific, suggesting enhanced precipitation, the increased strength of the subtropical jet streams, and the anomaly pattern over the United States, with more northerly winds over the eastern U.S. and more southerly winds over the western U.S., suggesting cooler than normal temperatures in the eastern U.S. and warmer than normal temperatures in the western U.S. Note also the pattern of alternating cells, which varies from one ENSO event to another depending upon the location of the anomalously warm ocean temperatures. What initiates the major ENSO events is still not known in full, but once begun there is an impressive consistency in their subsequent development, at least in the tropical regions. The ENSO demonstrates a very large-scale ocean/atmosphere coupling, with pronounced impacts of the atmosphere on the tropical and mid-latitude Pacific waters and vice versa.

Impacts of the ice

Just as the atmosphere and the oceans influence the sea ice and each other, so the presence of sea ice has numerous climatic consequences, influencing the temperature and circulation patterns of both the atmosphere and the oceans. Sea ice lessens the amount of solar radiation absorbed at the ocean's surface; it serves as a strong insulator, restricting exchanges of heat, mass, momentum, and chemical constituents between ocean and atmosphere; and its formation often results in a deepening of the oceanic mixed layer due to the salt rejection that occurs as the water freezes. Sometimes this mixed-layer deepening further leads to bottom water formation, thereby influencing the deep circulation of the entire world ocean. For reviews and details on some of the impacts of the ice, the reader is referred to Ackley (1981), Ackley and Keliher (1976), Allison (1981), Barry (1983), Budd (1975), Fletcher (1969), Gordon (1978, 1981), Killworth (1983), Parkinson and Cavalieri (1982), Parkinson and Gratz (1983), Walsh and Johnson (1979), and Walsh and Sater (1981).

One of the most prominent effects of the ice is the reduced absorption of solar radiation at the ocean's surface: With an ice cover, only 30–50% of the incident solar radiation is absorbed, the rest being reflected, while without the ice 85–95% is absorbed. (The percent of incident solar radiation reflected is referred to as the shortwave *albedo* of the surface. It is the high shortwave albedo of ice, 50–70%, relative to that of open water, 5–15%, which produces the lessened absorption.) Although part of the reflected radiation does help warm the atmosphere, the largest portion passes through to space and is therefore lost to the earth/atmosphere system. Hence the ice reduces the availability of absorbed shortwave radiative energy to the polar regions.

Another prominent effect of the ice is thermal insulation. The presence of sea ice significantly reduces heat transfer between ocean and atmosphere, the slow process of molecular conduction being the primary transfer mechanism for heat through the ice cover. In the Arctic winter, air temperatures vary roughly from $-20°C$ to $-40°C$, while water temperatures vary from $0°C$ to $-2°C$. This marked temperature contrast can lead to heat transfers of 10^2 to 10^3 W m^{-2} from the ocean to the atmosphere in locations where the two come in direct contact. Where direct contact is prevented by one or more meters of ice, however, the heat transfer is only about 10 to 20 W m^{-2}. The ice cover thereby strengthens the net polar atmospheric cooling in winter, which should create larger average temperature gradients from pole to equator and lead to a more intense atmospheric circulation than might otherwise exist. However, as with many of the coupled aspects of the ocean/atmosphere/sea ice system, there are complications: As the stronger circulation from the larger temperature gradients leads to greater advection of warm air into the polar regions, polar atmospheric temperatures are increased and equatorward gradients hence reduced, creating a negative feedback. Thus whether the equilibrium atmospheric circulation (if such exists) is more or less intense in the presence of a polar ice cover remains uncertain.

In summer, the thermal insulating effect of the ice reduces sensible heat transfer from atmosphere to ocean rather than vice versa. This combined with the shortwave albedo effect, whereby ice reduces the absorption of solar radiation, greatly suppresses summer heat gain by the polar oceans, although the area affected is much less than the area of suppressed ocean-to-atmosphere heat transfer in winter, because of the much reduced sea ice cover in summertime.

Another aspect of the insulation provided by sea ice is the lessened evaporative transfer to the atmosphere in the presence of an ice cover and the resultant reduction of moisture available for cloud formation, rain, or snow. Abnormally low ice concentrations in a given year can be expected to increase both heat and moisture fluxes from the ocean to the atmosphere. This is likely to result in increased winter storm activity. Should large regions that currently have winter ice become free of ice, the increased evaporation could reasonably be expected to increase snowfall not only over the immediate ocean area but over all surrounding land areas as well. The possibility that this would lead to enough additional snowfall to initiate large-scale glaciation is the basis of a cyclical theory of Northern Hemisphere glaciations advanced in the mid-1950s (Ewing and Donn, 1956).

On the seasonal scale, the freezing and melting of ice affect climate by releasing energy (during the change of state from liquid to solid) in the winter season and absorbing energy (during the change of state from solid to liquid) in the summer, thus lessening seasonal temperature extremes. The same mechanism lessens regional temperature contrasts as well, because of the net equatorward advection of sea ice in both polar regions: the net ice formation in the central Arctic and near the Antarctic continent releases heat to the polar oceans and atmosphere, while the net ice decay farther equatorward absorbs heat from the subpolar oceans and atmosphere. The result produces what has been termed a *negative heat transport* out of the polar regions.

These seasonal and regional effects on temperature are complemented by similar seasonal and regional effects on water salinity. As water freezes in winter, the discharge of salt to the ocean beneath increases the salinity of the ocean mixed layer, whereas in summer the melting of the ice adds nearly fresh water to the mixed layer, thereby decreasing its salinity. Regionally, since the salt rejection process results in increased salinities in regions of net ice formation and decreased salinities in regions of net ice decay, the overall ice dynamics produces what is termed a *negative salt transport* out of the polar regions just as it produces a negative heat transport.

The increased salinity of the mixed layer resulting from the formation of ice in turn affects the vertical density structure of the ocean. The higher salinity increases the density of the mixed layer and so increases the chances of an unstable density stratification and resulting convection with underlying water (Fig. 2.22). In instances when such convection occurs, the depth, temperature, and salinity of the mixed layer are all further modified. Even more importantly, if the original density stratification is such that the increased salinity at the surface leads to very deep convection or even bottom water formation, then the entire world ocean circulation could be affected in the long run. This is most likely where the amount of ice formation is the greatest, which tends to be at the open-ocean and shoreline edges of the ice pack. Offshore ice motion near a coast in winter increases the ice production rate near the coast, as the ice motion continually generates open water, which then quickly freezes. Although observations are limited, it is believed that a large part of the world's bottom water derives from the polar regions, near the landward and seaward edges of the sea ice cover.

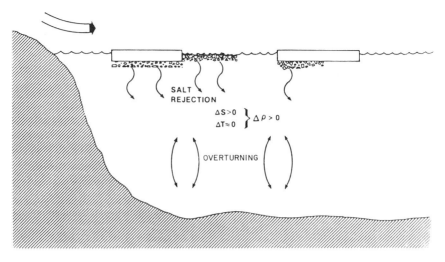

Fig. 2.22 Schematic diagram of the possible inducement of deep oceanic convection by the rejection of salt during sea ice formation. [From Parkinson (1985).]

In view of the varied impacts of atmosphere, oceans, and sea ice on each other, a full numerical treatment of the climate system eventually must entail a model that includes atmosphere, hydrosphere, and cryosphere interactively. However, much can and has been learned with noninteractive models as well as with interactive models, and one of the major issues in the development of a numerical model is the determination of which physical components and processes to include in the calculations and in how much detail. The best selection will depend heavily upon the purposes of the specific model. For instance, short-term atmospheric predictions over mid-continental areas would not benefit from oceanic calculations, but longer-term atmospheric predictions might. Simulations obtainable with both coupled and uncoupled atmosphere, ocean, and ice models are illustrated by a variety of examples in Chapter 5, following two chapters dealing with model details and numerics.

CHAPTER 3

Basic Model Equations

Fundamental Equations

The fundamental equations that govern the motions of the atmosphere, oceans, and sea ice are derived from the basic laws of physics, particularly the conservation laws for momentum, mass, and energy. The first of these comes from Isaac Newton's second law of motion, first explicitly stated in Newton's *Principia Mathematica* in 1687. The second and third, although more widely recognized now by the layman than the conservation of momentum, historically were formulated later, the law of conservation of mass in the late eighteenth century and the law of conservation of energy in the mid-nineteenth century. Albert Einstein in 1905 suggested that mass and energy can be converted from one to the other, replacing the two individual conservation laws by a single law of conservation of mass/energy. However, since conversions between mass and energy are not of relevance to normal atmosphere, ocean, and sea ice behavior, the two individual laws of conservation are used in numerical modeling. The various equations for the conservation of energy require incorporation of both internal and external energy sources, which vary with the climate component being considered.

In addition to the three conservation laws, climate models require an *equation of state* relating several of the parameters in the other equations, plus a moisture equation. For the atmosphere the equation of state relates the pressure, density, and temperature, and for the ocean the equation of state relates the pressure, temperature, density, and salinity. This equation, together with the moisture equation and the equations for the conservation of momentum, the conservation of energy, and the conservation of mass, constitute the basic equations used in modeling the climate system.

Conservation of momentum

The law of conservation of momentum asserts that the net force exerted upon a body is proportional to the change of motion induced. Often expressed

$$F = ma \qquad (3.1)$$

or more accurately

$$\boldsymbol{F} = d(m\boldsymbol{v})/dt \qquad (3.2)$$

where $\boldsymbol{F}$ symbolizes the net force, m mass, a acceleration, $\boldsymbol{v}$ velocity, and $d(\;\;)/dt$ the total derivative, this is considered the fundamental equation of classical mechanics. Applied to the atmosphere, oceans, and sea ice, this equation states that the net force on an imaginary particle or volume of air, water, or ice equals the time rate of change of the product of the mass of the particle or volume and its velocity. Here a particle is something greater than a single atom or molecule and retains its initial mass, so that the $F = ma$ form of Newton's law becomes appropriate and the prose translation can be reworded to: the net force on a particle or volume equals its mass times its acceleration. For convenience we will divide by m to obtain $a = F/m$, and consider forces per unit mass.

Since only large-scale models with three spatial dimensions will be considered in this book, the laws will be expressed in spherical form appropriate to the earth's nonplanar surface. The three primary directions are: longitude λ increasing to the east, latitude ϕ increasing to the north, and radial distance r measured from the center of the earth (Fig. 3.1). The third direction, r, equals the height z above the surface of the earth plus the radius of the earth, a. Furthermore, a is traditionally taken as a constant in meteorology and oceanography, so that $r = a + z$ and $dr = dz$. By convention, latitude is negative in the Southern Hemisphere and positive in the Northern Hemisphere, ranging from $-90°$ at the south pole to $+90°$ at the north pole. Labeling the linear velocities in the three primary directions as u, v, and w, respectively, u is then positive for a wind blowing from the west, v is positive for a wind blowing from the south, and w is positive for

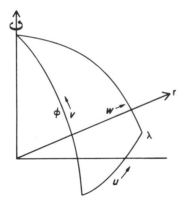

Fig. 3.1 Schematic of a spherical coordinate system.

an upward wind. Recalling from plane geometry that the length of a circular arc is the circle radius times the extended angle in radians, and noting, for instance, that the radius of a latitude circle at latitude ϕ is $a \cos \phi$, we can determine the curvilinear velocities along a latitude circle and along a meridian, and the linear velocity along a local vertical to be

$$u = r \cos \phi \frac{d\lambda}{dt} \qquad (3.3)$$

$$v = r \frac{d\phi}{dt} \qquad (3.4)$$

and

$$w = \frac{dr}{dt} \qquad (3.5)$$

respectively.

To expand Newton's second law, (3.1), for use with atmospheric, oceanic, and sea ice motions, expressions are required for the forces acting on a volume or mass of the atmosphere, ocean, or sea ice. The major such forces are the frictional forces, the gravitational force due to the attraction of the earth's mass, the pressure gradient force due to the pressure differences on the separate sides of the air, water, or sea ice volume, and the forces resulting from the earth's motion. For the time being, the frictional forces per unit mass in the three primary directions will be labeled F_λ, F_ϕ, and F_z. The gravitational force is essentially unidirectional toward the center of the earth, so its components in the λ and ϕ direction are 0 and in the w direction $-g$, the gravitational acceleration. The remaining forces will require more extensive discussion.

Using a spherical coordinate system with three space dimensions (λ, ϕ, z) at right angles to each other, Newton's second law of motion in each of the three directions can be written

$$a_\lambda = \frac{du}{dt} = \sum \text{forces}_\lambda \qquad (3.6)$$

$$a_\phi = \frac{dv}{dt} = \sum \text{forces}_\phi \qquad (3.7)$$

$$a_z = \frac{dw}{dt} = \sum \text{forces}_z \qquad (3.8)$$

where u, v, and w represent velocity components in the λ, ϕ, and z directions, t is time, and the forces in each of the three directions are forces per unit mass. [The subscripted a's in (3.6)–(3.8), representing accelerations, should not be confused with the mean radius of the earth, labeled a.] The total derivatives can be expanded from elementary calculus using the chain rule and recognizing the velocity components u, v, and w to be functions of the three curvilinear spherical space distances λ, ϕ, z, and time t. For instance, $u = u(\lambda, \phi, z, t)$ and

$$du = \frac{\partial u}{\partial t} dt + \frac{\partial u}{\partial \lambda} d\lambda + \frac{\partial u}{\partial \phi} d\phi + \frac{\partial u}{\partial z} dz \qquad (3.9)$$

After dividing by the total time increment dt and inserting the expressions of (3.3)–(3.5) plus $dr = dz$, (3.9) becomes

$$\frac{du}{dt} = \frac{\partial u}{\partial t} + \frac{u}{r \cos \phi} \frac{\partial u}{\partial \lambda} + \frac{v}{r} \frac{\partial u}{\partial \phi} + w \frac{\partial u}{\partial z} \qquad (3.10)$$

Corresponding expressions for dv/dt and dw/dt can be derived similarly.

Newton's equations apply to an inertial or fixed frame of reference. For many applications in which the motions of the stars are considered negligible, this fixed frame might appropriately be the stellar background. In our case, because the motions of interest are relative to the rotating earth, it is more convenient to refer velocities to the earth. A derivation of equations of motion relative to the rotating earth is given in Appendix A. Here it is important simply to note the major forces affecting the motion. These include, for each of the components of the climate system, the centripetal and Coriolis forces due to the rotation of the earth, gravitation, pressure forces, and frictional forces.

Since the basis for most considerations of weather and climate on time scales of a month or less lies largely in the atmosphere,

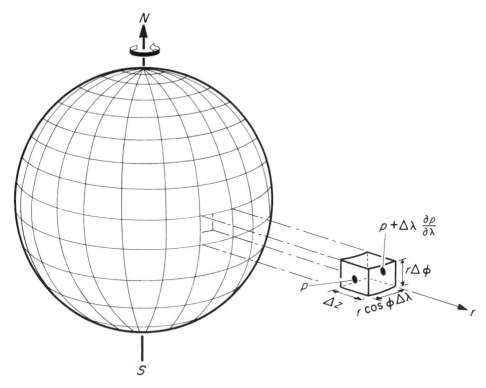

Fig. 3.2 Schematic of pressure forces on an imaginary volume of air.

we consider the atmospheric equations first. For the pressure gradient force, consider a volume of air as depicted in Fig. 3.2. This volume is assumed to be rectilinearly shaped with volume $(r \cos \phi \, \Delta \lambda)(r \, \Delta \phi) \Delta z$. Since pressure p is force per unit area and the area of the left-hand wall is $r \, \Delta \phi \, \Delta z$, the total force acting on the left-hand wall is

$$p \, r \Delta \phi \, \Delta z \tag{3.11}$$

The total force acting on the right-hand wall is

$$\left(p + r \cos \phi \, \Delta \lambda \frac{\partial p}{r \cos \phi \, \partial \lambda} \right) r \Delta \phi \, \Delta z \tag{3.12}$$

Thus the net force in the λ direction is

$$- \left(p + r \cos \phi \, \Delta \lambda \frac{\partial p}{r \cos \phi \, \partial \lambda} \right) r \Delta \phi \, \Delta z + p r \Delta \phi \, \Delta z$$

$$= -r^2 \cos \phi \, \Delta \lambda \, \Delta \phi \, \Delta z \frac{\partial p}{r \cos \phi \, \partial \lambda} \tag{3.13}$$

Dividing by mass to obtain force per unit mass as desired in $a = F/m$, we obtain

$$-\frac{r^2 \cos\phi \, \Delta\lambda \, \Delta\phi \, \Delta z}{m} \frac{\partial p}{r \cos\phi \, \partial\lambda} = -\frac{1}{\rho} \frac{\partial p}{r \cos\phi \, \partial\lambda} \qquad (3.14)$$

as the λ-component of the force caused by the gradient of pressure, or more simply, the pressure gradient force, where density $\rho = m/\text{volume}$ and volume $= r^2 \cos\phi \, \Delta\lambda \, \Delta\phi \, \Delta z$. The other two components of the pressure gradient force can be obtained similarly in the spherical coordinate system as

$$-\frac{1}{\rho} \frac{\partial p}{r \, \partial\phi} \qquad \text{and} \qquad -\frac{1}{\rho} \frac{\partial p}{\partial z} \qquad (3.15)$$

As mentioned earlier, other forces result due to the rotation of the earth. For simplicity, consider an earth whose only motion is its west-to-east rotation about its axis (Fig. 3.1). During one rotation, any object on the earth and stationary with respect to it will travel due to the earth's rotation a linear distance equal to the length of the latitude circle where it is located. Thus such an object at one location has a velocity component which exceeds that of any such object farther poleward and is less than that of any such object farther equatorward. Consider an object that artificially is given a velocity component toward the north or south. If the object is directed poleward, its west-to-east velocity will exceed that of the region it is approaching, so that it will appear to the observer on earth to be deflected to the east (that is, to the right in the Northern Hemisphere and to the left in the Southern Hemisphere). Conversely, if the object is directed equatorward, its west-to-east velocity will be less than that of the region it is approaching, so that it will appear to be deflected to the west. So again the apparent deflection is to the right in the Northern Hemisphere and to the left in the Southern Hemisphere. Therefore, whether an object is directed poleward or equatorward, in the Northern Hemisphere the apparent deflection is to the right when observed from the point of origin along the direction of motion, while in the Southern Hemisphere the apparent deflection is to the left. This provides an intuitive explanation of why the earth's rotation causes a v-component of velocity to impact du/dt (but not dv/dt or dw/dt). A w-component of velocity similarly impacts du/dt (but not dv/dt or dw/dt), since an object directed upward will approach a region with a greater linear eastward velocity and appear deflected to the west, while an object directed downward will approach a region with a lesser linear velocity and hence appear deflected to the east. An object deflected westward or eastward will not impact du/dt as a result of the earth's rotation but will impact dv/dt and dw/dt. Mathematically, the force per unit mass in the λ direction is

$$(2\Omega \sin \phi)v - (2\Omega \cos \phi)w \tag{3.16}$$

in the ϕ direction is

$$-(2\Omega \sin \phi)u \tag{3.17}$$

and in the z direction is

$$(2\Omega \cos \phi)u \tag{3.18}$$

where Ω is the earth's angular velocity. For convenience, definitions are made as follows: $f = 2\Omega \sin \phi$ and $\hat{f} = 2\Omega \cos \phi$, so that the above forces per unit mass are $fv - \hat{f}w$ in the λ direction, $-fu$ in the ϕ direction, and $\hat{f}u$ in the z direction. Note that $\hat{f}$ is everywhere positive, while f is positive in the Northern Hemisphere, negative in the Southern Hemisphere.

The terms involving $\hat{f}$ and f are called apparent force terms and arise from the rotation of the coordinate system. They are referred to as the Coriolis terms, and f as the Coriolis parameter, after the French mathematician G. G. Coriolis, who studied the physics of rotating systems in 1835. The important role of the Coriolis terms in the dynamics of the atmosphere, ocean, sea ice system will become apparent in many sections of this book.

Using the above expressions for the force terms, (3.6)–(3.8) can be expressed as follows for the u, v, and w components of the wind:

$$\frac{du}{dt} - \frac{uv \tan \phi}{r} + \frac{uw}{r} = -\frac{1}{\rho r \cos \phi} \frac{\partial p}{\partial \lambda} + fv - \hat{f}w + F_\lambda \tag{3.19}$$

$$\frac{dv}{dt} + \frac{u^2 \tan \phi}{r} + \frac{vw}{r} = -\frac{1}{\rho r} \frac{\partial p}{\partial \phi} - fu + F_\phi \tag{3.20}$$

$$\frac{dw}{dt} - \frac{u^2 + v^2}{r} = -\frac{1}{\rho} \frac{\partial p}{\partial z} - g + \hat{f}u + F_z \tag{3.21}$$

The additional terms on the left-hand sides of (3.19)–(3.21), beyond the total derivative terms, derive from the centripetal forces on a rotating sphere. Somewhat more rigorous derivations of (3.19)–(3.21) using vector notation can be found in Appendix A, and detailed derivations can be found in Holton (1979), Kasahara (1977), Haltiner and Williams (1980), and Marchuk et al.

(1984). Phillips (1966) and others have shown that (3.19)–(3.21) can be simplified further by energy consistency and scale considerations, and that many of the terms are generally small since the atmosphere and oceans are shallow envelopes on the earth, their 4–20 km depths paling in comparison to the 6371 km radius of the earth.

Recall that a is the earth's mean radius, z is height above the earth's surface, and $r = a + z$. Since $z \ll a$, r generally can be replaced by a as well as dr by dz. This is known as the *shallowness approximation*, because it results from the relatively shallow depth of the atmosphere compared to the earth's radius. Also, since in climate-oriented general circulation models the motions of primary interest have characteristic horizontal space scales of hundreds to thousands of km and velocities on the order of 10 m s^{-1}, we are able to simplify (3.19)–(3.21) by eliminating several of their generally less-significant terms, such as $\hat{f}w$, $\hat{f}u$, and uw/r, vw/r, and $(u^2 + v^2)/r$, as well as by replacing r by a and assuming g is constant. The resulting set of equations is

$$\frac{du}{dt} - \left(f + u\frac{\tan\phi}{a}\right)v = -\frac{1}{a\cos\phi}\frac{1}{\rho}\frac{\partial p}{\partial\lambda} + F_\lambda \qquad (3.22)$$

$$\frac{dv}{dt} + \left(f + u\frac{\tan\phi}{a}\right)u = -\frac{1}{\rho a}\frac{\partial p}{\partial\phi} + F_\phi \qquad (3.23)$$

$$\frac{dw}{dt} = -\frac{1}{\rho}\frac{\partial p}{\partial z} - g + F_z \qquad (3.24)$$

with

$$\frac{d}{dt} = \frac{\partial}{\partial t} + \frac{u}{a\cos\phi}\frac{\partial}{\partial\lambda} + \frac{v}{a}\frac{\partial}{\partial\phi} + w\frac{\partial}{\partial z} \qquad (3.25)$$

and

$$u = a\cos\phi\frac{d\lambda}{dt} \qquad (3.26)$$

$$v = a\frac{d\phi}{dt} \qquad (3.27)$$

$$w = \frac{dz}{dt} \qquad (3.28)$$

where (3.25) and (3.26)–(3.28) are similar to (3.10) and (3.3)–(3.5), although r has been replaced by a and dr has been replaced by dz. Equations (3.22)–(3.28) form the basis for predicting the three components of atmospheric motion or, as described later in this chapter, oceanic motion. The forces that drive the motion are the local gradients (or differences per unit distance) of pressure, gravity, the Coriolis terms, and the frictional terms F_λ, F_ϕ, and F_z.

The principal terms in (3.22)–(3.24) for large-scale atmospheric motions are generally the pressure gradient, Coriolis, and gravity terms. Reducing (3.22)–(3.24) to include only these terms and solving the first two equations for the horizontal wind components, we obtain the following:

$$v = \frac{1}{a \cos \phi} \frac{1}{f\rho} \frac{\partial p}{\partial \lambda} \qquad (3.29)$$

$$u = -\frac{1}{a} \frac{1}{f\rho} \frac{\partial p}{\partial \phi} \qquad (3.30)$$

$$g = -\frac{1}{\rho} \frac{\partial p}{\partial z} \qquad (3.31)$$

Equations (3.29)–(3.30) are termed the *geostrophic wind equations*, the *geostrophic wind* being the hypothetical wind resulting from a perfect balance between the Coriolis and pressure gradient forces. Note that such a wind is indeed defined uniquely by the pressure gradient and Coriolis parameter, assuming density is constant. This balance is often a very good approximation to large-scale motions of the atmosphere, and the corresponding approximation, with u and v referring to water velocities rather than winds, is even better over much of the oceans, where motions are typically slower. Because the Coriolis parameter, which equals 0 at the equator, is in the denominator of (3.29)–(3.30), the geostrophic approximation becomes invalid near the equator.

Figure 3.3 shows in schematic form the relationship between pressure and wind for the case of the geostrophic approximation (3.29)–(3.30). Starting north of the equator with the case of a low pressure system (in which pressure is lowest at the center), $\partial p/\partial \lambda$ is positive to the right of the center of the low and is negative to the left of the center. Hence, from (3.29), v is positive to the right and negative to the left of the center, since f is positive in the Northern Hemisphere. Similarly, $\partial p/\partial \phi$ is positive north of the low center and negative south of the low. From (3.30), u is negative to the north and positive to the south of the low. Note that a negative u wind is a wind blowing from the east and a negative v wind is a wind blowing from the north. The resulting schematic picture of the geostrophic wind flow around a low pressure system in the Northern Hemisphere is depicted in Fig. 3.3A.

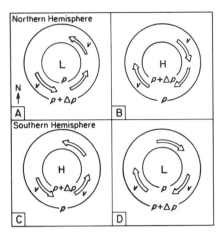

Fig. 3.3 Schematic of the relationship between the geostrophic wind and pressure patterns in the Northern and Southern Hemispheres. L denotes the center of a low pressure system and H denotes the center of a high pressure system.

The reader should go through a similar thought exercise for the corresponding points in the high pressure region to verify the direction of the arrows in Fig. 3.3B. South of the equator, the direction of the circulation reverses because the Coriolis parameter f in (3.29) and (3.30) is negative. Therefore v becomes negative to the right of the low pressure center and positive to the left, and u becomes negative to the south of the center and positive to the north. As a result, the circulation is in the opposite direction from that around a low pressure center to the north of the equator (see Fig. 3.3D versus Fig. 3.3A). A corresponding reversal in wind direction can be seen around a Southern Hemisphere high pressure center (Fig. 3.3C) compared to a Northern Hemisphere high pressure center (Fig. 3.3B). To summarize, in the Northern Hemisphere the geostrophic winds in low pressure regions rotate counterclockwise and in high pressure regions rotate clockwise, while in the Southern Hemisphere the rotation is in the opposite direction. Although observed winds are not exactly geostrophic, the geostrophic approximation is generally good in the large scale ($> 10^3$ km), so that Northern Hemisphere hurricanes (which are intense low pressure systems) always rotate counterclockwise, whereas Southern Hemisphere hurricanes always rotate clockwise.

The geostrophic wind tends to be a reasonably good first approximation to upper atmospheric motions in extra-tropical latitudes. This is evidenced by middle- and high-latitude weather maps of the winds and pressures, with the winds blowing approximately parallel to the isobars (lines of equal pressure). Very low in the atmosphere the geostrophic approximation is not valid, since the frictional terms in (3.19)–(3.20) cannot be ignored as the earth's surface is approached. Also, since f approaches 0 at low latitudes, the geostrophic balance is not applicable near the equator. In fact, in the tropics and for certain small-scale weather phenomena in other latitudes, such as dust devils, tornadoes, and water spouts, the dominating terms in (3.19) and (3.20) tend to be the centripetal acceleration (the second and third terms on the left-hand sides of the equations) and the pressure terms, leading to a balance termed the *cyclostrophic* balance. These systems do not have the strong tendency for a preferred direction of rotation that exists with the systems in which the Coriolis force is a dominant factor.

Returning to (3.31), this equation, generally termed the hydrostatic balance, reflects a balance between the force of gravity and the vertical gradient of pressure. It tends to be well maintained for large-scale atmospheric and oceanic motions. When the balance is temporarily upset, the tendency is for rapid accelerations of air (or water in the case of the ocean) to reestablish the hydrostatic equilibrium. The only frequent situation in which the hydrostatic balance is not approximately maintained is small-scale motion (10 km or less) such as convection, where accelerations can be a sizable fraction of the magnitude of the force of gravity. Rigorous discussions of the conditions when the hydrostatic balance is invalid can be found in numerous advanced theoretical texts such as those by Holton (1979), Haltiner and Williams (1980), and Gill (1982). For the scales of motion treated explicitly in climate models, the hydrostatic balance is generally assumed to be valid.

Returning to the more general equations (3.22)–(3.28), the large-scale motions of the atmosphere (including synoptic weather systems, which are high and low pressure systems) have typical horizontal velocities u and $v \simeq 10$ m s^{-1}, vertical velocity $w \simeq 0.01$ m s^{-1}, horizontal length scales $L \simeq 10^3$ km, vertical scale $H \simeq 10^4$ m, horizontal pressure differences $\Delta p \simeq 1.0$ kPa ($= 10$ mb), and a time scale $t \simeq 10^5$ s (Charney, 1948). [The pressure unit Pa is a Pascal, named after the French scientist Blaise Pascal (1623–1662), who performed experiments on the decrease of atmospheric pressure with height. 100 kPa is equivalent to 1000 mb (millibars), a more commonly used unit.] The Coriolis parameter is $\simeq 10^{-4}$ s^{-1} in mid-latitudes (defined as f_0).

If one forms a ratio of the characteristic scales of the acceleration and the Coriolis terms in mid-latitudes, the following nondimensional number results:

$$R_0 = \frac{u^2/L}{f_0 u}$$ (3.32)

which can be simplified to

$$R_0 = \frac{u}{f_0 L}$$ (3.33)

This nondimensional number was first used by I. A. Kibel (1940) in the Russian literature (see A. S. Monin, 1972), but generally is referred to in Western literature as the Rossby number, after the Swedish meteorologist C. G. Rossby, who made substantial contributions to understanding large-scale atmospheric motions. A typical value of R_0 for large-scale motions can be obtained by substituting the typical values of u, f_0, and L provided above into (3.33). Upon doing so, one obtains a value of 0.1, indicating that for motions of this scale the acceleration terms are about one order of magnitude smaller than the Coriolis terms. The winds approximate a geostrophic balance. However, the small difference from geostrophy is of crucial importance since exact balance implies a steady-state atmosphere. A crucial element in predicting the weather or ocean circulations is to evaluate this small imbalance appropriately for large-scale synoptic motions of the atmosphere or oceans rather than for waves that are more representative of smaller-scale high frequency motions. The latter include acoustical waves, which propagate through a medium by compression and rarefaction, and gravity waves, which propagate due to variations in hydrostatic pressure. The bulk of the energy in the atmosphere and oceans is manifested in the larger-scale lower frequency motions, not in the acoustical or gravity wave motions.

Conservation of mass

The second conservation law explicitly modeled in climate models is the conservation of mass, sometimes referred to as the equation of continuity. Again reference can be made to the imaginary volume of air or water shown schematically in Fig. 3.2, with the restriction that the mass times the velocity (called mass transport or flux) for each of the six sides of the indicated volume must balance in such a way that mass is conserved. The net mass flux difference across the longitudinal faces of the volume is

$$\rho u a \, \Delta\phi \, \Delta z - \left[\rho u + \frac{a \, \Delta\lambda \cos\phi}{a \cos\phi} \frac{\partial}{\partial\lambda}(\rho u)\right] a \, \Delta\phi \, \Delta z \qquad (3.34)$$

which can be reduced to

$$-\frac{1}{a \cos\phi} \frac{\partial}{\partial\lambda}(\rho u) a^2 \cos\phi \, \Delta\lambda \, \Delta\phi \, \Delta z \qquad (3.35)$$

by subtraction. Similarly the net mass fluxes across the other two pairs of opposite faces of the volume are

$$-\frac{1}{a \cos\phi} \frac{\partial}{\partial\phi}(\rho v \cos\phi) a^2 \cos\phi \, \Delta\lambda \, \Delta\phi \, \Delta z \qquad (3.36)$$

and

$$-\frac{\partial}{\partial z}(\rho w) a^2 \cos\phi \, \Delta\lambda \, \Delta\phi \, \Delta z \qquad (3.37)$$

Because the time change of the mass of the volume must equal the total mass flux difference across the faces,

$$\frac{\partial}{\partial t}(\rho a^2 \cos\phi \, \Delta\lambda \, \Delta\phi \, \Delta z) = - \left[\frac{1}{a \cos\phi} \frac{\partial}{\partial\lambda}(\rho u)\right.$$
$$+ \frac{1}{a \cos\phi} \frac{\partial}{\partial\phi}(\rho v \cos\phi)$$
$$\left. + \frac{\partial}{\partial z}(\rho w)\right] a^2 \cos\phi \, \Delta\lambda \, \Delta\phi \, \Delta z \quad (3.38)$$

Since the volume $(a^2 \cos\phi \, \Delta\lambda \, \Delta\phi \, \Delta z)$ is constant by definition, it can be factored out to yield

$$\frac{\partial}{\partial t}(\rho) = -\frac{1}{a \cos\phi} \left[\frac{\partial}{\partial\lambda}(\rho u) + \frac{\partial}{\partial\phi}(\rho v \cos\phi)\right] - \frac{\partial}{\partial z}(\rho w) \quad (3.39)$$

or, by use of (3.25),

$$\frac{d\rho}{dt} + \frac{\rho}{a \cos\phi} \left[\frac{\partial u}{\partial\lambda} + \frac{\partial}{\partial\phi}(v \cos\phi)\right] + \rho \frac{\partial w}{\partial z} = 0 \qquad (3.40)$$

Equation (3.39), or (3.40), is the fundamental equation of the conservation of mass, often termed the equation of mass continuity. Specifically, the net mass entering or leaving a volume results in the corresponding increase or decrease of mass within the volume. In the atmosphere the time derivative term is very important since the air is compressible for large-scale motions; however, water and sea ice are nearly incompressible, so that the total derivative term in (3.40) can generally be ignored for ocean and ice modeling.

First law of thermodynamics

The first law of thermodynamics for a gas is a statement of how the thermal energy of a system is related to the work done by compression or expansion (that is, by changing its volume). In mathematical form it can be written as

$$C_v \frac{dT}{dt} = -p \frac{d}{dt}\left(\frac{1}{\rho}\right) + Q \qquad (3.41)$$

This is a law of conservation of energy, in particular conservation of thermodynamic energy. The term on the left represents the time change of internal energy, where C_v is the specific heat for air at constant volume and T is temperature in kelvins. The first term on the right of (3.41) is work done upon a unit mass of air by compression or expansion of the volume, reflecting the elementary physics rule that if a gas is compressed without the addition or subtraction of external heat then its temperature will increase, and if a gas is expanded it will cool. The reciprocal of density, $1/\rho$, is termed the specific volume. In the ocean the $(d/dt)(1/\rho)$ term is essentially negligible since seawater is almost incompressible; and in sea ice studies the term is usually neglected relative to the term Q. In (3.41), Q is the net heat gain or loss to the system from external sources, such as heating due to solar insolation, heating or cooling due to longwave radiation, latent heating due to condensation of water vapor into liquid water, and sensible heating due to conduction and convection. The system being considered in the case of atmospheric models is a small particle or volume of gas. When there is no external gain or loss of heat, so that $Q = 0$, processes are termed *adiabatic*. For many purposes atmospheric and oceanic motions can be considered essentially adiabatic, especially over short time periods. However, for climate studies, the assumption of exclusively adiabatic processes is not appropriate since the amount of heat added or lost to a unit volume of air or water over a long period of time can be substantial. The term Q is called the nonadiabatic (or diabatic) term.

Equation of state

From the classical physics of J. Charles, J. Gay-Lussac, and R. Boyle, a relationship was developed between the density (ρ), pressure (p), and temperature (T) of a gas. This relationship has been termed the ideal gas law or equation of state. Again because of the incompressible nature of water, the equation of state is somewhat different for a fluid such as seawater, as will be shown below. The equation of state for dry air is

$$p = \rho R T \tag{3.42}$$

where R is the so-called gas constant for dry air. A further refinement to (3.42) can be made by modifying either R or T to take into account moisture. (The reader is referred to the section on precipitation for inclusion of moisture in the equation of state and first law of thermodynamics). By dividing by ρ and differentiating (3.42) with respect to time:

$$p \left(\frac{d(1/\rho)}{dt} \right) = R \frac{dT}{dt} - \frac{1}{\rho} \frac{dp}{dt} \tag{3.43}$$

Substituting (3.43) into (3.41) and setting $C_p = R + C_v$ yields a different form of the first law of thermodynamics:

$$C_p \frac{dT}{dt} - \frac{1}{\rho} \frac{dp}{dt} = Q \tag{3.44}$$

C_p is the specific heat at constant pressure, while C_v used in (3.41) is the specific heat at constant volume. For an adiabatic process, with $Q = 0$, division by $C_p T$, insertion of the equation of state (3.42), and definition of $\kappa = R/C_p$ yield a more convenient form of the first law of thermodynamics:

$$\frac{1}{T} \frac{dT}{dt} - \frac{\kappa}{p} \frac{dp}{dt} = 0 \tag{3.45}$$

The above equation, in conjunction with (3.31), allows the determination of the rate of temperature decrease with height in a hypothetical atmosphere in which all processes are adiabatic and in hydrostatic balance. Specifically, (3.45) and a total differential form of (3.31) can be used to obtain

$$\frac{dT}{dz} = -\frac{\kappa \rho g}{p} T \tag{3.46}$$

which reduces to

$$\frac{dT}{dz} = -\frac{g}{C_p} \qquad (3.47)$$

upon inserting (3.42) and the definition $\kappa = R/C_p$. This yields a temperature decrease with height of 9.8 K/km in the specified hypothetical atmosphere. Termed the *dry adiabatic lapse rate* and symbolized Γ, this value provides a standard against which to compare the actual lapse rate. Lapse rates exceeding this value (meaning the temperature decreases even more strongly with height) signify unstable atmospheric conditions with respect to the hydrostatic assumption, whereas lapse rates considerably smaller than this value signify stable conditions.

By using the standard formula for differentiating natural logarithms, $d(\ln A)/dt = (1/A)\,(dA/dt)$, (3.45) can be integrated to yield

$$\ln\left[\frac{T}{T_0}\left(\frac{p_0}{p}\right)^{\kappa}\right] = C \qquad (3.48)$$

or

$$\frac{d}{dt}\left[T\left(\frac{p_0}{p}\right)^{\kappa}\right] = 0 \qquad (3.49)$$

where p_0 usually is taken to be a standard sea level pressure. The above equation shows how the temperature and pressure are related in an adiabatic process. The bracketed term on the left-hand side of (3.49) is defined as potential temperature, θ; i.e., by definition,

$$\theta \equiv T\left(\frac{p_0}{p}\right)^{\kappa} \qquad (3.50)$$

From (3.49), θ must be constant in an adiabatic process. Another way to consider θ is that it is the temperature a gas would have if it were compressed or expanded adiabatically from a temperature T and pressure p to the standard sea level pressure p_0 of, say, 100 kPa (1000 mb). Because pressure necessarily must decrease with height and because θ is constant, inspection of (3.50) shows that temperature must decrease with height when an air parcel is raised vertically under adiabatic conditions, and conversely, the temperature will increase when an air parcel descends adiabatically. In a simplified sense, this and the fact that air becomes less able to hold moisture as it cools explain why condensation and precipitation occur when air rises. Conversely, when a particle descends it tends to compress, warm, and lessen its relative humidity. More on the role of moisture will be included later under discussion of atmospheric convection and precipitation.

When the processes of radiation and condensation are not negligible, the heating/cooling term, Q, in (3.41) and (3.44) cannot be set to zero, and therefore θ is not conserved. As mentioned earlier, the term Q is called the nonadiabatic term. For short-range forecasting out to a few days, this term is often neglected, although it can be quite important even on short time scales under certain conditions such as heavy precipitation. For climate models, the nonadiabatic term is crucial, and in fact much of climate modeling is devoted to refinements of the calculations involved in determining the component terms of Q. These will be discussed in greater detail later.

Summary of Basic Predictive Equations for the Atmosphere

Here we summarize the fundamental equations discussed above for atmospheric numerical modeling. The horizontal equations of motion (3.22)–(3.23) are predictive equations (meaning that they include a time derivative) for u and v:

$$\frac{du}{dt} - \left(f + u\frac{\tan\phi}{a}\right)v = -\frac{1}{a\cos\phi}\frac{1}{\rho}\frac{\partial p}{\partial\lambda} + F_\lambda \qquad (3.51)$$

$$\frac{dv}{dt} + \left(f + u\frac{\tan\phi}{a}\right)u = -\frac{1}{\rho a}\frac{\partial p}{\partial\phi} + F_\phi \qquad (3.52)$$

These derive from Newton's second law of motion, or the law of conservation of momentum. The hydrostatic equation (3.31) is an approximation relating density to pressure that is derived from the third equation of motion:

$$g = -\frac{1}{\rho}\frac{\partial p}{\partial z} \qquad (3.53)$$

The equation of continuity (3.39) and first law of thermodynamics (3.44) give predictive equations for density and temperature:

$$\frac{\partial\rho}{\partial t} = -\frac{1}{a\cos\phi}\left[\frac{\partial}{\partial\lambda}(\rho u) + \frac{\partial}{\partial\phi}(\rho v\cos\phi)\right] - \frac{\partial}{\partial z}(\rho w) \qquad (3.54)$$

and

$$C_p\frac{dT}{dt} - \frac{1}{\rho}\frac{dp}{dt} = Q \qquad (3.55)$$

These are the laws of conservation of mass and energy, respectively. Finally, the equation of state (3.42), or gas law, relates pressure, density, and temperature:

$$p = \rho RT \tag{3.56}$$

This system of six equations contains six unknowns $(u, v, w, \rho, p,$ and $T)$, making it in principle a solvable system, once the values of the constants and of Q, F_λ, and F_ϕ are given. However, the above form of the equations is not the most convenient for practical solution. Transforming these equations into a practical predictive system will be our next major task. The important point here is that the six variables $u, v, w, \rho, p,$ and T can be related in time and three-dimensional space in a compact system of six equations. An important seventh equation and a moisture variable will be added later (see (3.145)). This moisture equation relates the time rate of change of moisture in the atmosphere to the moisture inputs from the surface below and the change-of-state conversions within the atmosphere.

The total system (3.51)–(3.56) is not closed, because the frictional terms F_λ and F_ϕ and the heating/cooling term Q must be determined from knowledge of the other variables. As mentioned earlier, these terms are very important for climate simulation, although for the purposes of weather prediction out to a few days they often are ignored. Richardson (1922) used the above system of equations for his historic forecast. In the current literature this is called the primitive equation system because it is a return to a more basic set than the one used for the first computer modeling studies by Jule Charney and Norman Phillips in the early 1950s. More will be said later about early models. The word *primitive* is somewhat a misnomer since it does not imply a simple set of equations but rather refers to the full set of basic equations, prior to its simplification by various approximations.

Vertical Coordinate Systems

There are several alternative ways of treating the vertical coordinate in atmospheric models. Because of the hydrostatic nature of large-scale motions, the vertical coordinate can be converted to a pressure coordinate, as was done in early attempts at modeling the atmosphere in the 1940s (Eliassen, 1949; Starr, 1946). This scheme had some merit since it led to a simpler continuity equation than (3.54); however it suffered from the handicap that the treatment of mountains (orography) was somewhat awkward

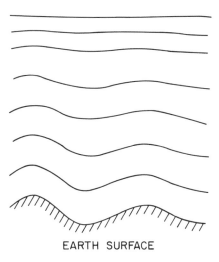

EARTH SURFACE

Fig. 3.4 Schematic of the transformed σ-coordinate system, approximately following mountain heights in the lower atmosphere and standardized pressure in the upper atmosphere. [From Arakawa and Lamb (1977).]

because the vertical coordinate at a given point on a mountain would change with time as the height of the pressure surface changed. More specifically, since a particular constant pressure surface might intersect a mountain at certain times but not others, defining the vertical coordinate as a pressure coordinate gave the vertical coordinate this property also. In early atmospheric computer models this caused troublesome computational problems, especially near the earth's surface (Leith, 1965).

Phillips (1957) devised a transformed coordinate system that avoided the orographical difficulties of the pressure coordinate system. In Phillips's system, termed the sigma-coordinate system, the vertical coordinate, σ, is defined as

$$\sigma = \frac{p}{p_s} \tag{3.57}$$

instead of z or p. In (3.57) p signifies the atmospheric pressure at the point in question and p_s, which is a function of λ, ϕ, and t, signifies the atmospheric pressure at the earth's surface vertically below the point in question. Note that $\sigma = 0$ at the top of the atmosphere, where $p = 0$, and $\sigma = 1$ at the earth's surface, where $p = p_s$. Hence there is no longer the problem of having a vertical coordinate level intersect mountains, because the $\sigma = 1$ level by definition follows the model's orography precisely (Fig. 3.4).

In order to convert the basic predictive equations, (3.51)–(3.56), to an appropriate representation for a σ-coordinate system, we need the following relationship between the partial derivatives of a function A along a constant σ surface and along a constant z surface:

$$\frac{\partial A}{\partial c}\bigg|_\sigma = \frac{\partial A}{\partial c}\bigg|_z + \frac{\partial \sigma}{\partial z}\frac{\partial z}{\partial c}\bigg|_\sigma \frac{\partial A}{\partial \sigma} \qquad (3.58)$$

where c can be λ, ϕ, or t. For example, the longitudinal pressure gradient term in (3.51) can be taken along a σ surface and expanded:

$$\frac{1}{\rho}\frac{1}{a\cos\phi}\frac{\partial p}{\partial \lambda}\bigg|_\sigma = \frac{1}{\rho}\frac{1}{a\cos\phi}\left[\frac{\partial p}{\partial \lambda}\bigg|_z + \frac{\partial p}{\partial \sigma}\frac{\partial \sigma}{\partial z}\frac{\partial z}{\partial \lambda}\bigg|_\sigma\right] \qquad (3.59)$$

The last term arises because σ is a function of z and λ. A similar expression can be obtained for the pressure gradient in the latitudinal direction, as needed in (3.52). Note that instead of one term for the horizontal pressure gradient there are two terms in (3.59). The term on the left of (3.59) represents the pressure gradient along the σ surface, which is sloped in mountain areas, and the second term on the right represents the hydrostatic pressure term that must be subtracted from the term on the left-hand side in order to yield the pressure gradient along a constant z surface. Because vertical pressure gradients are always much larger than horizontal pressure gradients, (3.59) reflects two large terms (the first and last) being subtracted from each other to yield the smaller horizontal gradient. Experience has shown that smoothed mountains must be used in atmospheric models that incorporate the σ-system in order to avoid excessive numerical approximation errors. Models that use the z-system ((3.51)–(3.56)) do not suffer this particular problem but have other problems when mountains intersect the constant-z layers (see Kasahara and Washington, 1971, and Washington and Williamson, 1977). Many present atmospheric climate models use the σ-system with the proviso that the mountains be smoothed in a way consistent with the horizontal and vertical resolution of the model.

The total derivative (3.25) can be converted to the σ-system as follows:

$$\frac{d}{dt}\bigg|_\sigma = \frac{\partial}{\partial t}\bigg|_\sigma + \frac{u}{a\cos\phi}\frac{\partial}{\partial \lambda}\bigg|_\sigma + \frac{v}{a}\frac{\partial}{\partial \phi}\bigg|_\sigma + \dot{\sigma}\frac{\partial}{\partial \sigma}\bigg|_\sigma \qquad (3.60)$$

where $\partial/\partial\lambda$ and $\partial/\partial\phi$ are taken along a constant σ surface instead of along a constant z surface, and $\dot{\sigma}$ is the vertical velocity,

$d\sigma/dt$, in the σ-system. Using (3.58), the horizontal equations of motion (3.51) and (3.52) can be written in the σ-system as

$$\left.\frac{du}{dt}\right|_\sigma - \left(f + \frac{u}{a}\tan\phi\right)v = -\frac{1}{\rho a\cos\phi}\left[\left.\frac{\partial p}{\partial\lambda}\right|_\sigma - \frac{\partial p}{\partial\sigma}\frac{\partial\sigma}{\partial z}\left(\frac{\partial z}{\partial\lambda}\right)\bigg|_\sigma\right]$$

$$+ \left.F_\lambda\right|_\sigma \qquad (3.61)$$

$$\left.\frac{dv}{dt}\right|_\sigma + \left(f + \frac{u}{a}\tan\phi\right)u = -\frac{1}{\rho a}\left[\left.\frac{\partial p}{\partial\phi}\right|_\sigma - \frac{\partial p}{\partial\sigma}\frac{\partial\sigma}{\partial z}\left(\frac{\partial z}{\partial\phi}\right)\bigg|_\sigma\right]$$

$$+ \left.F_\phi\right|_\sigma \qquad (3.62)$$

The hydrostatic equation (3.53) can be written in the σ-system as

$$\frac{\partial p}{\partial\sigma} = -\rho g\frac{\partial z}{\partial\sigma} \qquad (3.63)$$

Substitution of (3.63) into (3.61) and (3.62) leads to a slightly different form of the equations of motion, which the reader might encounter elsewhere.

In models using the σ-system, the equation of continuity must also be converted. This can be done by first dividing the continuity equation (3.40) by ρ to yield

$$\frac{d}{dt}\ln\rho + \frac{1}{a\cos\phi}\left[\frac{\partial u}{\partial\lambda} + \frac{\partial}{\partial\phi}(v\cos\phi)\right] + \frac{\partial w}{\partial z} = 0 \qquad (3.64)$$

Using (3.58) on each of the partial derivatives in brackets, the middle term of (3.64) can be converted from the z-system to the σ-system by

$$\frac{1}{a\cos\phi}\left[\frac{\partial u}{\partial\lambda} + \frac{\partial}{\partial\phi}(v\cos\phi)\right]\bigg|_z =$$

$$\frac{1}{a\cos\phi}\left\{\left[\frac{\partial u}{\partial\lambda} + \frac{\partial}{\partial\phi}(v\cos\phi)\right]\bigg|_\sigma\right.$$

$$\left.-\frac{\partial\sigma}{\partial z}\frac{\partial z}{\partial\lambda}\frac{\partial u}{\partial\sigma} - \frac{\partial\sigma}{\partial z}\frac{\partial z}{\partial\phi}\frac{\partial(v\cos\phi)}{\partial\sigma}\right\} \qquad (3.65)$$

Also, from (3.60) and the definition of w as dz/dt,

$$w = \left.\frac{\partial z}{\partial t}\right|_\sigma + \frac{u}{a\cos\phi}\left.\frac{\partial z}{\partial\lambda}\right|_\sigma + \frac{v}{a}\left.\frac{\partial z}{\partial\phi}\right|_\sigma + \dot\sigma\left.\frac{\partial z}{\partial\sigma}\right|_\sigma \qquad (3.66)$$

Calculation of $\partial w/\partial \sigma$ from (3.66) and then substitution of that result along with $\partial w/\partial z = (\partial \sigma/\partial z) \cdot (\partial w/\partial \sigma)$ and (3.65) into (3.64), after some manipulation, yield,

$$\frac{d}{dt}\ln \rho \bigg|_\sigma + \frac{d}{dt}\ln\left(\frac{\partial z}{\partial \sigma}\right)\bigg|_\sigma + \frac{1}{a\cos\phi}\left[\frac{\partial u}{\partial \lambda} + \frac{\partial}{\partial \phi}(v\cos\phi)\right]\bigg|_\sigma$$

$$+ \frac{\partial \dot\sigma}{\partial \sigma} = 0 \qquad (3.67)$$

Since

$$\ln \rho + \ln\left(\frac{\partial z}{\partial \sigma}\right) = \ln\left(\rho\,\frac{\partial z}{\partial \sigma}\right) \qquad (3.68)$$

and $\partial \sigma = \partial p/p_s$ from (3.57), then using (3.60) and (3.63), the continuity equation in the σ-system can be written as

$$\frac{\partial p_s}{\partial t} + \frac{1}{a\cos\phi}\left[\frac{\partial p_s u}{\partial \lambda} + \frac{\partial}{\partial \phi}(p_s v\cos\phi)\right]\bigg|_\sigma + p_s\frac{\partial \dot\sigma}{\partial \sigma} = 0 \quad (3.69)$$

To transform (3.69) into a form more commonly used in atmospheric modeling, vertically integrate the equation from $\sigma = 0$ to $\sigma = 1$ with the boundary conditions of $\dot\sigma = 0$ at $\sigma = 0$ and $\sigma = 1$:

$$\int_0^1 \frac{\partial p_s}{\partial t}d\sigma + \int_0^1 \frac{1}{a\cos\phi}\left[\frac{\partial p_s u}{\partial \lambda} + \frac{\partial}{\partial \phi}(p_s v\cos\phi)\right]d\sigma$$

$$+ \int_0^1 p_s\frac{\partial \dot\sigma}{\partial \sigma}d\sigma = 0 \qquad (3.70)$$

Integration of two of the terms of (3.70) yields

$$\frac{\partial p_s}{\partial t} = -\int_0^1 \frac{1}{a\cos\phi}\left[\frac{\partial p_s u}{\partial \lambda} + \frac{\partial}{\partial \phi}(p_s v\cos\phi)\right]d\sigma \qquad (3.71)$$

Integration of (3.69) instead from $\sigma = 0$ to $\sigma = \sigma$ yields

$$p_s\dot\sigma = -\sigma\frac{\partial p_s}{\partial t} - \int_0^\sigma \frac{1}{a\cos\phi}\left[\frac{\partial p_s u}{\partial \lambda} + \frac{\partial}{\partial \phi}(p_s v\cos\phi)\right]d\sigma \quad (3.72)$$

which allows the vertical velocity, $\dot\sigma$, to be computed at any vertical level in the atmosphere.

The equation of state (3.56) remains identical in the σ-system to its form in the z-system, while the first law of thermodynamics (3.55) retains its earlier form, although requiring that the derivatives be taken along constant-σ surfaces:

$$C_p\frac{dT}{dt}\bigg|_\sigma - \frac{1}{\rho}\frac{dp}{dt}\bigg|_\sigma = Q \qquad (3.73)$$

Some modeling groups have defined the top of the model to be at some nonzero pressure such as 200 mb, thus ignoring the stratosphere and the various processes taking place within and above it. If the top of the model is not at $\sigma = 0$, then additional terms appear in (3.51) and (3.52).

There are several other types of vertical coordinate systems used in the atmospheric sciences, for instance those based on pressure or on potential temperature θ. However, for large-scale general circulation models, the σ scheme with its many variants is currently the most widely accepted, even though it has drawbacks (see Kasahara, 1977). Kasahara (1974, 1977) gives details of the transform coordinate system of any generalized variable that is monotonic with respect to the vertical coordinate, i.e., that either increases or decreases with the coordinate, but not both.

Atmospheric and Ocean Dynamics

Vorticity and divergence equations

Another way of writing the equations of motion is particularly convenient for understanding the dynamics of the atmosphere and ocean and for some forms of numerical solution. As mentioned earlier, the equations of motion yield predictive equations for u and v, the two perpendicular components of horizontal velocity. These components together make up the horizontal velocity vector, which gives the magnitude and direction of flow. Vector calculus is the traditional method used in fluid dynamics for presenting many of the concepts of fluid flow. Consequently, some elementary aspects of vector calculus, along with a few examples of its use, will be presented.

A generalized, three-dimensional velocity vector, V_3, is defined as

$$V_3 = u\boldsymbol{i} + v\boldsymbol{j} + w\boldsymbol{k} \tag{3.74}$$

where $\boldsymbol{i}$, $\boldsymbol{j}$, and $\boldsymbol{k}$ are unit vectors (meaning that they have unit magnitude) in the three primary directions, in this case east-west, north-south, and vertical, respectively, and u, v, and w are the corresponding velocity components. Readers who are unfamiliar with vector concepts are referred to Appendix A for an introduction to such concepts and in particular for definitions of various vector operations in generalized coordinates and specifically in the spherical coordinate system used here. One of the important operators is the three-dimensional gradient operator, $\nabla_3(\ )$, defined as

$$\nabla_3(\ ) = \frac{i}{a \cos \phi}\frac{\partial}{\partial \lambda}(\ ) + \frac{j}{a}\frac{\partial}{\partial \phi}(\ ) + k\frac{\partial}{\partial z}(\ ) \qquad (3.75)$$

for the case of a latitude-longitude-height coordinate system. In much of what follows, for both the velocity vector and the gradient operator, the horizontal components will be the only portions needed. Hence we also define the horizontal velocity vector, V, and the horizontal gradient (or derivative) operator, ∇:

$$V = ui + vj \qquad (3.76)$$

and

$$\nabla(\ ) = \frac{i}{a \cos \phi}\frac{\partial}{\partial \lambda}(\ ) + \frac{j}{a}\frac{\partial}{\partial \phi}(\ ) \qquad (3.77)$$

Upon making appropriate assumptions, as presented in Appendix A, the dot product of ∇_3 and V_3 is

$$\nabla_3 \cdot V_3 = \nabla \cdot V + \frac{\partial w}{\partial z} = \frac{1}{a \cos \phi}\left[\frac{\partial u}{\partial \lambda} + \frac{\partial}{\partial \phi}(v \cos \phi)\right]$$

$$+ \frac{\partial w}{\partial z} \qquad (3.78)$$

and the cross product, $\nabla_3 \times V_3$, is

$$\nabla_3 \times V_3 = \left(\frac{1}{a}\frac{\partial w}{\partial \phi} - \frac{\partial v}{\partial z}\right)i + \left(\frac{\partial u}{\partial z} - \frac{1}{a \cos \phi}\frac{\partial w}{\partial \lambda}\right)j$$

$$+ \left(\frac{1}{a \cos \phi}\frac{\partial v}{\partial \lambda} - \frac{1}{a \cos \phi}\frac{\partial(u \cos \phi)}{\partial \phi}\right)k \qquad (3.79)$$

The horizontal wind vector, V, can be separated into two scalar terms by the Helmholtz theorem:

$$V = k \times \nabla \psi + \nabla \chi \qquad (3.80)$$

where ψ is the streamfunction (a parameter of two-dimensional nondivergent flow whose value is constant along a streamline, which is a line following the flow of the fluid) and χ is the velocity potential (a scalar function whose gradient equals the velocity vector of an irrotational flow). Figure 3.5 shows the relationship between V, ψ, and χ in schematic form, where Fig. 3.5B and

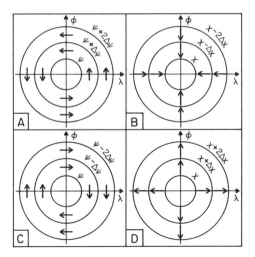

Fig. 3.5 Schematic of the relationship between velocity, V, indicated by bold arrows, streamfunction, ψ, and velocity potential, χ.

Fig. 3.5D show convergent and divergent velocity fields, respectively. Lines of concentric χ are highest at the center for convergence (case B) and lowest at the center for divergence (case D). In a similar fashion in A, the concentric circles represent increasing streamfunction values outward, and the opposite is true in C, where the motion is clockwise. The streamfunction is similar to the pressure field in that wind speed associated with it is proportional to the gradient. It turns out that, except in the tropics, the first term on the right-hand side of (3.80), involving the streamfunction, is generally the largest contributor to the velocity, while the velocity potential, or divergent component term $\nabla \chi$, is usually much smaller for large-scale motions.

By performing the $k \cdot \nabla \times (\ )$ and $\nabla \cdot (\ )$ vector operations on the horizontal equations of motion (3.51)–(3.52) we obtain the vorticity equation

$$
\overbrace{\frac{\partial \varsigma}{\partial t}}^{\text{large}} = \overbrace{-V \cdot \nabla(\varsigma + f)}^{\text{large}} \overbrace{- w \frac{\partial \varsigma}{\partial z}}^{\text{small}} \overbrace{- (\varsigma + f)D}^{\text{medium}} + \overbrace{k \cdot \nabla w \times \frac{\partial V}{\partial z}}^{\text{small}}
$$

$$
+ \underbrace{k \cdot \nabla p \times \nabla \left(\tfrac{1}{\rho}\right)}_{\text{small}} + \underbrace{k \cdot \nabla \times F}_{\text{small}} \tag{3.81}
$$

where $\varsigma \equiv k \cdot \nabla \times V$ is the vertical component of vorticity, and the divergence equation

$$\overbrace{\frac{\partial D}{\partial t}}^{\text{small}} = -\overbrace{\boldsymbol{\nabla}\cdot[(\boldsymbol{V}\cdot\boldsymbol{\nabla})\boldsymbol{V}]}^{\text{small}} - \overbrace{\boldsymbol{\nabla}\cdot(f\boldsymbol{k}\times\boldsymbol{V})}^{\text{large}} - \overbrace{\boldsymbol{\nabla}w\cdot\frac{\partial\boldsymbol{V}}{\partial z}}^{\text{small}} - \overbrace{w\frac{\partial D}{\partial z}}^{\text{small}}$$

$$\overbrace{-\boldsymbol{\nabla}\cdot\left(\tfrac{1}{\rho}\boldsymbol{\nabla}p\right)}^{\text{large}} + \overbrace{\boldsymbol{\nabla}\cdot\boldsymbol{F}}^{\text{small}} \tag{3.82}$$

where D is the horizontal divergence $\boldsymbol{\nabla}\cdot\boldsymbol{V}$. The size designations above the equations indicate typical relative magnitudes of the terms when considering large-scale weather and ocean circulation patterns. A systematic scale analysis shows that for mid-latitude large-scale motions, terms denoted as "large" in (3.81)–(3.82) have a scaling V^2/L^2, terms denoted as "medium" have a scaling $R_0 V^2/L^2$, and terms denoted as "small" are approximately an order of magnitude smaller than those denoted as "medium." Slightly different derivations of (3.81) and (3.82) can be found in numerous texts on dynamical meteorology (e.g., Haltiner and Williams, 1980, and Gill, 1982).

By performing the same curl and dot operations on (3.80), the vorticity and divergence can be expressed in terms of stream-function and velocity potential

$$\varsigma = \boldsymbol{k}\cdot\boldsymbol{\nabla}\times\boldsymbol{V} = \frac{1}{a\cos\phi}\frac{\partial v}{\partial\lambda} - \frac{1}{a}\frac{\partial u}{\partial\phi} + \frac{u\tan\phi}{a} = \boldsymbol{\nabla}^2\psi \tag{3.83}$$

$$D = \boldsymbol{\nabla}\cdot\boldsymbol{V} = \frac{1}{a\cos\phi}\frac{\partial u}{\partial\lambda} + \frac{1}{a}\frac{\partial v}{\partial\phi} - \frac{v\tan\phi}{a} = \boldsymbol{\nabla}^2\chi \tag{3.84}$$

where the $\boldsymbol{\nabla}^2$ on the right-hand side is called the Laplacian operator and is defined as

$$\boldsymbol{\nabla}^2(\ ) \equiv \frac{1}{a^2\cos\phi}\frac{\partial}{\partial\phi}\left(\cos\phi\frac{\partial(\ )}{\partial\phi}\right) + \frac{1}{a^2\cos^2\phi}\frac{\partial^2(\ )}{\partial\lambda^2} \tag{3.85}$$

(see Appendix A). For most nontropical large-scale motions, $\varsigma \gg D$. As will be discussed later, this is another way of saying the motions are nearly geostrophic.

Rossby wave equation

It may be helpful to consider the information the above equations reveal about the atmosphere and oceans. First, in the vorticity equation (3.81), the dominant terms for large-scale motions are those involving ς and the latitudinal change of the Coriolis parameter. Thus (3.81) can be simplified by retaining only the largest terms to yield the following approximation:

$$\frac{\partial \varsigma}{\partial t} + \boldsymbol{V} \cdot \boldsymbol{\nabla} \varsigma + \frac{v}{a}\frac{\partial f}{\partial \phi} = 0 \tag{3.86}$$

Although this equation had been known earlier, C. G. Rossby in the 1930s was the first to identify it as a key equation for describing large-scale motions in the atmosphere.

By turning briefly to Cartesian coordinates and assuming

$$\frac{1}{a\cos\phi}\frac{\partial}{\partial \lambda} = \frac{\partial}{\partial x} \tag{3.87}$$

$$\frac{1}{a}\frac{\partial}{\partial \phi} = \frac{\partial}{\partial y} \tag{3.88}$$

$$f = f_0 + \beta y \tag{3.89}$$

$$\beta \equiv \frac{df}{dy} \tag{3.90}$$

(3.83) and (3.86) become

$$\varsigma = \frac{\partial v}{\partial x} - \frac{\partial u}{\partial y} \tag{3.91}$$

$$\frac{\partial \varsigma}{\partial t} + u\frac{\partial \varsigma}{\partial x} + v\frac{\partial \varsigma}{\partial y} + v\beta = 0 \tag{3.92}$$

where $\boldsymbol{V} \cdot \boldsymbol{\nabla}\varsigma$ is expanded into its two horizontal components. Equation (3.92) is further simplified by substituting (3.91), allowing variations of u and v in the x direction only, and subdividing u into a constant part, U, and a smaller component, u', called a perturbation, such that $u = U + u'$. Further assuming that $U \gg u'$, products of the perturbation can be ignored and (3.92) can be simplified to

$$\frac{\partial^2 v}{\partial x \partial t} + U\frac{\partial^2 v}{\partial x^2} + \beta v = 0 \tag{3.93}$$

where U is a constant horizontal advection (wind) speed. Let us further assume a wave solution to (3.93) of the form

$$v = A\sin[k(x - ct)] \tag{3.94}$$

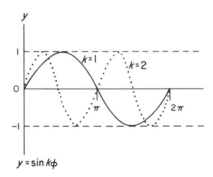

Fig. 3.6 Simple sine waves of amplitude 1 and wave number $k = 1$ and $k = 2$.

where A is some arbitrary constant amplitude, the wave number k is the number of waves between 0 and 2π (so that, on a spherical domain, k is the number of waves around the globe), and c is the speed of the wave. The wavelength $L = 2\pi/k$ is the distance between successive crests or troughs (see Fig. 3.6), and L/c is the period of the wave. Substituting (3.94) into (3.93) yields

$$c = U - \frac{\beta}{k^2} \tag{3.95}$$

From (3.95), C. G. Rossby (1939) succeeded in explaining a great deal about the behavior of large-scale atmospheric patterns. Rossby knew from the early upper-air charts made from upper-air sounding instruments that large eastward-moving weather patterns existed. What was needed was a theoretical underpinning to explain the movement of these waves, and (3.95) helps to provide that. Rossby applied (3.95) to large-scale atmospheric patterns to show that the speed of movement of the large-scale meteorological waves depends upon the mean speed of the wind, U, and the wave number, k, or wavelength, L. Typical values of the variables are $L \approx 3000$ km, $U \approx 10$ m s^{-1}, and $\beta \approx 10^{-11}$ m^{-1}s^{-1}, which yield a wave speed of about 8 m s^{-1}. Waves satisfying (3.95) are termed Rossby waves. From (3.95), the Rossby waves propagate westward relative to the mean speed of the wind. Relative to the earth, they propagate westward or eastward depending upon whether U is greater than or less than β/k^2. If the wave has a wavelength of

$$k = \sqrt{\beta/U} \tag{3.96}$$

then $c = 0$ and the wave is stationary, moving neither westward nor eastward.

Equation (3.95) cannot be used for quantitative prediction because the assumptions and simplifications used in deriving it are met only occasionally in the real atmosphere. The main point to be made is that, to a first approximation, for large-scale motions of the atmosphere there is a balance in the vorticity equation between time change and horizontal advection of vorticity, ς, and the advection of planetary vorticity, β (see either (3.86) or (3.92)). Strictly speaking (3.86) applies to a vertically averaged flow where wind does not change with height. This type of flow is called *barotropic* and does not allow for horizontal temperature gradients. In a more exact sense, barotropic means pressure is constant on constant density surfaces; as a result, temperature is not connected to pressure or density changes. Since barotropic models do not make use of the first law of thermodynamics (because temperature is not involved), they have very limited usefulness for climate research; however, they do allow for the studies of the propagation of Rossby waves in the atmosphere and oceans. Thus, to a limited extent they are used by researchers to understand the basic idealized properties of large-scale dynamical motions rather than the details of the actual atmosphere or ocean circulations.

Baroclinic models

Baroclinic models allow for horizontal temperature variation, which in turn requires vertical variation of wind with height. In contrast to barotropic, baroclinic implies that density surfaces are inclined with respect to pressure surfaces, so that pressure is *not* constant on a constant density surface but instead varies because of temperature variations. This variation can be obtained by combining three relationships obtained earlier. Specifically, substitution of the equation of state (3.56) into both the hydrostatic approximation (3.53) and the geostrophic relationships (3.29)–(3.30) yields

$$\frac{g}{T} = -R\frac{\partial \ln p}{\partial z} \tag{3.97}$$

$$\frac{u}{T} = -\frac{R}{af}\frac{\partial \ln p}{\partial \phi} \tag{3.98}$$

$$\frac{v}{T} = \frac{R}{af\cos\phi}\frac{\partial \ln p}{\partial \lambda} \tag{3.99}$$

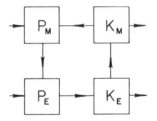

Fig. 3.7 Schematic of the atmospheric energy cycle. [Abbreviated from Peixóto and Oort (1984).]

Upon differentiating (3.98) and (3.99) with respect to z and substituting (3.97), the following pair of equations is obtained:

$$\frac{\partial}{\partial z}\left(\frac{u}{T}\right) = \frac{1}{af}\frac{\partial}{\partial\phi}\left(\frac{g}{T}\right) \tag{3.100}$$

$$\frac{\partial}{\partial z}\left(\frac{v}{T}\right) = -\frac{1}{af\cos\phi}\frac{\partial}{\partial\lambda}\left(\frac{g}{T}\right) \tag{3.101}$$

or

$$\frac{\partial}{\partial z}\left(\frac{u}{T}\right) = -\frac{g}{afT^2}\frac{\partial T}{\partial\phi} \tag{3.102}$$

$$\frac{\partial}{\partial z}\left(\frac{v}{T}\right) = \frac{g}{afT^2\cos\phi}\frac{\partial T}{\partial\lambda} \tag{3.103}$$

Following Haurwitz (1941), these (either (3.100)–(3.101) or (3.102)–(3.103)) are termed the *thermal wind equations* and reflect the essential point that in a baroclinic atmosphere or ocean, horizontal temperature variations imply vertical wind shear. In order to model a baroclinic fluid, at least two vertical layers of u and v must be included in the model to estimate the vertical shear, $\partial V/\partial z$. Essentially all modern climate models have these two or more layers.

A great deal has been learned over the last three decades about how the general circulation of the atmosphere works in terms of the flow of energy from one form to another (see Appendix C for definitions of the different forms of energy). The general pattern of flow is illustrated in Fig. 3.7. The symbol P denotes potential energy and K kinetic energy, while the subscripts M and E denote the zonal mean (i.e., averaged around a latitude circle) and deviations from that mean, respectively. The deviations from the mean are generally referred to as the eddy component. The net radiational heating, including latent heating from precipitation, described in Chapter 2, results in generation of mean potential energy, P_M. Due to the baroclinic nature of the atmosphere, some of P_M is converted to eddy potential energy, P_E, (not all is available for conversion) by the growth of baroclinic eddies (mid-latitude large-scale storm systems). Generally there are between five and eight baroclinic waves around the hemisphere at mid-latitudes on a given day. The reason for this many waves involves certain basic properties of the atmosphere such as the baroclinic nature of the observed flow. This is explained in many meteorological texts (e.g., Holton, 1979, and Haltiner and Williams, 1980). P_E is converted to eddy kinetic energy, K_E, by the sinking of cold air and rising of warm air, and some of K_E is dissipated at the ground or in the atmosphere by frictional loss. Another part of the K_E is converted back to zonal kinetic energy, K_M, some of which is also dissipated and the remainder of which is converted to P_M. In Chapter 5 an example of the generation of potential energy is given from a computer simulation, with breakdown of the potential energy into the baroclinic eddies. This example illustrates the sequence of energy conversions $P_M \rightarrow P_E \rightarrow K_E$.

Early General Circulation Model of the Atmosphere

In the first computer forecast, Charney, Fjørtoft, and von Neumann (1950) used a barotropic equation similar to (3.86), which is a balance of the largest terms in the vorticity equation (3.81). There are several consistent sets of simpler equations that have been used since 1950. These sets have the common characteristic that they are appropriate for larger-scale and slower motions of the atmosphere or oceans. The faster moving gravity and acoustic waves are excluded automatically. The vorticity equation is the basic predictive equation in these sets of equations, with terms involving time and vorticity advection $(V \cdot \nabla (\varsigma + f))$ being dominant (see Phillips, 1973). Several of these sets of simpler equations are called quasi-geostrophic because they use the geostrophic approximation in a consistent manner with only large-scale motions included (e.g., see Charney and Phillips, 1953).

A set of quasi-geostrophic equations was used by Phillips (1956) for the first climate general circulation experiment with a two-level baroclinic system in Cartesian coordinates. Phillips included the effects of heating and cooling, along with a simple treatment of frictional dissipation. The vorticity equation he used was of the form

$$\frac{\partial \varsigma}{\partial t} = -\boldsymbol{V} \cdot \boldsymbol{\nabla}(\varsigma + f) - f_0 D + A\left(\frac{\partial^2 \varsigma}{\partial x^2} + \frac{\partial^2 \varsigma}{\partial y^2}\right) \qquad (3.104)$$

where A is the lateral kinematic viscosity coefficient, which is a crude form of frictional dissipation, and f_0 is a constant Coriolis parameter. This vorticity equation (3.104) is obtained from the more general vorticity equation (3.81) by taking only the larger terms and then adding a dissipation term. Phillips also simplified the largest terms of the divergence equation (3.82), obtaining

$$f_0\varsigma = g\boldsymbol{\nabla}^2 h \quad \text{or} \quad f_0\boldsymbol{\nabla}^2\psi = g\boldsymbol{\nabla}^2 h \qquad (3.105)$$

where h is the height of a pressure surface and the second equation follows immediately from the first because of the relationship between streamfunction and vorticity given in (3.83). Equation (3.105) can be simplified further to yield

$$\psi \approx \frac{g}{f_0}h \qquad (3.106)$$

which can be used in the first law of thermodynamics (3.55). Proceeding in this manner, an energy and scale-consistent set of quasi-geostrophic equations can be constructed.

Unfortunately for calculation purposes, the set of quasi-geostrophic equations (which is simpler than the primitive equations) is not strictly valid near the equator, where the Coriolis parameter goes to zero. Also, it is a poor approximation for the very long waves in the atmosphere (one, two, or three waves around a latitude circle), so that most *global* atmospheric models have returned to the primitive set of equations. However, for special classes of problems where the quasi-geostrophic equations are valid, such as for individual ocean basins away from the equator, these equations are still used and can be more appropriate computationally than the primitive equations because of their exclusion of gravity and acoustical waves, which allows them to be solved more efficiently on modern computers. This will be elaborated on when numerical methods are considered in Chapter 4. The main point to be made here is that the early experiment by Phillips clearly demonstrated many observed features of the atmospheric climate system and led the way toward more refined and improved climate models. Phillips was able to simulate fairly realistic unstable baroclinic waves from an initial atmosphere with a uniform temperature distribution. The simulation developed a westerly jet stream and a meridional flow pattern with the basic three-cell structure discussed in Chapter 2, i.e., containing direct cells at low and high latitudes and an indirect cell at mid-latitudes. Thus the model captured several basic aspects of atmospheric flow.

Radiative and Cloud Processes

There is a wide variety in the methods by which different climate models incorporate the myriad of complex physical processes such as radiation, convection, precipitation, boundary exchanges of heat, moisture, and momentum with the earth's surface, and frictional dissipation. The attempt here is to indicate some of the basic methods currently being employed. The reader should consult individual research articles to find how particular modeling groups have implemented these methods. Very useful references on the state of parameterization of physical processes in atmospheric models are given in Kasahara (1977). In addition, Stephens (1984) has provided an up-to-date review of radiation methods, Mintz (1984) has reviewed the boundary treatments in use, and Anthes (1984) has surveyed the current status of cumulus convection schemes.

Figure 3.8 is a schematic of the relationship between the dynamics of climate models and physical processes. The dynamics involve interactions between motions, thermodynamics, and atmospheric water content. Radiation and clouds are links between the thermodynamics and moisture processes. At the bottom of Fig. 3.8, surface fluxes of sensible and latent heat are exchanged between the atmosphere and the various ground surfaces such as land, snow, ice, and oceans. Although it is not shown in Fig. 3.8, the momentum transfer between the atmosphere and the earth's surface also must be considered. Many of the relevant physical processes will be considered in subsequent sections with emphasis on the basic principles rather than on detailed explanations of the specific equations incorporated in particular models.

Radiation: basic principles

As described in Chapter 2, the radiation that heats and cools the climate system can be broken into two parts: (1) solar radiation, for which the electromagnetic wavelength λ is predominantly less than 4 μm (4 micrometers), and (2) terrestrial radiation, which is largely in the infrared region of the electromagnetic spectrum, with wavelengths ≥ 4 μm. This separation sometimes is referred

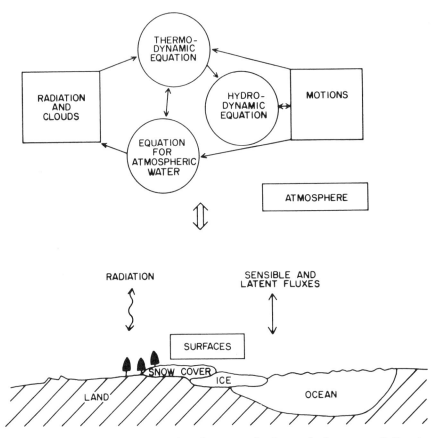

Fig. 3.8 Schematic of interactions between the dynamical aspects of climate models and various physical processes such as radiation, clouds, sensible and latent heat fluxes, and surface types. [From Dickinson (1985).]

to as the *shortwave* and *longwave* partitioning of the radiation involved in the climate system. The numerical treatment of the two types of radiation is generally quite distinct.

A more detailed breakdown of the familiar types of radiation in the electromagnetic spectrum is as follows:

Wavelength	Type of radiation
>1000 m	Longwave radio
100 m–1000 m	AM radio
5 cm–100 m	FM, television, shortwave radio
1 mm–5 cm	microwave-radar
0.7 μm–1 mm	infrared (terrestrial)
0.7 μm–4 μm	near infrared (terrestrial)
0.4 μm–0.7 μm	visible
0.01 μm–0.4 μm	ultraviolet
0.0001 μm–0.01 μm	x-ray
<0.0001 μm	gamma ray

The visible part of the spectrum can be subdivided further into the colors of the rainbow, with red centered at 0.7 μm, orange-yellow at 0.6 μm, green at 0.5 μm, and blue-violet at 0.4 μm. Most radiation that affects the climate system is in the infrared, visible, and ultraviolet parts of the electromagnetic spectrum.

All gases emit and absorb radiation. Each chemical element or combination of elements has a distinct spectrum indicating at which wavelengths in the electromagnetic spectrum it emits radiation. In some cases the emission may be confined to narrow parts of the electromagnetic spectrum rather than being a smoothly varying function of wavelength. The spectra produced by such discrete emissions are called line spectra. This phenomenon of distinct, discrete emission spectra results from fundamental principles of atomic physics and particularly from the emission of radiation as an electron moves from one orbit around the atomic nucleus to another orbit closer to the nucleus. For example, a hydrogen atom has just one electron revolving about a nucleus of one proton. This electron can occupy only certain definite or discrete orbits, determined by the balance of the electrostatic attraction and the centripetal forces. This gives rise to quantum mechanical selection rules that are beyond the scope of this book. Taking into account the angular momentum of the electrons based upon these selection rules, if the atom is excited by absorbing energy, then the electron can go to a higher (outer) discrete orbit. When the electrons return to their original orbit or base state, they emit electromagnetic energy of a specific frequency determined by the two orbits. For example, the hydrogen atom has three series of radiation emissions. As electrons return to the inner orbits, emission occurs in the ultraviolet part of the spectrum, producing the emission spectra called the Lyman series. As electrons return instead from higher orbits to intermediate orbits, the emission is in the visible part of the spectrum, producing what is called the Balmer series; for the outer orbits, the emission is in the infrared, producing the Paschen series. The particular series are not as important as the fact that each atomic gas has a discrete line spectrum based upon the forces involved and the energy required to move the electrons from one orbit to another.

The emission spectra of molecules can have more degrees of freedom than those of individual atoms. For example, the electrons of a molecule with two nuclei can rotate about some common axis, exhibiting rotational energy. Another mode in which molecular energy can be manifested is if the two or more atoms vibrate towards and away from each other. Again based upon the laws of physics, there are certain preferred modes of behavior for each gas that give specific emission characteristics for that gas. Not surprisingly, the emission spectrum of a molecule usually is much more complex than that of an atom. In fact, for many of the molecular gases that involve complex structures, the emission characteristics are determined experimentally instead of being derived theoretically as is done with simple atoms and molecules.

So far we have discussed only the emission of energy from a gas, not the absorption of energy. If radiational energy enters an atomic or molecular gas and is absorbed, then it can increase the atomic or molecular energy levels by the same amount of energy involved in emission. This absorption can be just as discrete or selective as the emission. Thus, if radiation entering a gas cannot excite the atoms or molecules in any of the energy forms, then the radiational energy will not be absorbed or emitted in the gas. In the earth's atmosphere, ozone (O_3), carbon dioxide (CO_2), and water vapor (H_2O) are very important triatomic molecules that both emit and absorb radiation in certain parts of the electromagnetic spectrum that affect the climate system. Ozone is a very strong absorber in the 9–10 μm region, carbon dioxide has absorption maxima in the 2, 3, 4, and 13-17 μm regions, and water vapor has several absorption regions in the 1–8 μm range and at wavelengths greater than 13 μm (Fig. 3.9). An enlarged section in the upper left-hand corner of Fig. 3.9 shows a very detailed line spectrum for water vapor between 1.98 and 2.0 μm. If other parts of the spectrum were similarly enlarged, the same sort of fine structure would be found. The bottom curve in Fig. 3.9 shows that the atmosphere is almost transparent in the 8–12 μm range with the exception of absorption near 9.6 μm by ozone. This 8–12 μm range, known as the 8–12 μm *atmospheric window* due to its transparency, is also the range where the infrared radiation is a maximum for the atmosphere (see Fig. 2.2). An

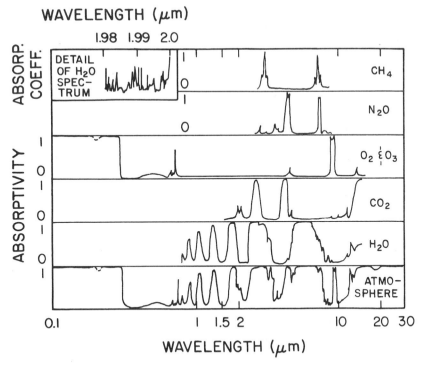

Fig. 3.9 The absorption spectra for H_2O, CO_2, O_2 and O_3, N_2O, CH_4, and the total atmosphere. Absorptivity ranges between 0 and 1. The bulk of infrared radiation within the earth's atmosphere is in the range 5–30 μm (see Fig. 2.2). [Redrawn from Fleagle and Businger (1980).]

important consideration for climate studies is that if this atmospheric 8–12 μm window is contaminated by small increases of CO_2 or trace gases like the chlorocarbons (ClC), large climatic impacts may result. More of this will be discussed in Chapter 6, since several numerical modeling studies have been performed on the potential climatic impacts of increases in atmospheric CO_2.

The other major gases in the earth's atmosphere, besides those indicated in Fig. 3.9, are not radiatively active in the temperature ranges experienced in the earth's environment except those at the very top of the atmosphere. Generally it is felt that such interactions at the fringes of the atmosphere are not of primary importance to the climate system, so that most climate models consider only the emission and absorption of the gases presented in Fig. 3.9. As a further simplification, climate model radiation treatments do not account for the structure of individual spectral lines but instead consider broader regions, or bands, of strong absorption. The various models differ in how this is done, with some models using relatively narrow bands and others using very wide bands. The wider the bands, the greater the number of assumptions, implied or otherwise, that have to be made.

If the shape of an individual spectral line is examined carefully, it can be shown to approximate a Lorentzian distribution function, where there is a slight falling off of intensity on both sides of the maximum emission. The width of the lines is mostly a result of molecular collisions. Narrow linewidths usually are associated with fewer intra-molecular collisions, and, conversely, wide linewidths are associated with high density gases, in which there are many collisions. This effect is called pressure broadening. Obviously, pressure broadening will generally be far less in the upper atmosphere, where the pressures are small compared to those near sea level.

The reason the absorption bands are located where they are in the electromagnetic spectrum is beyond the scope of this book; however, it is reasonably correct to say that a molecule has certain "natural" periods based on its structure, much as a pendulum has a natural period based upon its physical characteristics, such as the length of its arm. In an analogous manner a molecule can have many natural periods, or frequencies. It is at these natural frequencies that molecules absorb and emit most of their energy.

Radiation: physical laws

The computation of radiational heating and cooling in climate models usually is done by calculations of upward and downward fluxes through unit horizontal areas, taking into account the vertical distributions of temperature, water vapor, and other radiative absorbing gases such as carbon dioxide and ozone. The radiation treatment shown here relies on derivations found in Houghton (1979), Paltridge and Platt (1976), and Liou (1980), as well as the review papers of Ramanathan and Coakley (1978) and Stephens (1984). The reader is referred to these sources for more rigorous derivations.

As was shown in Fig. 2.2, the overlap of the blackbody radiation curves for the sun and the earth is not great, and consequently solar and terrestrial radiation can be separated on the basis of wavelength. The blackbody radiation curve was determined theoretically by Max Planck in 1900. A corollary of Planck's equation is the earlier displacement law of Wien, that

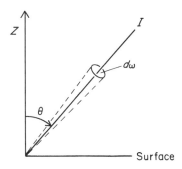

Fig. 3.10 Relationship between the vertical, z, intensity, I, element of solid angle, $d\omega$, and angle, θ, to the vertical. The bottom line refers to the horizontal surface.

the wavelength of maximum intensity, λ_m, is

$$\lambda_m = \frac{2897.8 \ \mu\text{m K}}{T} \tag{3.107}$$

For the earth, the approximate global mean surface temperature is 293 K, which yields a maximum intensity near 9.9 μm, whereas for the sun the surface temperature is approximately $T = 6110$ K, which yields a maximum intensity at 0.474 μm. This contrast is reflected in Fig. 2.2.

Much of the fundamental physics governing radiation transfer is embodied in the following two laws: (1) Lambert's law, which provides a formulation for the decrease in intensity of radiation of a given wavelength as the radiation passes through a given amount of absorbing gas; (2) Kirchhoff's law, which states that there is a proportionality between radiative absorptivity and emissivity of a gas at the same temperature for any wavelength. This can be paraphrased as: A good absorber of radiation at some wavelength is also a good emitter of radiation at the same wavelength. (*Absorptivity* is the ratio of the amount of radiative energy absorbed to the total incident radiation. *Emissivity* is the ratio of the emitted radiation to the maximum possible emitted radiation at the same temperature.)

Lambert's law expresses the change in radiation intensity, I, upon absorption in the vertical due to an atmospheric vertical layer of thickness dz:

$$dI = -Ik\rho \, dz \tag{3.108}$$

where k is the absorption coefficient, ρ is the density of the layer, and I is the radiance, defined as the energy per unit time per unit area per unit solid angle $d\omega$. Figure 3.10 shows the relationship between intensity from some given direction, unit solid angle $d\omega$, and the surface, such that integration over θ of the hemisphere above the horizontal surface yields the flux, F, arriving from all angles, i.e.,

$$F = \int I \cos \theta \, d\theta \qquad (3.109)$$

Dividing (3.108) by I and forming integrals on either side yield

$$\int \frac{dI}{I} = -\int k\rho \, dz \qquad (3.110)$$

which becomes

$$\ln \frac{I}{I_0} = -\int k\rho \, dz \qquad (3.111)$$

upon integrating the left-hand side. This can be rewritten exponentially as

$$I = I_0 e^{-\chi} \qquad (3.112)$$

where χ is $\int k\rho \, dz$, termed the *optical depth* or *optical pathlength*, and $\tau = e^{-\chi}$ is the fractional transmission, indicating the fraction of the original radiance I_0 that gets transmitted to optical pathlength χ within the attenuating gas. If $\chi = 1$, then $\tau = 0.37$, which means the initial intensity I is decreased by a factor of 0.63, and if $\chi = 2$ then $\tau = 0.14$, which yields a decrease of intensity of 0.86. Optical depths at least this great are frequently observed in the atmosphere with very thick dark clouds $(\chi \simeq 10)$, but for normal atmospheric conditions χ typically is much less than 1.

Again using results from basic physics, by applying Kirchhoff's law, the change of intensity dI due to blackbody emission can be found to be

$$dI = k\rho \, dz \, B(T) \qquad (3.113)$$

where $B(T)$ is the blackbody emission per unit solid angle per unit area for a given temperature T. The Stefan-Boltzmann law gives a relationship between the emission and the temperature. This emission $B(T)$ can be integrated over all angles contained in a hemisphere covering a horizontal surface, which gives the flux arriving from *all* angles at a parallel plane surface normal to the z-direction:

$$\int_0^{2\pi} B(T) \, dS \cos \theta \, d\omega = \sigma T^4 \, dS \qquad (3.114)$$

where σ is the Stefan-Boltzmann constant (5.67×10^{-8} W m^{-2} K^{-4}), θ is the angle between the vertical and an incoming ray arriving at the surface, and dS is an element of surface area. Integration of the left-hand side over the entire hemisphere above the parallel surface yields

$$\pi B(T) = \sigma T^4 \qquad (3.115)$$

Thus far it has been assumed that the radiation behaves as if the radiating body were a perfect blackbody with an emissivity equal to 1.0. If the emissivity, ϵ, is less than 1.0, then ϵ should be inserted as a factor on the right-hand side of (3.115).

Since infrared radiation in the earth/atmosphere system involves both absorption and emission, the change of intensity involves two terms. The equation for the change of intensity, referred to as Schwarzchild's equation, makes use of Lambert's and Kirchhoff's laws:

$$dI = -Ik\rho \, dz + B(T)k\rho \, dz \qquad (3.116)$$

The first term in (3.116) is the diminution of intensity by absorption and the second term is the change in intensity caused by emission in a layer or slab with thickness dz.

From (3.112) and the definition of τ following it, the fractional transmission of the radiative flux between the bottom of the layer at z and the top of the layer at z_1 is given by

$$\tau(z, z_1) = e^{-\int_z^{z_1} k\rho \, dz'} \qquad z \leq z' \leq z_1 \qquad (3.117)$$

Next considering the second term of (3.116), the emission portion of Schwarzchild's equation through a layer from z to z_1 becomes

$$dI_1 = k\rho \, dz \, B(z) e^{-\int_z^{z_1} k\rho \, dz'} \qquad (3.118)$$

or by differentiating (3.117) with respect to z, (3.118) becomes

$$dI_1 = B(z) \, d\tau(z, z_1) \qquad (3.119)$$

Equation (3.119) can be integrated from $z = z_0$ to z_1 to yield

$$I_1 = I_0 \tau(z_0, z_1) + \int_{\tau(z_0, z_1)}^{1} B(z) \, d\tau(z, z_1) \qquad (3.120)$$

Following Houghton (1977), (3.120) can be generalized by integrating the two terms on the right over frequency as

$$I_1 = \int_0^\infty I_0 \tau_{\tilde\nu}(z_0, z_1)\, d\tilde\nu$$

$$+ \int_0^\infty \int_{\tau_{\tilde\nu}(z_0, z_1)}^1 B_{\tilde\nu}(z)\, d\tau_{\tilde\nu}(z, z_1)\, d\tilde\nu \qquad (3.121)$$

where $B_{\tilde\nu}$ is the Planck function

$$B_{\tilde\nu} = c_1 \tilde\nu^3 [\exp(c_2 \tilde\nu/T) - 1]^{-1} \qquad (3.122)$$

which is a function of T alone for each wave number $\tilde\nu$. The conversion between wavelength and frequency can be accomplished by using the speed of light, c; i.e., $\nu = c/\lambda$. The constants in (3.122) are $c_1 = 1.1911 \times 10^{-8}$ W m^{-2} sr^{-1} cm^4 and $c_2 = 1.439$ K cm. (See Appendix in Houghton, 1977, and Paltridge and Platt, 1976, for details on how the constants are obtained.)

Integration over all angles of the hemisphere above a horizontal surface transforms the directional intensities I in Fig. 3.10 to upward fluxes $F^\uparrow$ and downward fluxes $F^\downarrow$. The difference of the upward and downward fluxes is the net radiative flux, $F = F^\uparrow - F^\downarrow$, and the intensity equation (3.121) is transformed to an equation for the net flux over a wave number interval $\Delta\nu$ as

$$F_i(z) = \pi B_i(0)\tau_i(z, 0) + \int_0^{\tau_z} \pi B_i(z')\, d\tau_i^*(z', z) \qquad (3.123)$$

Again the π comes from hemispheric integration over all angles above the horizontal surface; τ^* is the spectrally averaged flux transmissivity; and $\pi B_i(z)$ is the spectrally averaged value of the Planck function defined by

$$\pi B_i(z) = \frac{1}{\Delta\nu_i} \int_{\Delta\nu_i} \pi B_\nu(z)\, d\nu \qquad (3.124)$$

where $\pi B_\nu(z)$ is the Planck function for wave number ν and $\Delta\nu_i$ is a wave number interval. The hemispheric mean transmissivity from all angles over or under the surface is $\tau_\nu^*(|z - z'|) = 1.66\tau_\nu(|z - z'|)$, where 1.66 is a diffusivity factor.

The total net flux for all wave numbers is obtained by summing all wave number intervals:

$$F_T(z) = \sum_i F_i(z)\, \Delta\nu_i \qquad (3.125)$$

Obviously a summation over each line in the spectrum shown in Fig. 3.9 would lead to a cumbersome and lengthy computation. To avoid this, wave number intervals of relatively large width typically are used in climate-oriented models. As pointed out by Ramanathan and Coakley (1978), there are two methods in use for computing the mean or spectrally averaged transmissivities. In the first, theoretical concepts involving mean spectral line strength, mean pressure or density broadening of lines, and mean spectral line spacing or bandwidth are used in the evaluation of transmissivities. In the other, empirical methods are used to fit spectral regions to laboratory measurements. Ramanathan and Coakley (1978) review several specific formulations by Rodgers and Walshaw (1966), Rodgers (1968), and Cess and Ramanathan (1972). Also, the review article by Stephens (1984) discusses several methods for obtaining averaged transmissivities.

As an example of the band model approach, let us consider for most of the lower atmosphere closely spaced spectral lines that strongly overlap each other so that lines can be grouped together into bands. Figure 3.9 suggests several regions where the spectral lines can be grouped this way. The methods used in band treatments are approximate; however, they can be compared with very detailed line-by-line computations to see the approximate magnitude of the errors introduced due exclusively to the band approximation. The most commonly used band parameterization is based upon Goody (1964) and Malkmus (1967), where randomly spaced lines are assumed within a spectral region that has a statistical distribution of line strengths. Typically band models make use of averaged values for such properties as the strength of lines and the separation of lines. The transmission of radiation within a band is related to the mean line intensity, $\bar{S}$, the mean half-width, $\bar{\alpha}$, of the lines, and the mean separation or spacing of lines, $\bar{d}$. As an example, Goody (1952) assumed that randomly spaced lines with these mean properties lead to a transmission function over the interval $\Delta\nu$ of the form

$$\tau^* = \exp\left[-\frac{\bar{S}u\beta}{\bar{d}}\left(1 + \frac{\bar{S}u\beta}{\bar{\alpha}}\right)^{-1/2}\right] \tag{3.126}$$

where $\beta = 1.66$ and u is the amount of gas. The mean transmission function parameters in (3.126) have been defined slightly differently by several authors for the various individual absorbing gases (e.g., see Stephens, 1984).

In some infrared models the calculation of fluxes is simplified further. This is true particularly for water vapor, since it absorbs over a large spectral region (Fig. 3.9). The simplification normally is done by using the so-called emissivity method, in which an integration over relatively broad spectral intervals results in a more efficient computation (for an example, see the treatment of Ramanathan and Coakley, 1978).

By using (3.123) and subdividing into upward and downward fluxes, respectively, the flux equations can be written as

$$F_i^\uparrow(z) = \pi B_i(0)\tau_i^*(z,0) + \int_0^z \pi B_i(z')\, d\tau_i^*(z',z) \quad (3.127)$$

and

$$F_i^\downarrow(z) = \pi B_i(z_0)[\tau_i^*(z,z_0) - \tau_i^*(z,\infty)]$$
$$+ \int_{z_0}^z \pi B_i(z')\, d\tau_i^*(z',z) \quad (3.128)$$

The Planck function B_i in (3.127) and (3.128) can safely be assumed to be constant if the spectral interval is narrow; however, if the interval is large, then an assumption of constant B_i causes large errors. By defining the absorptivity as unity minus the transmissivity, that is,

$$A_i(z,z') = 1 - \tau_i^*(z,z') \quad (3.129)$$

then the two preceding equations can be written as

$$F_i^\uparrow(z) = \pi B_i(0) + \int_0^z A_i(z,z')\, d[\pi B_i(z')] \quad (3.130)$$

and

$$F_i^\downarrow(z) = \pi B_i(z_0)A_i(z,\infty) + \int_{z_0}^z A_i(z,z')\, d[\pi B_i(z')] \quad (3.131)$$

The above two equations can be summed over i to yield the total upward and downward longwave fluxes

$$F^\uparrow(z) = \pi B(0) + \int_0^z \tilde{\epsilon}(z,z')\, d[\pi B(z')] \quad (3.132)$$

and

$$F^\downarrow(z) = \pi B(z_0)\epsilon(z,\infty) + \int_{z_0}^z \tilde{\epsilon}(z,z')\, d[\pi B(z')] \quad (3.133)$$

where the emissivities are

$$\tilde{\epsilon}(z,z') = \sum A_i(z,z') \frac{dB_i(z')}{dB(z')} \quad (3.134)$$

and

$$\epsilon(z, z') = \sum_i A_i(z, z') \frac{B_i(z')}{B(z')} \qquad (3.135)$$

and $\pi B(z)$ is integrated over all wave numbers, i.e.,

$$\pi B(z) = \sigma T^4(z) \qquad (3.136)$$

For ease in numerical modeling, $\tilde{\epsilon}$ and ϵ must be made simple functions of water vapor content, pressure, and temperature for the path through which the radiation passes. The summations in (3.134) and (3.135) replace integration over wavelength by subdividing into broad wave number intervals $\Delta\nu_i$. Examples of the emissivity approach are given by Elsasser and Culbertson (1960), Sasamori (1968), Rodgers (1967), and Stone and Manabe (1968). In order to use (3.132)–(3.136), the transmission function τ^* is used to determine the absorptivities, A_i, from (3.129), after which the A_i are used to compute the emissivities $\tilde{\epsilon}$ and ϵ for each layer from (3.134) and (3.135), followed by calculation of the upward and downward longwave fluxes through the individual layers from (3.132)–(3.133). The numerical approximation of the integrals in (3.132) and (3.133) can be very important, especially wherever the temperature or moisture vertical profiles undergo sharp changes, which is the case near the earth's surface and at the tropopause.

Figure 3.11 shows from several authors the emissivities $\tilde{\epsilon}$ and ϵ as functions of water vapor path length u. The curves at the top, from Staley and Jurica (1970), show that emissivities increase substantially with longer water vapor path length and have a small dependence on temperature.

Solar radiation

Before deriving the radiation equations for solar heating and determining the effect of solar radiation on the atmosphere, oceans, and sea ice, it is necessary to compute the solar intensity at the top of the atmosphere. The distribution of the radiative flux arriving at the top of the atmosphere can be determined from spherical trigonometry, taking into account the date, time of day, and latitude. The *zenith angle*, Z, at any location is the angle between the vector perpendicular to the earth's surface at that location and the incoming solar rays. This angle can be seen to equal the angle between the plane tangent to the surface of the earth and the plane normal to the sun's rays. By the use of standard trigonometric equations (see Sellers, 1965), the cosine

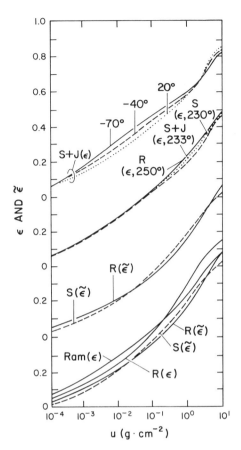

Fig. 3.11 Emissivity, ϵ, and modified emissivity, $\tilde{\epsilon}$, as functions of water vapor content (i.e., path length u) from several authors: R–Rodgers (1967), S&J–Staley and Jurica (1970), Ram–Ramanathan et al. (1983), S–Sasamori (1968). [From Stephens (1984).]

of this zenith angle can be calculated from

$$\cos Z = \sin \phi \sin \delta + \cos \phi \cos \delta \cos H \qquad (3.137)$$

where ϕ is latitude, δ is solar declination, which is the angular distance of the sun north of the equator and varies from about 23.5° on June 22 (summer solstice) to -23.5° on December 22 (winter solstice), and H is the hour angle, which is the longitudinal distance from the point in question to the meridian of solar noon and therefore is 0 at any point experiencing solar noon. At sunset and sunrise $\cos Z = 0$, and hence from (3.137)

$$\cos H = - \tan \phi \tan \delta \qquad (3.138)$$

From this the length of time between sunrise and sunset ($=$ $2 \times \cos^{-1}(-\tan\phi\tan\delta) \times$ hours$/15°$) is known uniquely as a function of time of year and latitude.

The solar flux, S, arriving at the top of the atmosphere is a function of $\cos Z$ and the distance from sun to earth, d, such that

$$S = S_0 f(d) \cos Z \qquad (3.139)$$

where S_0 is the *solar constant* and the factor $f(d)$ is 1.0344 in early January and 0.9674 in July for the present astronomical conditions. In paleoclimate studies where the orbital parameters are quite different from those of the present, both δ and $f(d)$ must be changed. This is an important point since there is increasing evidence that orbital parameters are a major cause of climate change, such as occurred during the Pleistocene ice ages. Paltridge and Platt (1976) give a discussion of these orbital variations. The solar constant S_0 as measured by observational satellite studies is thought to be about 1367 W m^{-2}. However, observational evidence indicates that S_0 changes, although by relatively small amounts compared to the δ and $f(d)$ variations.

The parameterization of the absorption of solar radiation in the climate system is critical for a proper simulation of the climate and its sensitivity. One of the principal absorbers in the atmosphere is stratospheric ozone, which absorbs very effectively in the ultraviolet ($\lambda < 0.4~\mu$m) and in the visible ($0.4~\mu$m $< \lambda < 0.7~\mu$m) portions of the electromagnetic spectrum. Water vapor is the primary absorber in the troposphere in the near-infrared ($0.7~\mu$m $< \lambda < 4~\mu$m). Figure 3.12 shows the spectral energy distributions of a blackbody with $T = 6000$ K, the solar energy at the top of the earth's atmosphere, and, for clear-sky conditions, the solar energy reaching the earth's surface. As shown, the most effective atmospheric absorber of solar energy in the shorter wavelengths is ozone, whereas water vapor and carbon dioxide are important absorbers for the longer wavelengths.

In climate models, the exact absorption, scatter, and transmission of solar energy are never used because of the enormous numerical calculations involved. Usually the detailed calculations are simplified by fitting the more complicated results to simpler analytical expressions, as done by Lacis and Hansen (1974). Thereby the amount of absorption of solar energy in each model layer can be computed based on clouds, ozone, water vapor, zenith angle of the sun, and the local albedo of the earth's surface. This simplified computation must account for multiple-beam reflections from other layers as well as the earth's

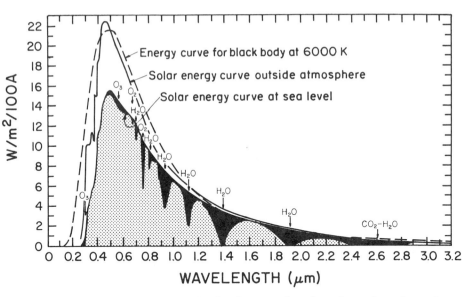

Fig. 3.12 Spectral energy distribution as a function of wavelength, at the top of the atmosphere and at sea level. The figure includes absorption by various atmospheric gases. Also shown is the shape of the blackbody spectral energy distribution for a temperature of 6000 K, although the vertical scale in the figure is not appropriate for this latter (dashed) curve. [From Stephens (1984), who modified an earlier version from Lacis and Hansen (1974).]

surface. For example, Rayleigh scattering of solar radiation by atmospheric molecules is an imporant part of the shortwave treatment in climate models.

The absorption of solar radiation by water vapor is a very strong factor in the heating of the lower troposphere, especially in the tropics. There are many formulations for this absorption, as reviewed by Stephens (1984). Figure 3.13 shows from various authors several absorption curves for water vapor. Note that the path length scale is logarithmic. Lacis and Hansen (1974) have approximated the function suggested by these curves with the following formula:

$$\overline{A} = \frac{2.9\,u}{(1 + 141.5\,u)^{0.635} + 5.925\,u} \tag{3.140}$$

where $\overline{A}$ is the absorptivity and u is the water vapor path length. Similar expressions have been obtained for ozone (O_3), carbon dioxide (CO_2), and oxygen (O_2).

The computation of diffuse versus direct solar radiation must be taken into account, as well as reflections and absorption by

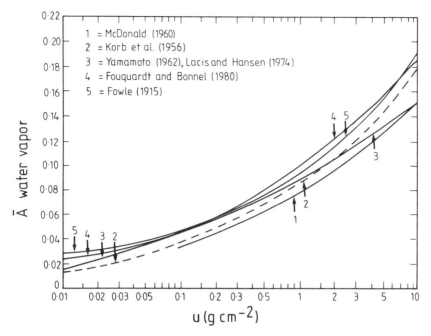

Fig. 3.13 Absorptivity of solar radiation due to water vapor, plotted as a function of water vapor path length, u. The individual curves are from various authors as indicated. [From Stephens (1984).]

clouds and the earth's surface. Stephens (1984) gives several tables showing which radiative aspects are included in various climate and weather prediction models used throughout the world. There are many uncertainties about the accuracy of the parameterizations used. As yet most climate models do not accurately take into account varying amounts of aerosols such as dust (volcanic and otherwise), one of the reasons being that little is known about aerosol size distribution and composition. However, it is expected that in the future radiation treatments will be given high emphasis because of their effect on the modeled sensitivity of climate to changes in the atmospheric constituents.

Net heating/cooling rates

The net heating or cooling of an atmospheric layer due to solar and infrared radiation is given by

$$\rho C_p \frac{dT}{dt} = \frac{d(F_N)}{dz} \qquad (3.141)$$

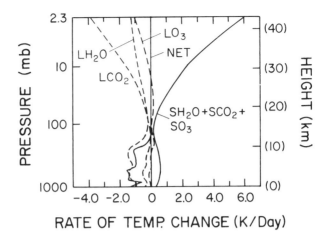

Fig. 3.14 Heating rates in the atmosphere due to the absorption of solar radiation (S) by atmospheric H_2O, CO_2, and O_3, and heating and cooling rates due to the absorption and emission of longwave or infrared radiation (L) by H_2O, CO_2, and O_3. [From Manabe and Strickler (1964).]

where ρ is air density, C_p is the specific heat of air at constant pressure, and F_N is the total net solar and infrared radiative flux.

The largest contributors to the heating/cooling rates in K/day are water vapor, CO_2, and O_3, as shown in Fig. 3.14 from a simplified one-dimensional climate model of Manabe and Strickler (1964). The longwave contributions are denoted by an L preceding the gas, the solar contribution by an S. The net heating/cooling is zero in the stratosphere since the stratosphere in the one-dimensional model is in radiative equilibrium. The troposphere shows a net radiative cooling that is compensated by vertical transfer of sensible and latent heat from below by moist adiabatic convection. As seen from Fig. 3.14, H_2O is the strongest contributor to tropospheric cooling. In the stratosphere, cooling due to CO_2, H_2O, and O_3 generally is compensated by heating due to absorption of solar radiation by ozone. Ozone's strong absorption band centered at a wavelength of about 9.6 μm (Fig. 3.9) helps to warm the 15 and 25 km regions where ozone exists in sizable quantities, by absorption of radiation from the earth's surface. This heating of the upper atmospheric levels is possible because of the relative lack of ozone in the troposphere and the radiation window (low absorptivity by water vapor near 10 μm; see Fig. 3.9) in the troposphere.

Recently Ramanathan et al. (1985) have shown that small amounts of trace gases, such as nitrous oxide (N_2O) and methane (CH_4), can "dirty" the radiation window near 10 μm. Like CO_2,

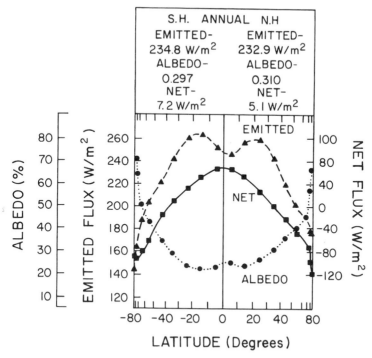

Fig. 3.15 Latitudinal distribution of the annually averaged emitted (outgoing) infrared flux at the top of the atmosphere, the net flux (absorbed solar minus emitted) at the top of the atmosphere, and the net albedo of the earth/atmosphere system. [From Stephens et al. (1981).]

these trace gases in the troposphere can trap upward infrared radiation, which is theorized to result in a warming of the troposphere and surface and a cooling of the stratosphere. Although several of the trace gases have undergone a dramatic increase over the last few decades that will likely have important consequences for the earth's radiation budget in the future, most present climate models do not include them as yet. The reader is referred to Ramanathan et al. (1985) for more details on trace gases and their potential impact on the climate system.

The latitudinal distributions of the annually averaged emitted infrared flux, planetary albedo, and net flux at the top of the atmosphere are shown in Fig. 3.15. The planetary albedos are far greater in high latitudes than elsewhere, in large part due to more ice and snow, along with moderate-to-heavy cloud covers. The albedo also undergoes large temporal changes because of high sensitivity to the nature of the surface and clouds. The emitted longwave flux $F^{\uparrow}$ is determined largely from temperature and moisture distributions, both of which vary greatly between tropics and poles. The net flux, which is the difference between the absorbed solar flux and the emitted infrared flux, is the energy source for atmospheric and oceanic circulations that carry heat from the tropics to the poles. In the absence of these circulations, the net radiation curve in Fig. 3.15 would imply increasingly warmer tropics and colder polar regions, as discussed in Chapter 2.

Clouds

Clouds play a very important role in the radiation characteristics of the earth's climate, but their radiative properties are not fully understood. In present climate models clouds either are specified as a function of space and time or are computed. In the latter case the precipitation and convective aspects of the model are used to predict where clouds exist based on whether the atmosphere is near saturation and whether convection is occurring. If the air is saturated but not convective then a stratiform (layer) cloud or fog exists. Usually it is assumed that stratiform clouds occupy all or nearly all of any grid cell where they appear. However, if a cloud is of convective origin, then usually it is assumed to occur over some small fraction of a grid cell. The reason for the difference is that convective clouds normally occur over horizontal areas that are smaller than the typical horizontal grid areas used in present climate models, whereas stratiform clouds normally cover large horizontal regions and are fairly uniform. Often climate models make use of either relative humidity (defined in the next section) or some other moisture parameter in conjunction with empirical factors to relate cloud amount to moisture. Cloud albedo is generally determined as a function of zenith angle and optical depth (e.g., Ramanathan et al., 1983).

Often the radiation treatment in a model assumes that clouds at different levels are overlapped randomly so that the cloud cover C_i at each layer i can be combined as follows to give the total cloud cover C_T:

$$C_T = 1 - \prod_{i=1}^{N} (1 - C_i) \qquad (3.142)$$

For example, if there is a two cloud-layer system $(N = 2)$ with 0.5 cloud cover in each layer, then the total cloud cover C_T is 0.75.

In the infrared part of the spectrum, clouds usually are treated as blackbodies, meaning that their emissivity is assumed to be unity, although there is observational evidence that thin ice and water clouds such as cirrus clouds have emissivities considerably less than 1. Most climate models have some difficulty in properly treating small amounts of moisture, and thus optically thin cirrus-type clouds often are treated somewhat arbitrarily. There remains a great deal of uncertainty in parameterizing clouds properly in numerical models. In order to improve this important aspect of climate modeling, continuing research needs to be conducted on cloud/radiation interactions.

Precipitation and cloud processes

The treatment of moisture processes is one of the more difficult aspects of constructing climate models. There are several reasons for this: First, the precipitation-cloud physics, specifically how condensation, sublimation, and freezing of water drops and ice particles occur, is not fully understood; second, the several hundred km horizontal and the km or so vertical resolution of climate models are much larger than the scales in which most clouds are formed; and third, there are some unique problems in approximating the precipitation-cloud aspects numerically, such as the wide range of moisture values, which can extend over several orders of magnitude. Small amounts of moisture tend to be difficult to handle in climate models.

In order to discuss the precipitation and cloud physics used in climate models, several basic concepts need to be introduced. The ratio of the density of water vapor, ρ_w, to the density of dry air, ρ, is defined as the mixing ratio, q:

$$q = \frac{\rho_w}{\rho} \qquad (3.143)$$

In some models a slightly different ratio is used, where ρ_w is divided by the total density of the air, $\rho_w + \rho$. This quantity $(\rho_w/(\rho_w + \rho))$ is called specific humidity. The difference between mixing ratio and specific humidity is small except near the earth's surface in the tropics, where ρ_w can be a sizable fraction of ρ. A third measure of water vapor content in the atmosphere is the relative humidity, which is the ratio of the amount of water vapor present to the amount which would be present if air of the same temperature were fully saturated with water vapor. Formally, the relative humidity, r, is defined as the ratio of actual mixing ratio, q, to saturation mixing ratio, q_s:

$$r \equiv \frac{q}{q_s} \qquad (3.144)$$

If $r = 1$, then the atmosphere is saturated. Equation (3.144) is approximately equivalent to $r = e/e_s$, where e is the partial pressure due to water vapor and e_s is the saturated partial pressure due to water vapor.

In a manner analogous to the continuity of mass treatment introduced earlier in this chapter, the changes in the amount of water vapor must be balanced by the moisture sources and sinks. An equation of continuity for water vapor mixing ratio can be written

$$\frac{dq}{dt} = \frac{1}{\rho}M + E \tag{3.145}$$

where M is the time rate of change of water vapor per unit volume due to condensation or freezing and E is the time rate of change of water vapor content per unit mass due to evaporation from the surface and subgrid-scale vertical and horizontal diffusion of moisture within the atmosphere. The first term on the right-hand side of (3.145) is a sink of moisture, while the second term is often a source of moisture. Often (3.145) is written in flux form by combining with the equation of continuity to obtain

$$\frac{\partial(\rho q)}{\partial t} + \nabla \cdot (\rho q V) + \frac{\partial}{\partial z}(\rho q w) = M + \rho E \tag{3.146}$$

If (3.146) is integrated over the entire volume of the atmosphere, the second and third terms on the left drop out, so that the sources and sinks of moisture must balance in order to have no secular change in atmospheric moisture over the globe.

If the atmosphere is saturated with moisture (i.e., the relative humidity is 100%), then sensible heat can be added to the atmospheric system from latent heat by the conversion of water vapor to liquid water or ice particles. Conversely, if water droplets evaporate or ice sublimates, sensible heat is taken out of the atmosphere and put into a latent form.

The first law of thermodynamics presented earlier (3.41) must incorporate in the nonadiabatic term Q this energy conversion process due to phase changes between liquid, solid, and gas. If the nonadiabatic process of conversion of water vapor to liquid water is the only energy source incorporated in Q, then (3.41) becomes

$$C_v dT + p\, d\left(\tfrac{1}{\rho}\right) = -L\, dq_s \tag{3.147}$$

where L is the latent heat of condensation or fusion, depending upon whether the condensation is from water vapor to liquid (rain) or from water vapor to ice (snow). Introducing the equation of state (3.42), (3.147) can be converted to

$$C_p dT - \frac{1}{\rho} dp = -L dq_s \tag{3.148}$$

as in the conversion of (3.41) to (3.44). Inserting a total differential form of the hydrostatic equation (3.53) gives

$$\frac{dT}{dz} = -\frac{g}{C_p} - \frac{L}{C_p} \frac{dq_s}{dz} \tag{3.149}$$

Considering q_s as a function of T and p, the derivative in the last term of the above equation can be expanded to

$$\frac{dq_s}{dz} = \frac{\partial q_s}{\partial T} \frac{dT}{dz} + \frac{\partial q_s}{\partial p} \frac{dp}{dz} \tag{3.150}$$

The equation of state for dry air was given in (3.42). For water vapor, and still assuming an ideal gas, the equation of state can be written as

$$e = \rho_w R_m T \tag{3.151}$$

where e is the partial pressure for water vapor, ρ_w is the density of water vapor, and R_m is the gas constant for water vapor. Recognizing that the total pressure p is the sum of e and the dry-air pressure p_d, (3.151) and (3.42) as modified to $p - e = p_d = \rho R T$ can be inserted into (3.143) to yield

$$q = \frac{R}{R_m} \frac{e}{p - e} \tag{3.152}$$

The ratio R/R_m of the gas constants for dry air and water vapor can be shown to be the ratio of the molecular masses of water vapor and dry air, which is 0.622. Since p is usually much larger than e, the mixing ratio generally can be approximated as

$$q \simeq 0.622 \frac{e}{p} \tag{3.153}$$

and the saturation mixing ratio as

$$q_s \simeq 0.622 \frac{e_s}{p} \tag{3.154}$$

With this approximation, converted to $e_s \simeq 1/0.622\, pq_s$, the total differential of e_s can be approximated as

$$de_s \simeq \frac{1}{0.622} (p dq_s + q_s dp) \tag{3.155}$$

or, dividing by $e_s \simeq 1/0.622\, pq_s$ and changing the order of terms,

$$\frac{dq_s}{q_s} \simeq \frac{de_s}{e_s} - \frac{dp}{p} \tag{3.156}$$

The Clausius-Clapeyron equation is a well-known equation from classical thermodynamics that relates the change in saturation vapor pressure to the latent heat involved in a phase change from water vapor to liquid water or from liquid water to ice (for more details see standard texts on atmospheric thermodynamics, such as Fleagle and Businger, 1980). The most convenient form of the equation for our purposes is

$$\frac{de_s}{e_s} = \frac{L}{R_m} \frac{dT}{T^2} \tag{3.157}$$

which shows e_s as a function of T alone. Multiplication of (3.156) by q_s/dz and substitution of (3.157), a total differential form of the hydrostatic equation (3.53), and the equation of state (3.56) yield

$$\frac{dq_s}{dz} \simeq \frac{Lq_s}{R_m T^2} \frac{dT}{dz} + \frac{q_s g}{RT} \tag{3.158}$$

Using (3.149) gives a new lapse rate

$$\Gamma_m = -\frac{dT}{dz} \simeq \frac{g}{C_p} \left(1 + \frac{Lq_s}{RT}\right) \left(1 + \frac{L^2 q_s}{C_p R_m T^2}\right)^{-1} \tag{3.159}$$

which is the so-called moist or saturation adiabatic lapse rate and is less than the adiabatic lapse rate defined earlier (as the negative of dT/dz in (3.47)). The reason the moist lapse rate is less than the dry lapse rate is that the latent heat released by condensation or freezing is added to the air, so that a parcel of air that is ascending does not cool as much as it would otherwise. Also, in an analogous manner the potential temperature in an atmosphere undergoing precipitation can be determined to be

$$\theta_e = \theta e^{Lq_s/C_p T} \tag{3.160}$$

Termed the equivalent potential temperature, θ_e is approximately constant in an air mass as it undergoes precipitation. This can be compared to the potential temperature, θ, under dry conditions, which is approximately constant when the atmospheric processes are approximately adiabatic.

If air is supersaturated and not undergoing convection, then latent heating can be computed from the first law of thermodynamics in the form (3.44) with $dp = 0$ and the nonadiabatic term Q being the stable (without convection) latent heating Qs_L:

$$Qs_L = \frac{C_p dT}{dt} \tag{3.161}$$

The temperature change in (3.161) is the temperature change obtained by the conversion of water vapor into liquid water, i.e.,

$$\frac{L}{C_p}(q - q_s) \qquad (3.162)$$

Note from (3.156) that the saturation mixing ratio q_s is only a function of e_s and p, which in turn are functions of T. In fact, the dependence of q_s or e_s on T is approximately exponential. The term e_s can be approximated by an empirical formula (e.g., as in Murray, 1967) or by tabulated values (List, 1951). In summary, stable or nonconvective latent heating can be computed in an atmospheric model whenever the atmosphere is supersaturated (i.e., $q > q_s$). Usually the moisture is reset to saturation ($q = q_s$). Furthermore, often the relative humidity criterion for assuming nonconvective latent heating is less than 100%; for example, some models use 80%. The justification for a criterion less than 100% is that precipitation is known observationally to occur often on smaller scales than the grid scale of most climate models and to take place even if the large-scale environment has a relative humidity less than 100%.

The manner in which convective latent heat release is included in atmospheric models is much less uniform. Anthes (1984) has recently reviewed the current understanding of this type of latent heating, often referred to as a cumulus parameterization. Quoting from GARP Publication Series No. 8 (WMO, 1972), the key to successful parameterization is "formulation of quantitative rules for expressing the location, frequency of occurrence, and intensity of the subgrid-scale processes in terms of the resolvable scale." Although some methods of treating small-scale convective or cumulus parameterization are more physically sound than others, it is still uncertain which of the many methods in use is the best overall.

Before discussing particular schemes it may be useful to review in schematic form how convective clouds form. If a parcel of air near the earth's surface (at an atmospheric pressure of approximately 1000 mb) starts to rise, say from strong heating, it will cool at approximately the dry adiabatic lapse rate of 9.8 K km^{-1}. Assume that the parcel of air has a relative humidity of 50% at the surface and that it does not mix with the environmental air as it ascends and cools, so that its water vapor content remains the same. The assumed lack of mixing is termed *no entrainment*, and is not a realistic assumption under many situations. As the parcel cools, its relative humidity increases, since cool air is unable to hold as much water as warm air. When the

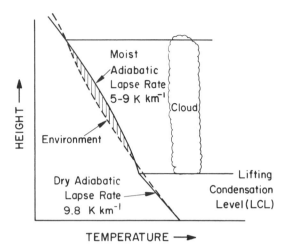

Fig. 3.16 Schematic of cumulus cloud formation showing the lifting condensation level (LCL), dry and moist adiabatic lapse rates, and lapse rate for the air outside the cloud (labeled "environment").

parcel cools sufficiently to have a relative humidity of 100%, the parcel is saturated. The level at which this occurs is termed the lifting condensation level, LCL, and is typically at the cloud base (Fig. 3.16). If the air near the earth's surface is moist, then the LCL is low in altitude, perhaps 100 meters or so, while if the surface air is dry, the LCL can be quite high. Obviously, if the parcel does not rise high enough to reach a LCL, then a cloud should not form. From the LCL upward, the parcel will cool less rapidly as it continues to rise, since the latent heat of condensation or fusion releases heat. This lessened rate of temperature decrease with height is the moist adiabatic lapse rate referred to earlier. Typically in the lower tropical troposphere it is 5–6 K km^{-1}. If the temperature during a model simulation is greater than 0°C, then the latent heat of condensation is invoked, and if it is less than 0°C, the latent heat of fusion is invoked. Also, as depicted in Fig. 3.16, the temperature/height curve for the parcel of air crosses the curve for environmental air at some colder temperature and higher altitude. This limits the upper reaches of the cloud. In reality, because of buoyancy forces and due to its upward momentum, the cloud extends past the crossover point, so that the cloud air eventually becomes cooler and heavier than the environmental air. It will then sink back to near the crossover point.

This simple model of the convective process ignores many important physical processes such as entrainment of environmental air into the clouds, especially through the sides. This mixing will dilute the temperature and moisture differences between the cloud and the surrounding environmental air. Rising motion in a cloud core and compensating sinking motion outside the cloud (with the sinking air warming at the dry adiabatic lapse rate) can lead to a large difference in temperature and moisture between cloud and environment. This also leads to a large differential buoyancy, which can accelerate the growth of clouds. The larger clouds have less mixing between cloud and environment, and thus they will have a preferential growth compared to small clouds that are rapidly mixed with the environmental air. The challenge for cloud researchers is to determine how nature decides which process dominates under different large-scale environmental conditions. If the entrainment mixing is strong, then the buoyancy forces and the liquid water content will be small and so cloud growth will be inhibited. The proper parameterization of this process in atmospheric models requires an improvement of our knowledge of cloud processes.

Before considering specific cumulus parameterization schemes it will be useful to examine some simple aspects of stable and unstable lapse rates and how they relate to potential temperatures θ and θ_e, since the schemes are based in part on understanding this concept. Figure 3.17 shows dry and moist lapse rates in schematic form. The dotted line is the curve for the environmental air, and the solid line is either the dry or the moist adiabatic temperature line along which a parcel will ascend from point A to B. There is a point B' on the curve for the environmental air that is colder in both cases than the point B for the parcel at the same height. Since the air at B' is colder than that at B, the parcel will have positive buoyancy. Thus in principle it will continue ascending. This suggests that the environmental air is unstable. If on the other hand a parcel is at C and ascends to point D, it will be colder than the surrounding environmental air at D'. Thus the parcel will have negative buoyancy, which will cause it to sink to its original position. Hence environmental conditions corresponding to CD' are stable, whereas environmental conditions corresponding to AB' are unstable. This relatively simple concept for dry and moist convection has been well verified by both theoretical discussion (Fleagle and Businger, 1980) and laboratory experiments. Theoretically, the conditions for a stable atmosphere with respect to dry and moist convection are

$$\frac{\partial \theta}{\partial z} \geq 0 \qquad (3.163)$$

for dry convection, and

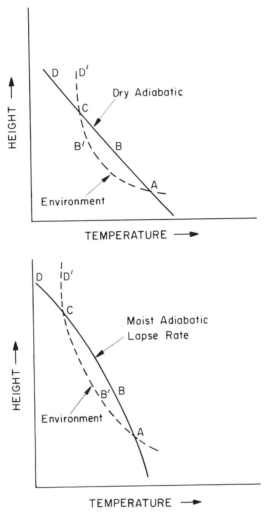

Fig. 3.17 Diagram of stable and unstable environmental conditions with respect to dry and moist adiabatic lapse rates. In each case the air at point A is unstable, while that at point C is stable.

$$\frac{\partial \theta_e}{\partial z} \geq 0 \qquad (3.164)$$

for moist convection.

Convective adjustment parameterization

In many of the present atmospheric models a scheme of convective adjustment is used to simulate convective overturning in the atmosphere, even though it oversimplifies the physical process (see Anthes, 1984, for a review of the uses and adaptations of convective adjustment). Manabe, Smagorinsky, and Strickler (1965) were the first to use this method in a full-scale atmospheric general circulation model, although it had been used in limited atmospheric modeling studies before then. In fact early researchers in atmospheric modeling recognized that the lapse rate in primitive equation models must be kept stable in order not to have the model generate small-scale unresolvable motions. This meant at the very least that the lapse rate must always be less than the dry adiabatic lapse rate, and possibly less than the moist lapse rate if the atmosphere is saturated. Taking the derivative of the natural logarithm of the potential temperature and assuming the hydrostatic approximation (3.53) and the equation of state (3.56), one obtains from (3.50) the following:

$$\frac{1}{\theta}\frac{\partial \theta}{\partial z} = \frac{1}{T}\left(\Gamma + \frac{\partial T}{\partial z}\right) \qquad (3.165)$$

where $\Gamma = g\, C_p^{-1}$ is the dry adiabatic lapse rate defined earlier (after (3.47)). The corresponding equation for the moist adiabatic lapse rate Γ_m is

$$\frac{1}{\theta_e}\frac{\partial \theta_e}{\partial z} = \frac{1}{T}\left(\Gamma_m + \frac{\partial T}{\partial z}\right) \qquad (3.166)$$

In early atmospheric models the right-hand sides of (3.165) and (3.166) were forced to be ≥ 0 in order to maintain stable atmospheric conditions. Depending upon whether the model atmosphere is unsaturated or saturated, $-\partial T/\partial z$ is set $\leq \Gamma$ or Γ_m. Also, the moisture distribution is adjusted to prevent a supersaturated condition. If there is excess moisture, it is allowed to fall out of the column with a resultant addition of latent heat to the atmosphere. In order to start the adjustment process the atmosphere is first examined with the conditions (3.165) and (3.166). If no layers are unstable, then the lapse rate is not modified. If, on the other hand, the model vertical temperature profile is unstable, i.e., the right-hand side of (3.165) or (3.166) is less than 0, then these equations are used in conjunction with the constraint that the total energy is conserved. This constraint says in effect that the sum of internal, potential, and latent energies does not change during the adjustment process. A convenient way to do this is to define quantities called dry and moist static energies, h_{dry} and h_{moist}:

$$h_{dry} = C_p T + gz \qquad (3.167)$$

and

$$h_{moist} = C_p T + gz + Lq \qquad (3.168)$$

where the internal energy is $C_p T$, the potential energy is gz, and the latent energy is Lq. The vertical integrals of (3.167) and (3.168) are constant before and after the adjustment. In order to solve the system one must use (3.154) relating q_s and e_s and (3.157) relating e_s and T.

The convective adjustment scheme often produces excessive precipitation, especially in the lower troposphere, and also produces temperatures in the lower troposphere that are colder than the observed temperatures. The major merit of the scheme is that it is simple and well understood. As Anthes (1984) points out, there are many variations of this scheme, such as using relative humidities less than 100% or assuming that convective adjustment occurs only over a fraction of the area, the latter assumption having been suggested by Krishnamurti et al. (1980) and Corby et al. (1977). These modifications usually make the lapse rates closer to the observed and improve the precipitation rates; however, they must be regarded as semiempirical in that the modifications are fits to the observations and are not derivable from basic physical laws.

Convective adjustments to reference lapse rates other than Γ or Γ_m have been suggested by Betts and Miller (1986), motivated largely by empirical studies of deep cumulus convection and by an idealized parameterization of the mixing of cloudy and clear air during shallow cumulus convection.

More refined schemes for cumulus convection

Whereas the convective adjustment scheme examines the vertical temperature and moisture distribution to determine whether convection occurs, an alternative scheme developed by Kuo (1974; also see Haltiner and Williams, 1980, for more details) uses the low-level convergence of moisture into a region and low-level lifting as determining conditions of whether cumulus convection is taking place. There have been numerous experimental studies showing that in areas of active precipitation, especially in the tropics, the low-level wind patterns show convergence of moisture-laden air inflow into a restricted region. Kuo presents the rate of moisture accession per unit horizontal area, M_t, as

$$M_t = -\frac{1}{g} \int_{p=p_t}^{p=p_0} \nabla \cdot (qV)\, dp + EVAP_{\text{surface}} \qquad (3.169)$$

where the first term is the convergence of moisture (water vapor) into a vertical column from the top of the troposphere (where $p = p_t$) to the earth's surface (where $p = p_0$), and the second term, $EVAP_{\text{surface}}$, is the surface evaporation into that column. The integral is negative if there is low-level convergence taking place and positive if divergence is occurring.

Kuo (1974) assumes that a fraction, $(1 - b)$, of the total water vapor convergence, M_t, is condensed from water vapor and precipitated, while the remaining portion, b, is stored in the atmosphere, increasing the humidity of the vertical column. The heating due to cumulus convection, Q_{cu}, is

$$Q_{cu} = L\overline{C^*} - C_p \left[\frac{\partial \overline{F}_{VH}}{\partial p} + \nabla \cdot \overline{F}_{HH} \right] \qquad (3.170)$$

where $\overline{C^*}$ represents the condensation rate occurring in cumulus convection, and $\overline{F}_{VH}$ and $\overline{F}_{HH}$ represent, respectively, vertical and horizontal fluxes of sensible heat within the column. The overbar denotes a large-scale (100 km or more) average. The terms in the square brackets are vertical and horizontal convergence of sensible heat flux. The condensation rate is assumed to be proportional to the large-scale moisture convergence:

$$\overline{C^*} = \frac{g(1 - b)M_t}{(p_b - p_t)} N(p) \qquad (3.171)$$

where, following Kuo (1974) and Anthes (1977), $N(p)$ is the vertical density function for condensation. This function is usually made proportional to the difference between cloud temperature, T_c, and the temperature of the environmental air outside the cloud, T:

$$N(p) = \frac{(T_c - T)}{(p_b - p_t)^{-1} \int_{p_t}^{p_b} (T_c - T)\, dp} \qquad (3.172)$$

Note that heating is largest when $(T_c - T)$ is largest. The fraction of large-scale water vapor convergence, b, which is stored in the atmosphere is assumed to be a function of the relative humidity, r. Following Donner et al. (1982), this function is:

$$b = 1 - \frac{1}{0.6p_g} \int_{0.4p_g}^{P_g} r \, dp \qquad (3.173)$$

where p_g is the atmospheric pressure at ground level and the factors 0.6 and 0.4 limit the integration to be performed over the lowest part of the atmosphere only. Anthes (1977) presents a more general functional form with empirical parameters also dependent on relative humidity.

The time rate of change in moisture due to cumulus convection has an equation similar to that for heating, (3.170), namely:

$$\frac{dq}{dt} = -\overline{C^*} - \left[\frac{\partial \overline{F}_{VM}}{\partial p} + \nabla \cdot \overline{F}_{HM} \right] \qquad (3.174)$$

where $\overline{F}_{VM}$ and $\overline{F}_{HM}$ are the vertical and horizontal fluxes of moisture by cumulus convection. In order to obtain the fluxes $\overline{F}_{VH}$ and $\overline{F}_{VM}$ for (3.170) and (3.174), one-dimensional cloud models are used to obtain the in-cloud vertical motion and temperature.

Independent verification of the Kuo (1974) scheme is difficult because of limited observations; however, several attempts have been made (see Krishnamurti et al., 1983, and Anthes, 1984). Anthes (1984) shows several variations on the Kuo scheme used in modeling mesoscale or small storm systems that improve the moisture, temperature, and precipitation forecasts. There are still some semiempirical factors that have to be fitted to obtain good agreement with observations. Donner, Kuo, and Pitcher (1982) and Tiedtke (1984) have shown comparisons between the convective adjustment and the Kuo scheme in general circulation modeling studies.

Another scheme that is being employed by some climate modelers is the Arakawa-Schubert (AS) scheme (see Arakawa and Schubert, 1974; Haltiner and Williams, 1980; Lord, 1982; Anthes, 1984; Hack et al., 1984). The derivation of the method is beyond the scope of this book, but it may be useful to outline some of its major attributes. First, it allows for the coexistence of a spectrum of idealized model clouds ("subensembles"), such as a mixture of shallow and deep clouds, that interact with each other as well as with the large-scale environment. Each member of the spectrum has its own mass, heat, and moisture budget, and is uniquely characterized by a fractional entrainment rate. Second, the scheme interacts strongly with the boundary layer, thereby differing from the Kuo method, which uses moisture convergence and surface evaporation as the key variables. In order to determine the vertical transports accomplished by the spectrum of clouds, Arakawa and Schubert introduce the cloud work function, a measure of the efficiency of kinetic energy generation by buoyancy forces (a form of conversion of internal and potential energy to kinetic energy). This cloud work function is largely determined by the thermodynamic characteristics of the ensemble of cumulus clouds with respect to the thermodynamic properties of the large-scale environment, including processes such as rising motion inside the clouds, induced subsidence (sinking) between the clouds, and the evaporation of detrained cloud air. All these processes are innovatively linked by assuming an approximate equilibrium of the cloud work function with respect to time. Thus, the generation of moist instability by the large-scale flow is balanced with cumulus convective processes that destroy the instability.

The Arakawa-Schubert method is an attempt to make the interaction of parameterized cumulus convection with the explicitly predicted large-scale fields more realistic. As yet it has not gained widespread acceptance because many of the assumptions are based upon relatively poorly observed quantities. Perhaps with time and more verification of data it or one of its variants will become the preferred method of handling cumulus convection in numerical models since it is based upon more realistic cloud-process physics than the other methods described above. Its major drawback is that it can require substantially more computational time than other parameterization schemes. Tiedtke (1984) has carried out a general circulation model comparison study of the three types of cumulus convective schemes discussed in this section, showing some of their respective merits and shortcomings.

Surface Processes

Boundary fluxes at the earth's surface

The treatments of boundary fluxes of momentum, moisture, and sensible heat between the earth's surface and the atmosphere are very important for all components of the climate system. Most atmospheric modeling studies make use of simple bulk formulae for the transfer of quantities within the constant flux or Prandtl layer, which is 100 to 200 m above the surface. All assume that the transfer process is proportional to the local gradient between the surface and atmosphere, multiplied by the wind speed.

The boundary flux formulae for momentum are

$$\tau_\lambda = \rho C_D V_a u_a \tag{3.175}$$

$$\tau_\phi = \rho C_D V_a v_a \tag{3.176}$$

while that for sensible heat is

$$H = \rho C_p C_H V_a (\theta_s - \theta_a) \tag{3.177}$$

and that for evaporation is

$$LE = \rho L C_E V_a (q_s - q_a) \tag{3.178}$$

where ρ is air density, C_D is the drag coefficient or momentum transfer coefficient, C_H is the transfer coefficient for sensible heat, C_E is the transfer coefficient for moisture, L is the latent heat of evaporation, $V_a = (u_a, v_a)$ is air velocity, V_a is velocity magnitude,

$$V_a = |V_a| = \sqrt{u_a^2 + v_a^2} \, , \tag{3.179}$$

and θ_a and q_a are boundary layer potential temperature and mixing ratio. θ_s and q_s are potential temperature and mixing ratio at the earth's surface. Over the ocean θ_s is computed from the temperature of the ocean surface, and the surface is assumed saturated, so that q_s is a function of temperature alone and is computed from (3.154).

Approximate transfer coefficients have been determined from both theoretical considerations and observational studies. For ocean surfaces and smooth surfaces

$$C_D \simeq C_H \simeq C_E \simeq 10^{-3} \tag{3.180}$$

with larger values of about 3×10^{-3} over rough terrain or regions where there is considerable boundary convection. More detailed drag coefficient formulations have been devised that take into account whether the boundary layer is stable or unstable (see discussion following (3.165) and (3.166)).

Above the surface boundary layer adjacent to the earth's surface exists the planetary boundary layer, where the wind turns with height. Within this region the balance of forces can be approximated by the Coriolis, pressure gradient, and frictional forces through the following simplification of (3.51)–(3.52):

$$-fv = -\frac{1}{\rho a \cos \phi} \frac{\partial p}{\partial \lambda} + F_\lambda \tag{3.181}$$

$$+fu = -\frac{1}{\rho a}\frac{\partial p}{\partial \phi} + F_\phi \qquad (3.182)$$

The frictional term can be expressed as the vertical gradient of stress, so that

$$F_\lambda = \frac{1}{\rho}\frac{\partial \tau_\lambda}{\partial z} \qquad (3.183)$$

$$F_\phi = \frac{1}{\rho}\frac{\partial \tau_\phi}{\partial z} \qquad (3.184)$$

and the stress can in turn be related to the vertical gradient of wind shear

$$\tau_\lambda = \rho K_m \frac{\partial u}{\partial z} \qquad (3.185)$$

$$\tau_\phi = \rho K_m \frac{\partial v}{\partial z} \qquad (3.186)$$

where K_m is the vertical eddy or transfer coefficient for momentum. Note that if the vertical wind shear is large, as one would expect near the earth's surface since the wind must approach zero at the surface, then the momentum transfer is also large. On the molecular scale, K_m takes on values appropriate for those space scales; however, in atmosphere and ocean models the space scales are much larger. In fact many of the eddies have dimensions of the order of the spatial grid structure. Furthermore, the character of the eddies is strongly influenced by the vertical static stability defined in (3.163) and (3.164) and the preceding discussion. If the layer is unstable there is strong vertical mixing and K_m is large; if the layer is stable there is no strong coupling and K_m is small. A useful measure of this stability is the Richardson number (named after the same Richardson who performed the early numerical simulation discussed earlier), Ri, which is defined as

$$Ri = \frac{g}{\theta}\frac{\partial \theta/\partial z}{(\partial V/\partial z)^2} \qquad (3.187)$$

where $V = |\boldsymbol{V}|$ and $(\partial V/\partial z)$ is the vertical wind (or current) shear, or vertical gradient of wind (or current) velocity. An alternate form of the Richardson number often used in ocean models is

$$Ri = g\frac{(1/\rho)\,(\partial\rho/\partial z)}{(\partial V/\partial z)^2} \tag{3.188}$$

Richardson in 1925 determined the importance of this ratio through examination of energy transformations between turbulent and nonturbulent flow by balancing the effects of negative buoyancy with the production of shear turbulence. If the former were dominant then nonturbulence would be maintained, and if the latter were dominant then turbulence would prevail. An easy way to look at the Richardson number is that if the shear is large or the lapse rate is unstable, turbulence is likely.

Equations (3.187) and (3.188) can also be uniquely obtained from nondimensional considerations (e.g., Obukhov, 1971). There are three types of situations that can be determined from Ri alone. If $Ri > 0$, then $\partial\theta/\partial z$ or $\partial\rho/\partial z > 0$ and the stratification of the atmosphere (or ocean) is stable, with turbulence inhibited. For values of Ri near a critical value, Ri_c, there is a transition between turbulent and laminar (smooth) flow. If $Ri < 0$, then the flow is fully turbulent. Many investigators have estimated values of Ri_c. Richardson thought it should be 1; however, he neglected the effects of viscous dissipation. If such dissipation is taken into account, then a value of 0.25 is obtained from theoretical considerations.

Returning to boundary layer considerations, mixings of momentum, moisture, and sensible heat above the surface mixed layer are determined largely by transfer processes, which are dependent upon vertical eddy (or turbulent) transfer coefficients. Ideally these K values would take into account small-scale or subgrid-scale mixing and its dependence upon the local stratification and vertical wind shear (current shear for the ocean). In calculating the vertical transfer of sensible heat, the air temperature can be separated into an average value, $\overline{T}$, over the horizontal grid area and a fluctuating or eddy part, T', where $T' = T - \overline{T}$. Using a concept called Reynolds averaging, $\overline{T'} = 0$ since $\overline{T} - \overline{\overline{T}} = 0$. Similarly the vertical wind speed w can be separated to $w = \overline{w} + w'$, and $\overline{w'} = 0$. However, products of fluctuating quantities may not average to zero, so that although $\overline{T'} = 0$ and $\overline{w'} = 0$, $\overline{T'w'}$ may not be 0 if, for instance, upward moving air is warm and downward moving air is cold. Formally, $\overline{T'w'}$ can easily be calculated to be $\overline{Tw} - \overline{T}\,\overline{w}$ by expanding $\overline{Tw} = \overline{(\overline{T} + T')(\overline{w} + w')}$. Since the model layer represents $\overline{T}\,\overline{w}$ in the vertical advection terms of the first law of thermodynamics, the eddy or subgrid-scale vertical transfer of sensible heat can be represented by using a mixing coefficient:

$$\overline{\rho T' w'} = -\rho K \frac{\partial T}{\partial z} \tag{3.189}$$

The more general equations for subgrid-scale transfer of sensible heat and moisture are

$$h = \rho K_h \frac{\partial \theta}{\partial z} = \rho K_h \left(\Gamma + \frac{\partial T}{\partial z} \right) \tag{3.190}$$

and

$$r = \rho K_q \frac{\partial q}{\partial z} \tag{3.191}$$

where K_h and K_q, respectively, are the transfer coefficients for sensible heat and moisture by the subgrid-scale eddies. These are often assumed to have the same values as K_m. A Richardson dependence can be further assumed so that

$$K_h = f(Ri) \tag{3.192}$$

K_h is small when the atmosphere is stable and large when it is unstable. Deardorff (1966, 1972) formulates one such functional relationship, which is currently used in several climate models.

Computation of surface temperature and hydrology

In early atmospheric modeling studies the surface of the earth was treated very simply. However, as models have become more sophisticated the treatment has come closer to the true complexity of the actual earth, incorporating a variety of surface types: ocean, dry land, wet land, desert, snow-covered surfaces, sea ice, and vegetated surfaces. The latter is among the most crudely treated in present climate models (see Dickinson (1983) and Sud and Fennessy (1982) for details on the inclusion of vegetation effects). Since different modeling groups treat the computation of surface temperature and hydrology in somewhat different manners, the emphasis here will be to show the fundamental physics rather than providing a detailed description of the differences in each modeling study.

The temperature change at the surface results from a balance of energy fluxes to and from the atmosphere, the energy flux due to the conduction of heat between the surface and some point under the ground, and the energy flux due to the temperature change itself. Since the earth is a relatively poor conductor of heat, the surface temperature change in land regions is determined largely by the fluxes of heat to and from the atmosphere. The method for computing surface temperature shown below assumes that a thin layer of surface is being considered without heat conduction from subsurface layers. Depending upon time and location, this neglect of heat conduction can be a serious shortcoming.

The equation for predicting surface temperature for a simple surface layer without a snow cover is

$$\rho c \frac{\partial T_*}{\partial t} = \frac{[S_* + F^\downarrow - F^\uparrow - H - LE]}{\Delta z} \qquad (3.193)$$

where ρ is the density of the surface, c the specific heat of the surface type, T_* the temperature of the surface layer, S_* the absorbed solar flux at the earth's surface, $F^\downarrow$ the downward infrared flux, $F^\uparrow$ the upward infrared flux, H and LE the sensible and latent heats, respectively, and Δz the thickness of the surface layer.

The surface computation can be made more realistic by including the effects of a diurnal cycle in S_*. As mentioned earlier, the transfers of sensible heat, moisture, and momentum are highly dependent upon the local static stability of the atmosphere. If diurnally-averaged solar flux is imposed, this may provide a different atmospheric structure than one in which a more realistic diurnal cycle is used. Also, if the diurnal cycle is included, then heat conduction from subsurface layers should be included in order to obtain the proper diurnal change in surface temperature.

In atmospheric models uncoupled to ocean calculations, the surface temperature of the oceans usually is specified from climatological atlases as a function of location and time. The physical basis for using climatological values is that oceans have enormous heat capacity compared to land, so that over a time integration of less than several months the ocean temperature can be assumed to be constant. In reality this is not strictly valid since even small temperature changes can have a profound effect on the atmosphere, as shown in the El Niño events discussed in Chapter 2. More on this subject will be discussed when coupled atmosphere-ocean climate models and simulations of the impact of ocean temperature anomalies on the atmosphere are considered in Chapter 6.

As mentioned earlier the prediction of surface physics and its interaction with vegetation usually are treated crudely in present climate models. One aspect is the treatment of surface hydrology. The surface albedo and transfers of latent and sensible heat will be quite different over a snow-covered region than over dry land. The difference in albedo often leads to large differences in temperature and transfers of heat and moisture to the atmosphere. Likewise, the fact that one surface is saturated with moisture while another (such as a desert) is dry could have a large effect on the regional climate, again due to the differences in the transfer of moisture to the atmosphere. Dickinson (1983) reviews the evapotranspiration processes that should be included in climate models. Figure 3.18 shows a schematic indicating the basic processes. Precipitation falls on both the vegetation and the underlying surface. Some of the precipitation is intercepted by the vegetation and is reevaporated, while the rest drops to the ground. The surface moisture can infiltrate the surface and contribute to surface flow in streams, or percolate to a deeper layer, to say 1 m, or even become groundwater, depending on the porous properties of the ground. In the root zones of plants (10 cm to 1 m) some of the surface moisture can be taken up by the plant roots and then transpired by plant leaves. The amount of transpiration is strongly dependent upon the same quantities that determine surface energy fluxes discussed earlier. In particular, the wind, solar intensity, albedo, and moisture and temperature of the surface air are involved. Most models do not take into account all these complexities in the detail discussed by Dickinson; however, greater realism can be expected in the future, especially for detailed simulations of regional climate.

One of the widely used methods of incorporating snow cover and soil moisture in numerical climate models follows that first introduced by Manabe (1969a,b) resulting from Soviet observational studies by Romanova (1954). Romanova found that for plains and forest regions most of the moisture change takes place in the top 1 m of soil, which usually encompasses the root zone of most vegetation. Alpatev (1954) further found that if the soil is 70–80% saturated, then the surface evaporates moisture at the potential rate (i.e., as a saturated surface). Obviously, the surface in reality is much more complex than the treatment by Manabe, but the method at least has the virtue of incorporating the fundamental processes of precipitation, evaporation, and water storage in the soil.

The basic equation for soil moisture prediction is

$$\frac{\partial W}{\partial t} = P - E + S_m \qquad (3.194)$$

where W is total soil moisture in meters stored in a surface layer of soil, P is the rain precipitation rate, E is the surface

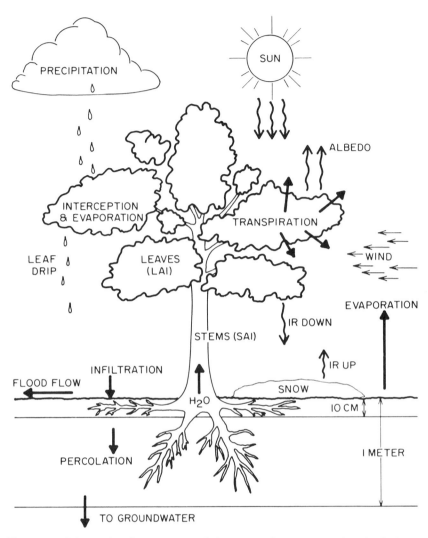

Fig. 3.18 Schematic of processes and features relevant to surface hydrology and evapotranspiration. [From Dickinson (1984).]

evaporation rate, and S_m is the snow melt rate, which is computed from the surface energy balance. The evaporation rate can be related to the soil moisture in the following manner:

$$\text{if} \quad W \geq W_c \qquad \text{then} \quad E = E_{ap} \qquad (3.195)$$

and

$$\text{if} \quad W < W_c \qquad \text{then} \quad E = E_{ap}\frac{W}{W_c} \qquad (3.196)$$

where W_c is a critical value of soil moisture and E_{ap} is the potential evaporation rate for a saturated surface. The above equations state that if the soil moisture is greater than W_c, then the evaporation rate is a maximum, E_{ap}, and if the soil moisture is less than W_c, then the evaporation rate is reduced linearly as a function of W. Manabe assumes that field capacity for soil moisture, W_{Fc}, is 0.15 m and W_c is 75% of that value.

In a similar approach Manabe uses the following snow cover prediction equation:

$$\frac{\partial S}{\partial t} = P - E - S_m \qquad (3.197)$$

where S is the snow cover amount in liquid water equivalent, P is snow precipitation rate, and E is rate of sublimation. It is possible for the snow cover to melt, thereby increasing soil moisture (3.194) and decreasing S (3.197). Also, if rain falls on snow, it is assumed that the snow does not hold the water but that the water instead seeps through to contribute to soil moisture. If snow covers the soil, no evaporation of soil moisture is allowed to take place, although sublimation is allowed. The determination of whether snow or rain is falling usually is based upon whether the lower tropospheric model layers have temperatures below or above freezing.

The existence of snow melt is determined by the sign of S_m, calculated as

$$S_m = \frac{1}{L_f}[S_* + F^\downarrow - F^\uparrow - H - LE] \qquad (3.198)$$

where L_f is the latent heat of fusion and the other terms are those of (3.193). If, in the presence of a snow cover, setting the surface temperature T_* at 0°C yields $S_m > 0$, then melting is inserted in (3.194) and (3.197). Otherwise, S_m is set equal to zero. The runoff, R_f, into the ocean is computed whenever $W > W_{Fc}$ as

$$R_f = P - E \qquad (3.199)$$

Usually the runoff is not translated into a river flow because that assumes a detailed description of the terrain, which most climate models do not yet incorporate.

Ocean Models

Much of the basis for ocean modeling has been described above in the discussion and equations for fluid flow, fluid thermodynamics, and the surface fluxes of radiation, evaporation, and sensible heat. The modeling of the world's oceans for climate purposes is similar in many respects to the modeling of the atmosphere in terms of the basic physical laws. There are significant differences, however, in that the time and space scales of relevance are different, the radiative flux considerations are more complex for the atmosphere, and water motions are greatly constrained by the ocean bottom topography. The dynamics that reflect the time and space scales of primary importance to climate in the oceans, as in the atmosphere, are those that respond to low frequencies and large-scale motions. However, the dominant time scales in the oceans are much longer than those in the atmosphere because of the greater thermal inertia. At the same time, the horizontal length scales of the relevant motions in the ocean are approximately one-tenth of those in the atmosphere.

One measure of the relevant horizontal length scales is the radius of deformation, L, defined as the ratio of the phase speed of gravity waves, c_{gr}, to the Coriolis parameter:

$$L = c_{gr}/f \tag{3.200}$$

This is a critical parameter in geophysical fluid dynamics. In the atmosphere $c_{gr} \simeq 100$ m s^{-1}, so that for $f \simeq 10^{-4}$ s^{-1}, $L \simeq 1000$ km, whereas in the ocean, where there are lesser density differentials, $c_{gr} \simeq 10$ m s^{-1}, which reduces L by an order of magnitude to 100 km. The importance of this arises because for scales smaller than L the waves disperse rapidly and eventually dissipate, while for scales larger than L the waves move slowly and are of lower frequency. It is important in atmosphere and ocean models to include motions with length scales larger than L, either explicitly or implicitly. Because of the smaller radius of deformation in the ocean, the horizontal grid size of an ocean model needs to be much smaller than that for an atmospheric model. As discussed in Chapter 2, the ocean has highly energetic mesoscale eddies. The role of these eddies must be considered in climate models, although not necessarily explicitly.

The first numerical ocean models were developed in the United States by Bryan (1963) and in the Soviet Union by Sarkisyan (see Sarkisyan, 1966, for a review of the early Soviet work, which took place in the 1950s). These models attempted to simulate the ocean dynamics with numerical rather than analytical methods because the geometry of ocean basins makes it impossible to find simple mathematical solutions to the governing equations. Since that time, ocean models have evolved along several fronts, with the most widely used climate-oriented ocean model in the mid-1980s being a model devised by Bryan (1969a) at the Geophysical Fluid Dynamics Laboratory/NOAA. A similar but computationally less efficient model was developed by Crowley (1968, 1970) using a different technique for handling gravity waves. Semtner (1974, 1986) and later Cox (1984) modified Bryan's model with respect to numerical details and wrote generalized codes that are extensively used throughout the ocean modeling community. It is their basic formulation of Bryan's model that will be described in some detail here.

As discussed earlier, the atmosphere is nearly always in hydrostatic balance with regard to large-scale motions. The hydrostatic approximation is an even better assumption in the oceans. Furthermore, since seawater is nearly incompressible, the equation of continuity (3.40) can be abbreviated by eliminating the time change of density, although density variations are generally included as far as buoyancy effects are concerned. By eliminating the derivative of density, (3.40) becomes

$$\frac{\partial w}{\partial z} = -\frac{1}{a \cos \phi} \frac{\partial u}{\partial \lambda} - \frac{1}{a \cos \phi} \frac{\partial}{\partial \phi} (v \cos \phi) \qquad (3.201)$$

Note that (3.201) is diagnostic in that if u and v are known, then w can be determined, so that a prognostic or prediction equation in time for w is not needed.

The equation of state for the ocean is significantly different from the ideal gas law (3.56), which is the equation of state for the atmosphere. For the ocean, the equation of state relates seawater density, ρ_w, to temperature, T, salinity, S, and pressure, p:

$$\rho_w = f(T, S, p) \qquad (3.202)$$

Several explicit approximations, of varying complexity, have been formulated for this relationship, with the one given by Bryan and Cox (1972) being convenient in terms of including the most important elements and yet remaining computationally efficient. Their approximation consists of a set of polynomial functions, one for each vertical level, of temperature, salinity, and pressure. The coefficients in these functions are determined by fitting to a standard formula of Knudsen (Fofonoff, 1962). As expected, cold saline waters have high densities, and warm, less saline waters have low densities.

A more elaborate equation of state has been defined by the Joint Panel on Oceanographic Tables and Standards (UNESCO, 1981). Density is obtained in a sequence of steps beginning with a calculation for the density of pure water as a fifth-order polynomial function of temperature. The second step adds to the density of pure water a nine-term nonpolynomial function of temperature and salinity to obtain the seawater density at 1 standard atmosphere. This is then divided by a separate nonpolynomial function of temperature, salinity, and pressure to obtain the generalized function for water density. The reader can find the details in Gill (1982), where it is also indicated that the resultant equation of state fits available measurements for pressures (beyond atmospheric pressures) of 0–1000 bars, temperatures of 0–40°C, and salinities of 0–42‰ to a standard error of 3.5 parts per million.

Following Semtner (1974, 1986), the remaining basic equations for ocean modeling will be presented in the spherical coordinate system, with depth, z, defined as negative downward from $z = 0$ at the surface. In this system the horizontal equations of motion are:

$$\frac{\partial u}{\partial t} + L(u) - \frac{uv \tan \phi}{a} - fv = -\frac{1}{\rho_0 a \cos \phi} \frac{\partial p}{\partial \lambda} + \mu \frac{\partial^2 u}{\partial z^2}$$

$$+ A_m \left\{ \nabla^2 u + \frac{(1 - \tan^2 \phi)u}{a^2} - \frac{2 \sin \phi}{a^2 \cos^2 \phi} \frac{\partial v}{\partial \lambda} \right\} \quad (3.203)$$

$$\frac{\partial v}{\partial t} + L(v) + \frac{u^2 \tan \phi}{a} + fu = -\frac{1}{\rho_0 a} \frac{\partial p}{\partial \phi} + \mu \frac{\partial^2 v}{\partial z^2}$$

$$+ A_m \left\{ \nabla^2 v + \frac{(1 - \tan^2 \phi)v}{a^2} + \frac{2 \sin \phi}{a^2 \cos^2 \phi} \frac{\partial u}{\partial \lambda} \right\} \quad (3.204)$$

where

$$L(\alpha) = \frac{1}{a \cos \phi} \left[\frac{\partial}{\partial \lambda}(u\alpha) + \frac{\partial}{\partial \phi}(\cos \phi \, v\alpha) \right] + \frac{\partial}{\partial z}(w\alpha) \quad (3.205)$$

and α is a dummy variable, ρ_0 is a constant approximation to the density of seawater, μ is the vertical eddy viscosity coefficient, and A_m is the horizontal eddy viscosity coefficient. The acceleration, Coriolis, and pressure gradient terms are similar to the corresponding terms in the equations of motion for the atmosphere,

(3.51) and (3.52), while the frictional term for the atmosphere is now replaced by vertical and horizontal viscous terms. The involved form of the horizontal viscous terms (curly brackets) is needed to conserve angular momentum properly (see Bryan, 1969a).

The hydrostatic equation for the ocean is identical in form to that for the atmosphere (3.53):

$$\frac{\partial p_w}{\partial z} = -\rho_w g \tag{3.206}$$

Here the water density ρ_w is no longer necessarily assumed constant, in contrast to ρ_0 in (3.203) and (3.204).

The first law of thermodynamics for the oceans differs from that for the atmosphere (3.55) because of the assumption of incompressibility, which allows elimination of the term involving the time rate of change of density, and the replacement of the nonadiabatic term Q by explicit terms for the vertical and horizontal eddy diffusions of heat:

$$\frac{\partial T}{\partial t} + L(T) = \kappa \frac{\partial^2 T}{\partial z^2} + A_H \nabla^2 T \tag{3.207}$$

where κ and A_H are the vertical and horizontal eddy diffusivity coefficients, respectively, and the function $L(\ )$ remains as defined in (3.205). The specific heat for seawater does not appear explicitly in (3.207) because it has been factored out of each term. Just as the latitudinal differential heating of the atmosphere can cause density differences in the atmosphere, which in large part drive the wind fields, the temperature and salinity differences in the ocean can cause density differences that drive ocean circulations. Such circulations are termed *thermohaline*, *thermo* referring to the temperature influence and *haline* to the salinity influence. The prediction equation for salinity is given by

$$\frac{\partial S}{\partial t} + L(S) = \kappa \frac{\partial^2 S}{\partial z^2} + A_H \nabla^2 S \tag{3.208}$$

Methods of solving (3.201)–(3.208) numerically will be discussed in Chapter 4.

The parameterization of vertical mixing has been improved recently by making the coefficients κ and μ functions of the local Richardson number (3.187 and 3.188). Pacanowski and Philander (1981) have found that if the vertical eddy coefficients are too large the mixing of warmer water in the mixed layer near the surface will be excessive, which results in the thermocline not having strong vertical temperature gradients. There is observational evidence, however, that indicates rapid vertical mixing when the local shear is large. Pacanowski and Philander (1981) suggest using the following formulae:

$$\mu = \frac{\mu_0}{(1 + \alpha Ri)^2} + \mu_b \qquad (3.209)$$

$$\kappa = \frac{\mu}{(1 + \alpha Ri)} + \kappa_b \qquad (3.210)$$

where μ_0, μ_b, κ_b, and α are of the order of 50 cm^2 s^{-1}, 1 cm^2 s^{-1}, 0.1 cm^2 s^{-1}, and 5, respectively. Since the ocean is generally approximately thermally stable except near the polar regions (where bottom water may be forming), the instability is primarily shear-generated, so that Ri becomes small (3.187) and thus the denominator in (3.209)–(3.210) becomes small as well. If Ri decreases to below about 0.2, vertical mixing increases dramatically. The values indicated above for the parameters μ_0, μ_b, κ_b, and α were used by Pacanowski and Philander (1981) in a model with relatively high resolution in the vicinity of the upper mixed layer. Coarser vertical resolution models may require different values.

Other parameters that often differ from one model to another include the horizontal diffusion coefficients A_m and A_H. In particular, in coarse resolution models, these coefficients are modified to incorporate the effects of subgrid-scale eddies. Although most modeling studies use constant values for A_m and A_H, a few introduce flow-dependent parameterizations. Further improvements in horizontal eddy parameterizations are expected from high-resolution eddy-resolving ocean modeling studies in limited areas.

Following Bryan (1969a) and Semtner (1974, 1986), the boundary conditions at the bottom of the ocean, where $z = -H(\lambda, \phi)$, are as follows:

$$\rho_0 \mu \frac{\partial}{\partial z}(u, v) = 0 \qquad (3.211)$$

$$\rho_0 \kappa \frac{\partial}{\partial z}(T, S) = 0 \qquad (3.212)$$

These equations imply that at the ocean bottom there is zero stress and no vertical flux of sensible heat or salinity. The first equation is sometimes replaced by a quadratic law for bottom friction. The bottom boundary condition on the vertical motion is

$$w = -\frac{u}{a\cos\phi}\frac{\partial H}{\partial \lambda} - \frac{v}{a}\frac{\partial H}{\partial \phi} \qquad (3.213)$$

so that vertical motions at the bottom are induced by horizontal motions (u, v) interacting with the bottom topography, H. Horizontal flow impinging on an upslope will generate upward motion, and horizontal flow over a downslope will generate downward motion.

At the top of the ocean, where $z = 0$, fluxes of momentum (wind stress), heat, and moisture are either specified from atmospheric data, as in Bryan and Lewis (1979) and Meehl et al. (1982), or calculated with the use of atmospheric models. In either case, for flux continuity across the boundary,

$$\rho_0\mu\frac{\partial}{\partial z}(u, v) = (\tau_\lambda, \tau_\phi) \qquad (3.214)$$

and

$$\rho_0\kappa\frac{\partial}{\partial z}(T, S) = \left(\frac{1}{c_{pw}}H_{\text{ocn}}, \nu_s(E - P)S\right) \qquad (3.215)$$

where τ_λ and τ_ϕ are the momentum fluxes as defined in (3.175) and (3.176), H_{ocn} is the net heat flux into the ocean (positive for heating and negative for cooling), P is precipitation rate, E is evaporation rate, S is surface salinity, c_{pw} is specific heat of water, and ν_s is an empirical conversion factor. H_{ocn} can be obtained from a straightforward surface energy balance:

$$H_{\text{ocn}} = S_* + F^\downarrow - F^\uparrow - H - LE \qquad (3.216)$$

where the terms on the right-hand side were defined after (3.193), in the discussion of surface temperature computation. The term $(P - E)S$ can be interpreted as a salinity flux; when $P > E$ the water becomes less saline and when $P < E$ the water becomes more saline.

The typical climate-oriented ocean model has a rigid lid, with no vertical motion allowed at the top. This boundary condition of $w = 0$ at the top of the ocean automatically eliminates high frequency motions, which are of little importance to the lower frequency motions of climate processes, and allows a longer time step. This device was originally used by Smagorinsky (1963) for the atmosphere in one of the first atmospheric general circulation models that employed the primitive equations. Smagorinsky assumed an incompressible atmosphere for this simulation. Crowley (1970), in an ocean simulation, allowed for a free upper boundary, limiting the time increments used in the forecast of new variables. (Some of the numerical aspects of solving such systems are discussed in Chapter 4.)

The lateral boundaries of the oceans usually are treated as nonslip, meaning that the normal and tangential horizontal velocities at these boundaries are zero. The horizontal fluxes of sensible heat and salinity are also typically set equal to zero at these boundaries. Following Semtner (1974), the vertical integration of the hydrostatic equation (3.206) yields

$$p(z) = p_s + \int_z^0 g\rho_w \, dz' \tag{3.217}$$

where ρ_w is computed as a function of T, S, and p (as in (3.202)). There is no prediction equation for the surface pressure, p_s; however, it can be computed as a diagnostic variable for analysis purposes.

If the horizontal momentum equations (3.203)–(3.204) are differentiated with respect to z, the following equations result:

$$\frac{\partial}{\partial t}\left(\frac{\partial u}{\partial z}\right) - f\frac{\partial v}{\partial z} = \frac{g}{\rho_0 a \cos\phi}\frac{\partial \rho_w}{\partial \lambda} + \frac{\partial}{\partial z}(G^\lambda) \tag{3.218}$$

$$\frac{\partial}{\partial t}\left(\frac{\partial v}{\partial z}\right) + f\frac{\partial u}{\partial z} = \frac{g}{\rho_0 a}\frac{\partial \rho_w}{\partial \phi} + \frac{\partial}{\partial z}(G^\phi) \tag{3.219}$$

where G^λ and G^ϕ combine all the nonlinear and viscous terms, and the hydrostatic approximation (3.206) has been used to replace the pressure change with a density change.

Equations (3.218) and (3.219) allow prediction of the vertical shear of velocity. An additional specification of a pair of two-dimensional quantities, such as the vertically averaged velocity components, $(\bar{u}, \bar{v})$, is needed. These averaged components can be written in terms of a volume transport function, ψ, for the vertically integrated flow:

$$\bar{u} = \frac{1}{H}\int_{-H}^0 u \, dz = -\frac{1}{Ha}\frac{\partial \psi}{\partial \phi} \tag{3.220}$$

$$\bar{v} = \frac{1}{H}\int_{-H}^0 v \, dz = \frac{1}{Ha\cos\phi}\frac{\partial \psi}{\partial \lambda} \tag{3.221}$$

The existence of ψ is guaranteed by the nondivergent nature of the vertically integrated flow, which is a result of the assumption of incompressibility and the rigid lid condition, i.e.,

$$\frac{1}{a\cos\phi}\frac{\partial}{\partial\lambda}\left(\int_{-H}^{0}u\,dz\right)+\frac{1}{a\cos\phi}\frac{\partial}{\partial\phi}\left(\cos\phi\int_{-H}^{0}v\,dz\right)=0$$

$$(3.222)$$

This relation can be verified by integrating the continuity equation (3.201) by parts and applying the boundary conditions $w=0$ at $z=0$ and (3.213) for w at $z=-H$.

A prediction equation for ψ that does not include the surface pressure, p_s, is obtained by forming the vertical averages of (3.203) and (3.204) and taking the curl ($\nabla\times$) of these. In terms of ψ and p_s, the vertically averaged equations are as follows:

$$-\frac{1}{Ha}\frac{\partial}{\partial\phi}\left(\frac{\partial\psi}{\partial t}\right)-\frac{f}{Ha\cos\phi}\frac{\partial\psi}{\partial\lambda}=-\frac{1}{a\rho_0\cos\phi}\frac{\partial p_s}{\partial\lambda}$$

$$-\frac{g}{Ha\rho_0\cos\phi}\int_{-H}^{0}\int_{z}^{0}\frac{\partial\rho}{\partial\lambda}\,dz'\,dz$$

$$+\frac{1}{H}\int_{-H}^{0}G^{\lambda}\,dz \qquad\qquad (3.223)$$

$$\frac{1}{Ha\cos\phi}\frac{\partial}{\partial\lambda}\left(\frac{\partial\psi}{\partial t}\right)-\frac{f}{Ha}\frac{\partial\psi}{\partial\phi}=-\frac{1}{a\rho_0}\frac{\partial p_s}{\partial\phi}$$

$$-\frac{g}{Ha\rho_0}\int_{-H}^{0}\int_{z}^{0}\frac{\partial\rho}{\partial\phi}\,dz'\,dz$$

$$+\frac{1}{H}\int_{-H}^{0}G^{\phi}\,dz \qquad\qquad (3.224)$$

where (3.217) has been used to expand p to two terms and G^{λ} and G^{ϕ} again combine the nonlinear and viscous terms of (3.203) and (3.204). By applying the $\mathbf{k}\cdot\nabla\times$ operator, given by

$$\mathbf{k}\cdot\nabla\times(q_1,q_2)=\frac{1}{a\cos\phi}\left[\frac{\partial q_2}{\partial\lambda}-\frac{\partial}{\partial\phi}(q_1\cos\phi)\right] \qquad (3.225)$$

on (3.223) as q_1 and (3.224) as q_2, we obtain, after multiplication by $a^2\cos\phi$:

$$\left[\frac{\partial}{\partial \lambda} \left(\frac{1}{H \cos \phi} \frac{\partial^2 \psi}{\partial \lambda \, \partial t} \right) + \frac{\partial}{\partial \phi} \left(\frac{\cos \phi}{H} \frac{\partial^2 \psi}{\partial \phi \, \partial t} \right) \right]$$

$$- \left[\frac{\partial}{\partial \lambda} \left(\frac{f}{H} \frac{\partial \psi}{\partial \phi} \right) - \frac{\partial}{\partial \phi} \left(\frac{f}{H} \frac{\partial \psi}{\partial \lambda} \right) \right]$$

$$= - \left[\frac{\partial}{\partial \lambda} \left(\frac{g}{\rho_0 H} \int_{-H}^{0} \int_{z}^{0} \frac{\partial \rho}{\partial \phi} \, dz' \, dz \right) \right.$$

$$\left. - \frac{\partial}{\partial \phi} \left(\frac{g}{\rho_0 H} \int_{-H}^{0} \int_{z}^{0} \frac{\partial \rho}{\partial \lambda} \, dz' \, dz \right) \right]$$

$$+ \left[\frac{\partial}{\partial \lambda} \left(\frac{a}{H} \int_{-H}^{0} G^{\phi} \, dz \right) \right.$$

$$\left. - \frac{\partial}{\partial \phi} \left(\frac{a \cos \phi}{H} \int_{-H}^{0} G^{\lambda} \, dz \right) \right] \quad (3.226)$$

Equation (3.226) is a prediction equation for ψ that requires solution of a second-order differential operator to obtain the function. It is necessary to specify boundary conditions for a solution in which the domain of ψ will be a multiply-connected region. For instance, the boundary could consist of a primary continent and several islands. The value of ψ must be spatially constant along each individual coastline in order to have vanishing normal velocities. On the primary continent, usually some connected region made up of a combination of existent continents, ψ is held constant in time as well; on an "island" ψ varies in response to the changing circulation. We require that p_s be a single-valued function, in the sense that a line integral of ∇p_s around the coastline of each island should vanish. By integrating (3.223) over λ and (3.224) over ϕ around each island, adding, and applying this line-integral condition, the following equation is obtained to predict the change of ψ on the island:

$$\int \left(- \frac{\cos \phi}{H} \frac{\partial^2 \psi}{\partial \phi \, \partial t} \, d\lambda + \frac{1}{H \cos \phi} \frac{\partial^2 \psi}{\partial \lambda \, \partial t} \, d\phi \right)$$

$$
= -\int \frac{g}{\rho_0 H} \left[\left(\int_{-H}^{0} \int_{z}^{0} \frac{\partial \rho}{\partial \lambda} \, dz' \, dz \right) d\lambda \right.
$$

$$
+ \left(\int_{-H}^{0} \int_{z}^{0} \frac{\partial \rho}{\partial \phi} \, dz' \, dz \right) d\phi \right]
$$

$$
+ \int \left[\left(\frac{a \cos \phi}{H} \int_{-H}^{0} G^{\lambda} \, dz \right) d\lambda \right.
$$

$$
+ \left(\frac{a}{H} \int_{-H}^{0} G^{\phi} \, dz \right) d\phi \right] \tag{3.227}
$$

Examples of simulations with this type of ocean model will be given in Chapter 5.

Quasi-geostrophic Ocean Circulation Model

The equations above can be considerably simplified by making a set of assumptions that tend to be adequate for many restricted ocean circulation studies. Since large-scale ocean currents approximate geostrophic conditions even more closely than do large-scale atmospheric flows (typical Rossby numbers R_0 for large-scale motions being 0.01–0.1 in the ocean and 0.1–10.0 in the atmosphere), the vorticity equation (3.81) can be simplified to

$$
\frac{\partial \varsigma}{\partial t} = -\boldsymbol{V} \cdot \boldsymbol{\nabla}(\varsigma + f) + f_0 \frac{\partial w}{\partial z} \tag{3.228}
$$

where the last term on the right derives from the $(\varsigma + f)D$ term in (3.81) by assuming $f \gg \varsigma$, f is a constant, labeled f_0, and the flow is incompressible, with continuity equation (3.201). The use of a constant f_0 in the final term of (3.228) is required to avoid spurious sources and sinks of energy, and thereby to obtain an energy-consistent set of equations (Gill, 1982). Note that in the middle term of (3.228) f is not assumed constant. The buoyancy changes due to density changes can be calculated based on a simplified density formulation whereby density is assumed, as a first approximation, to be a function of temperature alone, rather than of temperature, salinity, and pressure, as in (3.202):

$$
\rho_w = \rho_0(1 - \alpha T) \tag{3.229}
$$

Here ρ_0 is a constant reference density and α is a thermal expansion coefficient. Equation (3.229) can be modified easily to allow for haline (salinity-driven) circulations by adding another term, which is a function of the salinity content.

The hydrostatic equation (3.206) can now be written as

$$\frac{\partial p}{\partial z} = \alpha \rho_0 \, g \, T - \rho_0 g \qquad (3.230)$$

by inserting (3.229) into (3.206). If only the two largest terms in (3.82) are retained, the vorticity is expressed as $\nabla^2 \psi$, horizontal density variations are ignored, and the Coriolis parameter is assumed constant, then the divergence equation (3.82) can be approximated by

$$f_0 \nabla^2 \psi = \frac{1}{\rho_0} \nabla^2 p \qquad (3.231)$$

or further simplified by ignoring the Laplacian, in which case

$$\psi = \frac{p}{f_0 \rho_0} \qquad (3.232)$$

Furthermore, in view of the assumption of quasi-geostrophic flow, velocity V in (3.228) can be simplified from (3.80) by ignoring the velocity potential term, which in large-scale ocean circulations is typically at least an order of magnitude smaller than the streamfunction term:

$$V \simeq V_\psi = k \times \nabla \psi \qquad (3.233)$$

The thermodynamic equation for quasi-geostrophic motions in the ocean can be derived from (3.55) by first assuming an incompressible fluid, to obtain

$$c_{pw} \frac{dT}{dt} = Q \qquad (3.234)$$

then by ignoring the nonadiabatic term (i.e., ignoring the diffusivity terms of (3.207)), to obtain

$$\frac{dT}{dt} = 0 \qquad (3.235)$$

and finally by expanding the total derivative and inserting (into all except the vertical advection term) T as calculated from the hydrostatic equation (3.230) modified by (3.232). The resulting thermodynamic equation for oceanic quasi-geostrophic motions is:

$$\frac{\partial}{\partial t}\left(\frac{\partial \psi}{\partial z}\right) + V \cdot \nabla \left(\frac{\partial \psi}{\partial z}\right) + \frac{g \alpha}{f_0} \frac{\partial \overline{T}}{\partial z} w = 0 \qquad (3.236)$$

The last term in (3.234) involves the vertical temperature stratification and thereby the vertical stability of the fluid. The overbar implies that the vertical gradient of temperature is defined in an area-averaged sense. This term can be related to the Brunt-Väisälä frequency of small-scale vertical oscillations in a fluid having density or buoyancy variations (see Gill, 1982; Haltiner and Williams, 1980; or Charney and Flierl, 1981, for more thorough discussions of this derivation).

Another very useful concept in large-scale geophysical fluid dynamics of atmosphere and ocean circulations is potential vorticity. The quasi-geostrophic prediction equation for potential vorticity can be derived by substitution of w from (3.236) into (3.228) to obtain

$$\left[\frac{\partial}{\partial t} + V \cdot \nabla\right] [q + f] = 0 \qquad (3.237)$$

where

$$q = \varsigma + \frac{\partial}{\partial z}\left[\left(\frac{g\alpha}{f_0^2}\frac{\partial \overline{T}}{\partial z}\right)^{-1}\frac{\partial \psi}{\partial z}\right] \qquad (3.238)$$

is the potential vorticity and V is approximated by (3.233). Advection by the velocity potential component usually is ignored. Holland, Keffer, and Rhines (1984) and others have shown that the quasi-geostrophic equations (3.228)–(3.238) are capable of describing many of the observed features of ocean circulation. Nevertheless, most climate-oriented models of ocean circulation revert to the primitive equations since the quasi-geostrophic models are not valid under strong nonlinearity, when R_0 is on the order of 1, or near the equator, where (3.232) and (3.236) fail because the Coriolis parameter in the denominator approaches zero.

Results from several ocean modeling studies appear in Chapter 5, including examples of simulations of particular ocean basins, such as the Atlantic, and of the global ocean circulation.

Sea Ice Models

Sea ice models are of necessity somewhat different from ocean models and atmospheric models in that sea ice is a solid, not a fluid, is discontinuous in nature, with the ice cover complicated by multiple cracks, leads, and polynyas, and varies significantly in its distribution from one time period to another. In contrast to ocean and atmospheric models, whose calculations center on determining the *properties* of the water and air, sea ice models for use in large-scale climate simulations have primary goals of determining, at each location and each time step, whether ice exists and, if it does, then the amount of ice, including both ice thickness and areal ice concentration. Determination of the properties of the ice becomes secondary in importance, although necessary for the primary goals of determining ice distributions and amounts.

Calculations for climate-oriented sea ice models can be divided into two broad categories, both of which affect the determination of ice distributions. These categories are: (1) calculations concerned with the thermodynamics of the ice cover and (2) calculations concerned with ice dynamics. The thermodynamic calculations determine the thickness and temperature structure of the ice and are based on the principle of the conservation of energy. The dynamic calculations determine the motions of the ice and are based on the principle of the conservation of momentum. Both sets of calculations contribute to determining the open water or lead area within the ice cover. Depending partly on usage, some numerical models of the ice include only dynamics, some only thermodynamics, and some both dynamics and thermodynamics.

The thermodynamic calculations used in many sea ice models have derived in large part from the detailed and high-resolution one-dimensional (vertical) sea ice calculations of Maykut and Untersteiner (1971) for the central Arctic, the simplification of those calculations by Semtner (1976), and the three-dimensional representation, the flux formulations, and the lead parameterization presented in Washington et al. (1976) and in Parkinson and Washington (1979). These calculations center on balancing incoming and outgoing energy fluxes at the air/snow, snow/ice, and ice/water interfaces. The fluxes included are solar radiation, incoming and outgoing longwave radiation, sensible and latent heat, conduction through the ice and snow layers, an ocean heat flux, and the absorption and emission of energy due to the change of state between ice and water.

Calculations of ice dynamics have derived in large part from the work of Campbell (1964), the various modelers of the Arctic Ice Dynamics Joint Experiment (AIDJEX) (Coon, 1980), and Hibler (1979), all concentrating exclusively on the Arctic, and from the work of Parkinson and Washington (1979) simulating the ice conditions in both hemispheres. Basically, sea ice dynamics tend to be calculated by creating a momentum balance among the five major stresses acting on the ice: air stress, water stress, Coriolis force, the stress produced by dynamic topography (corresponding to the tilt of the sea surface), and internal ice resistance. The major difference among the ice dynamics in different sea ice models lies in the formulations of the five stresses, especially the internal ice stress.

Ice thermodynamics

The thermodynamic calculations for the sea ice cover center on the determination of ice thickness or ice and snow thicknesses, plus, for those models with lead parameterizations, the determination of ice concentrations. Such determinations require ascertaining additionally the temperature structure within the ice and snow layers. This can be done to varying degrees of precision, usually controlled by the number of levels of vertical resolution included and the accuracy of the energy flux parameterizations.

Figure 3.19 diagrams the general case of snow-covered ice floating on water. Either the snow layer alone or both the ice and snow layers could have zero thickness. The major energy fluxes between the upper snow surface and the layers above are the incoming solar radiation, $S^\downarrow$, the downward longwave radiation from the atmosphere, $F^\downarrow$, the emitted longwave radiation from the surface, $F^\uparrow$, the sensible and latent heat fluxes between atmosphere and surface, H and LE, the conductive flux through the snow layer, G_s, and the energy flux due to snow melt, M_s. (Later the adjustments will be indicated for the case of an ice surface without snow.) The equation for the conservation of energy at the upper snow surface is

$$H + LE + \epsilon_s F^\downarrow + (1 - \alpha_s)S^\downarrow - I_0 - F^\uparrow + (G_s)_0 - M_s = 0 \quad (3.239)$$

where ϵ_s is the infrared emissivity of snow, α_s is the shortwave albedo of snow, I_0 is the solar radiation that penetrates the snow surface, and $(G_s)_0$ denotes the surface value of G_s. The sign convention used on H and LE in (3.239) is that both are considered positive when the flux is downward, negative when the flux is upward. (The definitions differ somewhat from those used in (3.193) because the atmosphere and sea ice literature developed differently. The sea ice conventions are used here to make references easier to follow.)

The formulations for several of the terms in (3.239) vary among models, but it is common to use bulk aerodynamic formulae, as in (3.177) and (3.178), for the sensible and latent heat

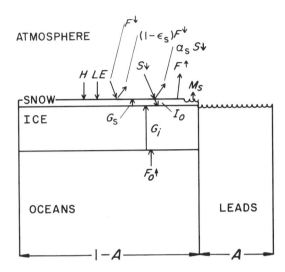

Fig. 3.19 Schematic of a grid square in a sea ice model, with ice and snow layers of uniform thickness, a variable lead fraction, A, and the relevant energy fluxes at the water/ice, ice/snow, and snow/air interfaces. [Modified from Parkinson and Washington (1979).]

fluxes and a graybody radiative flux formulation for the outgoing longwave radiation:

$$F^\uparrow = \epsilon_s \sigma \, (T_s)_0^4 \tag{3.240}$$

where σ is the Stefan-Boltzmann constant, T_s is the snow temperature, and $(T_s)_0$ is the snow temperature at the upper surface. (A graybody is an absorber or emitter that is independent of wavelength. In the more restrictive case of a blackbody, $\epsilon_s = 1$.) The surface conductive flux is

$$(G_s)_0 = k_s \left(\frac{\partial T_s}{\partial z}\right)_0 \tag{3.241}$$

where k_s is the thermal conductivity of snow, approximated as a constant, and T_s is the snow temperature. Setting h_s equal to the snow thickness and Q_s equal to the latent heat of fusion of snow (i.e., the amount of energy released per volume of snow upon conversion of snow to water), the snow melt term becomes

$$M_s = -Q_s \frac{dh_s}{dt} \tag{3.242}$$

When the sea ice calculations are carried out in conjunction with an atmospheric model, the incoming longwave and shortwave

fluxes to be used for the sea ice in (3.239) generally will be calculated from the atmospheric portion of the model. The reader
is referred to Parkinson and Washington (1979) for a complete
set of flux formulations used in a stand-alone sea ice model.

Since snow melt does not occur while the surface temperature
is below the freezing point, (3.239) can be expanded to

$$H + LE + \epsilon_s F^\downarrow + (1 - \alpha_s)S^\downarrow - I_0 - F^\uparrow$$

$$+ (G_s)_0 = \begin{cases} 0 & \text{if} \quad (T_s)_0 < 273 \text{ K} \\ -Q_s \, dh_s/dt & \text{if} \quad (T_s)_0 = 273 \text{ K} \end{cases} \quad (3.243)$$

$(T_s)_0$ is not allowed to exceed 273 K as long as snow is present.
The excess energy signified by a calculated $(T_s)_0 > 273$ K is used
in (3.243) first to deplete the snow layer through snow melt and
then, if h_s goes to 0, to deplete the ice layer through ice melt.

At the snow/ice interface, the equation for conservation of
energy is simply an equality of the conductive fluxes through the
ice and snow layers:

$$k_s \left(\frac{\partial T_s}{\partial z} \right)_{h_s} = k_i \left(\frac{\partial T_i}{\partial z} \right)_{h_s} \quad (3.244)$$

where the subscript h_s denotes the snow/ice interface, which occurs at depth h_s below the upper snow surface, k_i is the thermal
conductivity of the ice, and T_i is the ice temperature.

At the ice/water interface, the equation for the conservation
of energy balances the energy absorbed or released through melting or freezing at the interface (i.e., through a change of state)
with the difference between the energy flux, $F_0^\uparrow$, arriving from
the ocean and the upward conductive flux through the ice:

$$-Q_i \left(\frac{\partial h_i}{\partial t} \right)_{h_s + h_i} = F_0^\uparrow - k_i \left(\frac{\partial T_i}{\partial z} \right)_{h_s + h_i} \quad (3.245)$$

Q_i is the latent heat of fusion of ice, h_i is the ice thickness,
and the subscript $h_s + h_i$ denotes the ice/water interface, which
occurs at depth $h_s + h_i$ below the snow surface. The negative sign
on the left-hand side of (3.245) reflects the fact that if the net
flux to the interface (the right-hand side of (3.245)) is positive,
the ice will melt, leading to a negative $\partial h_i/\partial t$. $F_0^\uparrow$ is derived
from the ocean calculations when (3.245) is used in conjunction
with an ocean model. In stand-alone sea ice models, $F_0^\uparrow$ is often
set at a constant value, generally about 2 W m^{-2} in the Arctic

and 10–25 W m^{-2} in the Antarctic. The value is higher in the Antarctic because of the lesser stability of the pycnocline and the resultant greater mixing of surface waters with the warmer waters beneath.

Energy transfer through the snow layer, relevant for the determination of the temperature profiles needed for (3.241) and (3.244), is modeled with a standard heat conduction equation modified by the possibility of penetrating solar radiation:

$$\rho_s c_s \frac{\partial T_s}{\partial t} = \kappa_s \frac{\partial^2 T_s}{\partial z^2} + K_s I_0 e^{-K_s z} \qquad (3.246)$$

where ρ_s is the density of snow, c_s is the specific heat of snow, and K_s is a bulk extinction coefficient for snow. In the last term of (3.246) we follow Maykut and Untersteiner (1971) in assuming that the extinction of solar radiation in the snow is adequately approximated by Beer's extinction law, with a bulk coefficient. Energy transfer in the ice is modeled similarly:

$$\rho_i c_i \frac{\partial T_i}{\partial t} = \kappa_i \frac{\partial^2 T_i}{\partial z^2} + K_i I_0 e^{-K_i z} \qquad (3.247)$$

where ρ_i, c_i, and K_i are the density, specific heat, and bulk extinction coefficient for sea ice, respectively. A more complete formulation of the energy transfer approximated in (3.247) can be found in Untersteiner (1964), which includes an additional term for hydrostatic adjustment and two additional conductivity terms.

Equations (3.239)–(3.247) are the basic equations of energy conservation for a snow-covered ice layer and are the focus of the thermodynamic calculations in sea ice modeling. Although we have avoided discussion of numerical considerations throughout most of this chapter, leaving such discussions for Chapter 4, we will discuss the numerics of sea ice thermodynamics here, because the discussion can be made relatively brief and can add insight to the model formulations. In particular, when (3.239)–(3.247) are converted to finite difference form in order to use them in numerical models, a decision must be made on the vertical resolution, that is, on the number of levels to specify in the ice and snow layers. As the resolution becomes finer, the small-scale details of the actual temperature structure can be more closely simulated, although at the cost of additional computational expense. The reader is referred to Semtner (1976) for a comparison of the results of ice calculations at one point in the central Arctic from

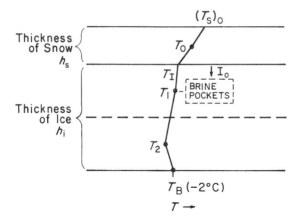

Fig. 3.20 Schematic of the temperature structure (T) in a sea ice model with two levels in the ice and one level in the snow. [Modified from Semtner (1976).]

(1) a very highly resolved model (that of Maykut and Unter-steiner, 1971), (2) a model with two levels in the ice and one in the snow, and (3) a model with one level in the ice and one in the snow. Each of the latter two models is computationally efficient enough to be used in large-scale simulations, and Semtner (1984b) has since argued for the use of the second model as an appropriate compromise between computational efficiency and simulation accuracy.

Whatever vertical resolution is selected, linear temperature profiles are generally assumed within the individual levels. Depending upon the model, the linearity holds either from the midpoint of one level to the midpoint of the adjacent level (Fig. 3.20) or from the bottom to the top of each level. The trend in the future will likely be toward two levels in the ice and one in the snow, as encouraged by Semtner (1984b). For simplicity, however, we present the equations for a single-level ice layer and a single-level snow layer, the temperature profile being modeled as linear from the bottom to the top of the ice and from the bottom to the top of the snow. With such a simplification, which is currently common in large-scale ice models,

$$\left(\frac{\partial T_s}{\partial z}\right)_0 = \left(\frac{\partial T_s}{\partial z}\right)_{h_s} = \frac{(T_s)_{h_s} - (T_s)_0}{h_s} \qquad (3.248)$$

$$\left(\frac{\partial T_i}{\partial z}\right)_{h_s} = \left(\frac{\partial T_i}{\partial z}\right)_{h_s+h_i} = \frac{(T_i)_{h_i+h_s} - (T_i)_{h_s}}{h_i} \qquad (3.249)$$

The temperatures of the ice and snow at the intervening interface are set equal:

$$(T_s)_{h_s} = (T_i)_{h_s} \qquad (3.250)$$

and the heat conduction equations (3.246) and (3.247) become unnecessary, as (3.248)–(3.250) can be inserted directly into (3.241) and (3.244)–(3.245). In models that make this simplification of having sea ice and snow unresolved in the vertical, including the zero-layer model of Semtner (1976) and the models of Bryan et al. (1975) and Parkinson and Washington (1979), the temperature at the bottom of the ice, $(T_i)_{h_i+h_s}$, generally is set precisely at the freezing point of seawater. This is realistic both conceptually and from observations.

As the sea ice calculations proceed, at every time step the snow melt term, M_s, is initially set equal to zero and (3.239)–(3.241) and (3.244) are solved for the temperatures at the upper snow surface and at the snow/ice interface. An adjustment is made at each grid square where the calculated temperature at the upper surface exceeds the melting point for snow, 273.15 K, because the energy is assumed to raise the snow temperature only to that point. The flux terms that are dependent upon $(T_s)_0$ are then recalculated with $(T_s)_0 = 273.15$ K, and M_s is determined as a residual from (3.239). The depth of melted snow is then calculated from (3.242).

A finite-difference form of (3.245) is used to calculate the change in ice thickness, Δh_i, caused by accretion or ablation at the bottom surface of the ice. Explicitly, the energy released per unit horizontal area when freezing a thickness Δh_i is $Q_i \times \Delta h_i$, and the energy absorbed when such a thickness is melted (Δh_i then being negative) is $-Q_i \times \Delta h_i$. Hence, whether melting or freezing is occurring, the energy flux E to the bottom surface of the ice needed to balance this flux due to the change of state must be

$$E = -Q_i \times \Delta h_i / \Delta t \qquad (3.251)$$

which is immediately solved for the change in ice thickness Δh_i once E has been determined from the right-hand side of (3.245).

In the event of ice without an overlying snow cover, any melt at the upper surface is ice melt rather than snow melt, so that h_i is affected by events at the upper surface as well as at the lower surface. With no snow cover, the fluxes at the upper surface are directly incident upon the ice, and the basic equation for the energy balance at the surface, (3.239), becomes

$$H + LE + \epsilon_i F^{\downarrow} + (1 - \alpha_i)S^{\downarrow} - I_0 - F^{\uparrow} + (G_i)_0 - M_i = 0 \quad (3.252)$$

while the supporting equations (3.240) and (3.241) become

$$F^{\uparrow} = \epsilon_i \sigma \, (T_i)_0^4 \qquad (3.253)$$

$$(G_i)_0 = k_i \left(\frac{\partial T_i}{\partial z} \right)_0 \qquad (3.254)$$

The incoming radiative fluxes $S^{\downarrow}$ and $F^{\downarrow}$ are identical to $S^{\downarrow}$ and $F^{\downarrow}$ in (3.239) and are, as before, calculated from empirical formulae or taken from the atmospheric model if the ice calculations are being carried out in conjunction with a coupled ice/atmosphere simulation; ϵ_i is the longwave emissivity of ice; α_i is the shortwave albedo of ice; $F^{\uparrow}$ is now the outgoing longwave radiative flux from the ice surface; $(G_i)_0$ is the conductive flux through the ice at the upper surface; and M_i is the energy flux due to ice melt at the upper surface. The penetration, I_0, of a portion of the radiation incident upon the surface usually is greater for an uncovered ice surface than for a snow surface and in fact is often ignored in (3.239) when a snow cover exists. I_0 varies depending on ice conditions, but in numerical models is generally set at a constant percentage, ranging from 0 to 10%, of the absorbed solar radiation. H and LE are calculated from (3.177) and (3.178).

As in the case of (3.239) for snow-covered ice, (3.252), in conjunction with the various flux formulations, is solved initially for surface temperature, assuming $M_i = 0$. When the calculated temperature exceeds the melting point of ice, the temperature is allowed to rise only to the melting point, the fluxes dependent upon the surface temperature are recalculated with $(T_i)_0$ set at the melting point, and (3.252) is used to determine M_i as a residual. This flux is then converted to a thickness change at the top surface of the ice using a finite difference approximation to

$$M_i = -Q_i \frac{dh_i}{dt} \qquad (3.255)$$

The melt and accretion at the bottom surface of the ice are determined from (3.245), which remains unchanged from the case of snow-covered ice. The other energy-balance equation for snow-covered ice, (3.244), is naturally eliminated for snow-free ice.

Finally, in the case of a grid square without ice, a stand-alone sea ice model still requires a temperature determination so that the timing of the onset of ice formation as the temperature reaches the freezing point can be determined. There are various methods of accomplishing this. One method artificially forces close correspondence with reality by using the observed sea ice edge as the outer boundary of freezing water temperatures. Another method determines the boundary on the basis of atmospheric temperature fields. Alternatively, the water temperature can be calculated with the following simple ocean mixed layer parameterization from Parkinson and Washington (1979).

Using an energy budget approach, the net energy flux, Q_N, into the mixed layer of the ocean is

$$Q_N = H + LE + \epsilon_w F^\downarrow + (1 - \alpha_w)S^\downarrow + F_w^\uparrow - \epsilon_w \sigma T_w^4 \quad (3.256)$$

where ϵ_w, α_w, and T_w are the longwave emissivity, shortwave albedo, and temperature of the water, respectively, and $F_w^\uparrow$ is the magnitude of the flux to the mixed layer from the deeper ocean. It is assumed that none of the absorbed solar radiation is transmitted through the mixed layer to the waters beneath. The net energy flux, Q_N, into the mixed layer must be balanced by the time rate of change of internal energy (per unit horizontal area) of the layer:

$$\frac{dI}{dt} = Q_N \quad (3.257)$$

Internal energy, I, is the product of the mixed-layer depth, d_{mix}, the volumetric heat capacity of water, c_w, and the water temperature, T_w. Assuming d_{mix} and c_w to be constants, as is generally done in stand-alone sea ice models, (3.257) can be rearranged to

$$\frac{dT_w}{dt} = \frac{Q_N}{d_{mix} \times c_w} \quad (3.258)$$

from which the change in water temperature is calculated, after replacing dT_w/dt by the finite difference approximation $\Delta T_w/\Delta t$, where Δt is the model time step. If the resulting T_w is below the freezing point of seawater (271.2 K), the negative energy flux Q_N is balanced instead by a term for cooling and a term for ice formation, so that (3.257) is expanded to

$$Q_N = \frac{dI}{dt} - Q_i \frac{d(C_i h_i)}{dt} \quad (3.259)$$

where dI/dt is calculated as the rate of change of internal energy resulting from a drop of temperature to the freezing point and C_i is the fractional concentration of sea ice. Through (3.259), the remaining energy deficit after cooling the water to the freezing point is used, along with the heat of fusion of ice, Q_i, to determine the amount of ice formed. Since the energy deficit determines only the ice volume (or $C_i h_i$, which is the ice volume divided by the grid cell area), the modeler has a free variable and can prescribe either the concentration or the thickness of the newly forming ice and then calculate the remaining variable. It is common in such a situation to prescribe an ice thickness somewhere between 0.01 and 0.50 m for the newly forming ice and then to calculate the fractional ice concentration C_i.

Treatments of leads or other open water areas within the ice cover range from a total denial of leads, with a uniform slab of ice simulated to cover the entire grid square, to full dynamic/thermodynamic formulations. The simplification of denying leads altogether probably goes too far if the ice model is to be used in conjunction with ocean and atmosphere calculations, since the leads are known to have a significant effect on the heat exchanges between ocean and atmosphere (e.g., Maykut, 1978). Of the models allowing leads, some treat them purely thermodynamically, some purely dynamically, and some with a combined thermodynamic/dynamic approach. The thermodynamic calculations within leads begin similarly to the calculations in grid squares free of ice. Equation (3.256) is used to calculate the net energy flux to the lead, and this flux is then balanced as in (3.259) by fluxes due to changes in water temperature and ice volume. The change in ice volume can occur laterally, altering the ice concentration, and/or vertically, altering the ice thickness. In the case of a negative Q_N, the energy deficit is made up entirely by cooling the water (using (3.258)) unless and until the water temperature reaches the freezing point, after which the deficit is made up entirely by ice formation (from (3.259)). In some models this is a strict lateral or vertical accretion to the existing ice; in other models it is the formation of a new category of very thin ice. In the case of a positive Q_N, the excess energy is proportioned between warming of the water and lateral or vertical melt of the ice, the exact proportioning varying among models. Often all the energy goes to melt until the grid square is free of ice, after which the excess energy goes to warming. Numerically, such models set all water temperatures in ice laden grid squares to be at the freezing point, so that $dI/dt = 0$ and the Q_N term becomes equated, in (3.259), exclusively with the energy flux from a change in ice volume, until the ice volume is reduced to zero.

The nonstandardized nature of such calculations reflects the uncertainty of sea ice scientists regarding the details of the physical mechanisms more than any numerical difficulties in distributing the Q_N flux.

Ice dynamics

Five major dynamic forces control the motions of sea ice floes: the air stress from above the ice, τ_a, steering the ice in the direction of the surface wind; the water stress from below the ice, τ_w, steering the ice in the direction of the water motion; the gravitational stress from the tilt of the sea surface, D, termed the *dynamic topography* and steering the ice downward from higher to lower surface levels; the stress induced by the earth's rotational motion, G, termed the *Coriolis force* (defined and discussed earlier) and steering the ice to the right of its otherwise-induced motion in the Northern Hemisphere and to the left in the Southern Hemisphere; and the pressure stresses, I, from within the ice cover, largely caused by the jostling of the individual ice floes. Thus the equation for the conservation of momentum becomes

$$m\frac{dV_i}{dt} = \tau_a + \tau_w + D + G + I \qquad (3.260)$$

where m is the ice mass per unit area and V_i is the ice velocity.

Equation (3.260) is often simplified to varying degrees in numerical models, with one or more of the terms set to 0 and others of the terms calculated with simplifying assumptions. In particular, the calculations are often done with the assumption of a steady-state ice velocity field, thereby reducing the left-hand side of the equation to 0. Partial justification of such a simplification derives from a scale analysis done by Rothrock (1973), who shows that the acceleration term, $m(dV_i/dt)$, generally is smaller than the wind and water stress terms by three orders of magnitude. However, the relative magnitudes of the various terms in (3.260) do vary greatly with the situation, and disagreement exists regarding which terms are essential and which are not under selected circumstances. Hence, in varied efforts to simplify the calculations, different researchers balance different subsets of the six terms in (3.260). For a complete model, to cover the full range of circumstances, all six terms should be included.

The parameterizations of the individual terms in (3.260) vary among researchers. For example, in dealing with the water stress, some assume a classical Ekman spiral under the ice. This theoretical spiraling and relaxing of ocean currents with depth, as well

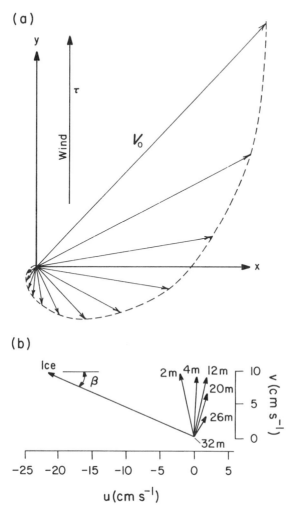

Fig. 3.21 (a) Theoretical Ekman spiral for wind-driven ocean currents, with V_0 the velocity vector for surface waters, directed 45° to the right of the wind in the Northern Hemisphere, and with each subsequent vector moving clockwise from V_0 representing the water velocity at a deeper level in the ocean. [Redrawn from Neumann and Pierson (1966).] (b) Velocity vectors measured for the water under a sea ice floe during a 1972 Arctic field study; numbers at the arrowheads of the vectors indicate water depths. [Redrawn and relabeled from McPhee (1986).]

as a corresponding theoretical Ekman spiral in the atmosphere, rarely are observed to agree closely with the theoretical predictions in either the open ocean or the atmosphere (Holton, 1979), but reasonable approximations have been observed in waters under the Arctic sea ice (McPhee and Smith, 1976; and Fig. 3.21).

In considering ice motions, Nansen (1902) assumes an Ekman spiral in the ocean reaching precisely to the ice boundary and fixes the angle between the ice velocity and the water stress at a constant 45°, while Shuleikin (1938) assumes a boundary layer between the Ekman spiral and the ice and then uses an empirical formula of Ekman (1905) that sets the water speed at the top of the Ekman spiral as a constant multiple of the wind speed (for a given latitude). As with Nansen's formulation, this results in a constant angle between the water stress and the ice velocity, but the angle is roughly 18° rather than 45°. Reed and Campbell (1960) instead calculate the magnitude of the water stress as

$$|\tau_w| = \rho_w K_w \frac{dr}{dz} \tag{3.261}$$

where ρ_w is water density, K_w the vertical eddy viscosity for water, and r the magnitude of the vector difference between the velocities of the ice and the water underneath. Due to numerical complications in (3.261) with a fully varying eddy viscosity, Reed and Campbell assume an eddy viscosity that is independent of depth in the Ekman spiral layer and decreases linearly in the boundary layer. As a result, water stress becomes

$$\tau_w = \rho_w \sqrt{|f| \, K_w} \, V_d \tag{3.262}$$

where f is the Coriolis parameter, and V_d is the vector difference between the ocean geostrophic velocity, obtained by balancing the dynamic topography and Coriolis forces, and the ice velocity. Equation (3.262) was first obtained by Ekman and later used by Campbell (1964) and by Hunkins (1966). A more detailed formulation of water stress instead incorporates nonlinear boundary layer theories, following McPhee (1975):

$$\tau_w = \rho_w c_w |V_d| (V_d \cos\theta + k \times V_d \sin\theta) \tag{3.263}$$

where c_w is the ocean drag coefficient, and θ is the turning angle in the ocean boundary layer. Although (3.263) is more sophisticated than the earlier formulations, users of this equation still simplify by assuming constant θ and still require a drag coefficient, c_w. The latter not only is not known with any precision, but is known to vary widely, further complicating its usage.

The wind stress term in (3.260) has a similar variety of formulations ranging from the simple forms discussed earlier in this chapter to nonlinear boundary-layer formulations, corresponding to (3.263) for the water stress. Such a nonlinear formulation, analogously assuming a constant turning angle into the atmosphere, is provided by Brown (1979):

$$\tau_a = \rho_a c_a |V_g|(V_g \cos \phi + k \times V_g \sin \phi) \qquad (3.264)$$

where ρ_a is air density, c_a is the atmospheric drag coefficient, V_g is the geostrophic wind, and ϕ is the turning angle in the atmospheric boundary layer. As with (3.263), (3.264) is often simplified by assuming constant, uncertain ϕ and c_a values. Furthermore, here there is the additional simplification of neglecting ice velocity, as the geostrophic wind is used directly in (3.264) rather than the difference, $V_g - V_i$. This can be partially justified on the basis of the generally much larger geostrophic wind velocity, but a more complete formulation would replace V_g in (3.264) by $V_g - V_i$.

In contrast to the wind and water stress terms, the Coriolis force has a precise formulation with no uncertain coefficients:

$$D = \rho_i h_i f V_i \times k \qquad (3.265)$$

where ρ_i is ice density, and the Coriolis parameter, $f = 2\Omega \sin \phi$, reflects the magnitude of the turning of the ice floes due to the earth's rotation.

The stress from the dynamic topography, G, depends on the gradient ∇ of sea surface height:

$$G = -\rho_i h_i g \nabla H \qquad (3.266)$$

where g is the acceleration due to gravity, H is the sea surface height field, and ∇H is the vector gradient of that field.

The final term in (3.260), the stress, I, due to internal ice resistance, in some senses is the most difficult to handle. With some justification, the difficulties can be avoided while simulating an open ice pack, especially under divergent conditions, simply by ignoring the internal stresses altogether and setting $I = 0$. In a concentrated ice pack such as occurs in the central Arctic Ocean, on the other hand, these internal stresses cannot readily be ignored. However, the proper method of modeling them remains uncertain, as the behavior of sea ice under compression is highly nonuniform. When thin floes collide, the ice often breaks and ridges. When thick floes collide, ridging is less likely, although it occurs if the convergent forces on the ice are sufficiently great. Since the strength of the ice depends on several factors, including thickness, salinity, and entrapped air content, modelers are faced with many uncertainties regarding the proper formulation of an ice stress term. In various works ice has been modeled alternatively as a viscous, plastic, or elastic material, sometimes being considered incompressible, sometimes partly compressible.

The stress/strain relationships for sea ice were modeled in the 1970s at the Arctic Ice Dynamics Joint Experiment (AIDJEX) according to an elastic/plastic formulation felt to be appropriate for large-scale (100 km) sea ice behavior and for events resolvable on a one-day time scale. Basically, with this formulation the modeled ice resists compression as a stiff elastic material until stress finally causes ridging, from which point the ice compresses as a plastic material. The plastic rheology provides a stress independent of deformation rate, in contrast to linear viscous rheologies in which the stress is proportional to deformation rate. As formulated in Coon et al. (1974) and tested in Coon et al. (1976) and Pritchard et al. (1977), the AIDJEX model calculations are based on an elastic/plastic model, with "strain hardening" incorporated by requiring more stress for ridging as the ice becomes thicker. Pritchard (1975) revised the model to make the strain more physically meaningful. In both formulations the model includes computation of ridging and lead widening.

Upon analyzing data from the Beaufort Sea for five days with relatively steady winds and motionless ice, Pritchard (1976) determined a lower bound of 1×10^5 N m^{-1} for yield strength under isotropic compression if a plastic model like the AIDJEX one is to simulate the ice properly.

In a model patterned after Coon et al. (1974), Hibler (1979) adjusted the constitutive law from an elastic/plastic formulation to a viscous/plastic formulation, having the ice behave as a linear viscous fluid for very small deformation rates and as a rigid plastic for larger deformation rates. This was done to eliminate certain numerical difficulties with the elastic/plastic formulation, in particular the necessity of indefinitely keeping track of the strain state (Hibler, 1979). The constitutive law used by Hibler provides the following formulation for the two-dimensional strain tensor, σ_{ij}:

$$\sigma_{ij} = 2\eta(\dot{\epsilon}_{ij}, P)\dot{\epsilon}_{ij} + [\varsigma(\dot{\epsilon}_{ij}, P) - \eta(\dot{\epsilon}_{ij}, P)]\dot{\epsilon}_{kk}\delta_{ij}$$
$$- P\delta_{ij}/2 \qquad (3.267)$$

where $\dot{\epsilon}_{ij}$ is the strain rate tensor, P is a pressure term dependent on the ice thickness, ς is the nonlinear bulk viscosity (not to be confused with the use of the same symbol for vorticity earlier in this chapter), η is the nonlinear shear viscosity, and δ_{ij} is the Kronecker delta:

$$\delta_{ij} = \begin{cases} 1 & \text{if } i = j \\ 0 & \text{if } i \neq j \end{cases} \tag{3.268}$$

The subscripts i and j refer to the x and y directions. [The reader is referred to Dutton (1976) for basic definitions of the strain tensor, tensor properties, shearing deformations, and stretching deformations. The strain rate tensor is defined as

$$\dot{\epsilon}_{ij} = \frac{1}{2} \left(\frac{\partial U_i}{\partial x_j} + \frac{\partial U_j}{\partial x_i} \right) \tag{3.269}$$

where U_i and U_j are velocity components and the terms on the right compose the divergence and deformation, the divergence terms being those for which $i = j$. Often the strain rate tensor is simplified to two dimensions.] Hibler calculates ς and η as functions of $\dot{\epsilon}_{ij}$ and P along an elliptical yield curve. The internal ice stress, $\boldsymbol{I} = (I_x, I_y)$, is then calculated as

$$I_x = \frac{\partial}{\partial x} \left[(\eta + \varsigma)\frac{\partial u}{\partial x} + (\varsigma - \eta)\frac{\partial v}{\partial y} - P/2 \right]$$

$$+ \frac{\partial}{\partial y} \left[\eta \left(\frac{\partial u}{\partial y} + \frac{\partial v}{\partial x} \right) \right] \tag{3.270}$$

$$I_y = \frac{\partial}{\partial y} \left[(\eta + \varsigma)\frac{\partial v}{\partial y} + (\varsigma - \eta)\frac{\partial u}{\partial x} - P/2 \right]$$

$$+ \frac{\partial}{\partial x} \left[\eta \left(\frac{\partial u}{\partial y} + \frac{\partial v}{\partial x} \right) \right] \tag{3.271}$$

Details can be found in Hibler (1979), along with a set of assumptions used to justify the constitutive laws and a mention of alternative constitutive laws used by others, some of which might be more realistic than the above, although more cumbersome numerically.

Although Hibler also simplified the treatment of continental boundaries from that in Coon et al. (1974) and increased the time step, the Hibler (1979) model remains dynamically more detailed than most other models suggested for climate studies. Recently Hibler (personal communication) proposed a considerable further simplification for use in certain restricted climate applications, although not for simulations of less than a few weeks. In this he suggests use of the following bulk viscous rheology:

$$\sigma_{ij} = \varsigma \left(\dot{\epsilon}_{11} + \dot{\epsilon}_{22} \right) \delta_{ij} \tag{3.272}$$

where

$$\varsigma \left(\dot{\epsilon}_{11} + \dot{\epsilon}_{22} \right) \begin{cases} = 0 & \text{for} \quad \dot{\epsilon}_{11} + \dot{\epsilon}_{22} \geq 0 \\ > 0 & \text{for} \quad \dot{\epsilon}_{11} + \dot{\epsilon}_{22} < 0 \end{cases} \tag{3.273}$$

and an upper limit is imposed on ς in the case of ice convergence $(\dot{\epsilon}_{11} + \dot{\epsilon}_{22} < 0)$. Semtner (1986a) has employed this method in a coupled ocean/ice model of the Arctic. Ling and Parkinson (1986) have also used a viscous rheology for an Arctic sea ice simulation, in which they have set the tensor σ_{ij} as follows:

$$\sigma_{ij} = (-p + \lambda \, \boldsymbol{\nabla} \cdot \boldsymbol{V}) \delta_{ij} + \mu \left(\frac{\partial u_i}{\partial x_j} + \frac{\partial u_j}{\partial x_i} \right) \tag{3.274}$$

where x_1 and x_2 are the two horizontal directions (x, y), u_1 and u_2 are the velocity components (u, v) in these directions, p is pressure, and λ and μ are the bulk viscosity and shear viscosity, respectively.

In view of the uncertainties regarding what might be the best constitutive law for sea ice, the uncertainties regarding the various terms within the constitutive laws, and the computer expense required for including an explicit constitutive law and a complete momentum equation (3.260), several efforts have been made to incorporate sea ice dynamics through less rigorous formulations. Such compromises between including a full momentum equation and omitting ice dynamics altogether range from including a momentum equation balancing the first four stress terms on the right-hand side of (3.260), followed by an adjustment for internal ice resistance, as done by Parkinson and Washington (1979), to replacing the momentum equation by a simple empirical relationship whereby approximate ice velocities are determined exclusively from either water velocities or wind velocities. For examples of the latter, Bryan (1969b) has sea ice of thickness less than 3 m move with the surface currents and disallows any motion for thicker sea ice, and several researchers are now considering instead setting sea ice speeds at a prescribed fraction of the wind speed and setting sea ice velocity directions at a prescribed turning angle from the wind direction. The latter suggestion derives historically from work of Sverdrup (1928) and Zubov (1944) and particularly from "Zubov's rule", an empirical rule determined by N. Zubov in 1939–1940 following analysis of the drift of the icebreaker *G. Sedov* in the Arctic ice pack. Zubov concluded that, to a reasonable approximation, sea ice drifts parallel to the atmospheric isobars and at a speed directly proportional to the pressure gradient. Subsequent Soviet analyses of the drift of various vessels have led to the rule of thumb that sea ice drifts at a speed of about 2% of the surface wind speed and at an angle of about 28–30° to the right of the wind direction in the Northern Hemisphere and to the left of the wind direction in the Southern Hemisphere (Gordienko, 1958). Although no such empirical approximation simulating ice velocities exclusively from winds or currents will be valid under all circumstances, for the purposes of specific individual simulations such formulations may be appropriate. For use in typical general circulation models, it remains to be seen which of the several formulations of sea ice dynamics will prove to have the best combination of adequate realism without excessive computer expense.

CHAPTER 4

Basic Methods of Solving Model Equations

The basic model equations for the atmosphere and oceans discussed in Chapter 3 involve spatial and temporal derivatives. In order to solve these equations in time, values for the variables first must be known in space at some starting instant. The full starting set of values in three-dimensional space is the set of *initial conditions*. Once this set is known, the time change or tendency terms in the equations can be solved to yield a new set of values at some future time. As mentioned in Chapter 1, this was first tried by L. F. Richardson (1922) in a prediction experiment for a portion of the European continent. Richardson did not know the numerical constraints on performing this type of calculation, although he did discuss numerical difficulties in solving a simple analog to the type of prediction equations discussed in Chapter 3. Richardson did a few experiments with an equation having a known analytical solution and found that solving the equation with smaller increments in time leads to a numerical solution closer to the analytical solution.

Courant, Friedrichs, and Lewy (1928) determined a numerical constraint to ensure a stable solution to a class of finite difference equations. Their constraint involves the ratio of the spatial to the temporal resolution and the speed of the fastest wave (or, more inclusively, the fastest signal propagation effect) being resolved, in whatever medium, whether atmosphere, ocean, or other.

As examples of the type of numerical constraints encountered in atmospheric and oceanic models, relatively simple equations are solved below to demonstrate the numerical methods. A more rigorous examination of the numerical properties can be found in advanced articles and texts, such as those by Haltiner and Williams (1980), Mesinger and Arakawa (1976), and Kreiss and Oliger (1973).

Consider an equation that governs the motion of a simple wave moving with constant speed c in the x direction. The time rate of change of the velocity u is related to the space rate of change of u as follows:

$$\frac{\partial u}{\partial t} + c\,\frac{\partial u}{\partial x} = 0 \qquad (4.1)$$

Note that this is similar to the prediction equations in Chapter 3. With c constant, the general solution to (4.1) is $u = f(x - ct)$. Our interest here is with wave solutions or harmonic solutions of the form

$$u = Ae^{ik(x-ct)} \qquad (4.2)$$

where A and k are constants and $i = \sqrt{-1}$. Substitution of (4.2) into (4.1) verifies that (4.2) is a solution to the partial differential equation. Equation (4.2) describes a wave moving with speed c, wave number k, wavelength $2\pi/k$, and wave amplitude A. The exponential $e^{ik(x-ct)}$ can be expressed in terms of trigonometric functions by the following identity due to Leonhard Euler and known as Euler's formula:

$$e^{iP} = \cos P + i\sin P \qquad (4.3)$$

where P is any value. When this is done, (4.2) becomes

$$u = A\{\cos[k(x - ct)] + i\sin[k(x - ct)]\} \qquad (4.4)$$

Although (4.1) itself has an analytical solution, many of the equations encountered in numerical modeling do not have analytical solutions, so that numerical, nonanalytical methods (often using finite differences) become essential. Some of these methods will be demonstrated below using (4.1), hence ensuring an analytical solution against which to compare the finite difference results.

Finite Differences

Equation (4.1) can be solved by approximating the spatial and temporal derivatives $\partial u/\partial t$ and $\partial u/\partial x$ by finite differences such as $\Delta u/\Delta t$ and $\Delta u/\Delta x$, where Δt and Δx are small, nonzero, but finite rather than infinitesimal increments. In order to have the differences centered on the point or moment of relevance, the form of our differences will often be

$$\frac{u(x + \Delta x) - u(x - \Delta x)}{2\Delta x} \quad \text{and} \quad \frac{u(t + \Delta t) - u(t - \Delta t)}{2\Delta t} \quad (4.5)$$

rather than

$$\frac{u(x + \Delta x) - u(x)}{\Delta x} \quad \text{and} \quad \frac{u(t + \Delta t) - u(t)}{\Delta t} \quad (4.6)$$

or

$$\frac{u(x) - u(x - \Delta x)}{\Delta x} \quad \text{and} \quad \frac{u(t) - u(t - \Delta t)}{\Delta t} \quad (4.7)$$

although for many purposes either of the latter two forms would be acceptable also. The form to be used here most frequently, (4.5), is termed *centered* differencing, the other two *forward* and *backward* differencing, respectively.

To examine the derivation of centered, forward, and backward differencing schemes, consider an arbitrary function $f(x)$ and expand it into a power series about the point x:

$$f(x + \Delta x) = f(x) + f'(x)\Delta x + f''(x)\frac{(\Delta x)^2}{2!}$$

$$+ f'''(x)\frac{(\Delta x)^3}{3!} + \cdots \quad (4.8)$$

where Δx is an increment in x, and f', f'', and f''' are first, second, and third derivatives of $f(x)$, respectively. Such a power series is termed a Taylor series expansion after Brook Taylor, who first published the expansion in 1715–1717. Its derivation and properties are discussed in most college calculus texts. Similarly, the Taylor expansion in the negative direction is

$$f(x - \Delta x) = f(x) - f'(x)\Delta x + f''(x)\frac{(\Delta x)^2}{2!}$$

$$- f'''(x)\frac{(\Delta x)^3}{3!} + \cdots \quad (4.9)$$

which can be obtained from (4.8) by replacing Δx by $-\Delta x$. Upon subtracting (4.9) from (4.8), alternate terms cancel, and a rearrangement of the remaining terms yields

$$f'(x) = \frac{f(x + \Delta x) - f(x - \Delta x)}{2\Delta x} - \frac{f'''(x)}{\Delta x} \frac{(\Delta x)^3}{3!} - \cdots \quad (4.10)$$

In many applications, second- and higher-ordered powers of Δx are ignored, in which case (4.10) reduces to the approximation

$$f'(x) \simeq \frac{f(x + \Delta x) - f(x - \Delta x)}{2\Delta x} \quad (4.11)$$

which is a second-order finite difference scheme centered on the point x. This is called "second-order" because the terms eliminated in moving from (4.10) to its approximation are the third- and higher-order terms. If, on the other hand, either (4.8) or (4.9) is used alone for calculating the derivative, generally the second- and higher-order terms are eliminated, yielding the "first-order" forward and backward differencing schemes mentioned above:

$$f'(x) \simeq \frac{f(x + \Delta x) - f(x)}{\Delta x} \quad (4.12)$$

and

$$f'(x) \simeq \frac{f(x) - f(x - \Delta x)}{\Delta x} \quad (4.13)$$

from (4.8) and (4.9), respectively. These first-order schemes, uncentered with respect to point x, generally are less accurate than the second-order, centered scheme of (4.11).

Many numerical atmosphere and ocean models using finite difference formulations make use of either first- or second-order approximations to derivatives in time and space. Higher-order schemes are used occasionally but have some drawbacks, since they can be more prone to inaccuracies in cases where there are scales of interest near the grid resolution (see Haltiner and Williams, 1980). Figure 4.1 illustrates a grid structure in space and time for one spatial dimension, x. The centered second-order approximation to the spatial derivative at point n uses the values of the variable being differentiated at the two adjacent points, $n-1$ and $n+1$, so that the differencing is done across a distance of $2\Delta x$ centered at point n. Similarly the second-order approximation to the temporal derivative at time m uses values at $m-1$ and $m+1$. By contrast, the first-order spatial forward and backward schemes use the value of the variable at the point n itself in

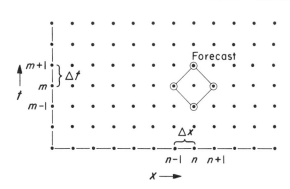

Fig. 4.1 Grid structure in space and time for one spatial dimension, x.

the differencing plus the value at one adjacent point (point $n+1$ for forward differencing and $n-1$ for backward differencing). These schemes are hence uncentered. Fourth-order schemes for approximating a derivative, like second-order schemes, are often centered, although they use two points on each side of the center point rather than only one point on each side.

In (4.1) there are partial rather than total derivatives. Centered, forward, and backward differencing will be done similarly to (4.11)–(4.13) although with a notational change to account for the time and space dimensions. Using subscripts on the function u to denote the space coordinate and superscripts on u to denote the time coordinate, the partial derivative of u with respect to x at time m can be approximated with a centered approximation as

$$\left(\frac{\partial u}{\partial x}\right)_m \simeq \frac{1}{2\Delta x}\left(u_{n+1}^m - u_{n-1}^m\right) \tag{4.14}$$

The corresponding second-order finite difference approximation to the partial derivative with respect to time is

$$\left(\frac{\partial u}{\partial t}\right)_n \simeq \frac{u_n^{m+1} - u_n^{m-1}}{2\Delta t} \tag{4.15}$$

where Δt is the time increment. Substituting (4.14) and (4.15) into (4.1) and rearranging terms yield

$$u_n^{m+1} \simeq u_n^{m-1} - c\,\frac{\Delta t}{\Delta x}\left(u_{n+1}^m - u_{n-1}^m\right) \tag{4.16}$$

Equation (4.16) can be solved for the velocities at time 2 by setting $m = 1$ and inserting the velocity conditions at times 0 and 1 in the appropriate terms on the right-hand side of (4.16). The velocity values at time 2 then can be substituted into (4.16), using $m = 2$, along with the velocities at time 1 to obtain a new set of u values for time 3, and so forth.

In practice, starting the computation to solve for u values from (4.16) requires only the values at time 0, not at both 0 and 1, because a forward difference in time can be used for the first time step before proceeding to centered differences for the remainder of the calculation. Thus, still using centered differences in space, u_n^1 is approximated as

$$u_n^1 \simeq u_n^0 - \frac{c\Delta t}{2\Delta x}\left(u_{n+1}^0 - u_{n-1}^0\right) \qquad (4.17)$$

This forward differencing is used to calculate a full set of u values at time 1, so that for any spatial position n, the u_{n+1}^1 and u_{n-1}^1 values are available for insertion into (4.16), along with the initial condition u_n^0, for calculating the u values for time 2. The u values for time 3 then are obtained through (4.16) from the values for times 1 and 2, and (4.16) then can be used repeatedly for as many time steps as desired. A crucial issue, however, is how closely the numerical solution obtained from using (4.16) and (4.17) will approximate the exact solution in (4.2).

Following Thompson (1961), the numerical wave solution to (4.16) can be written

$$u_n^m = Ae^{i(km\Delta t + \alpha n\Delta x)} \qquad (4.18)$$

Here the time, t, is expressed in terms of increments Δt, with the initial time at $t = 0$ for $m = 0$. In general, $t = m\Delta t$. Likewise the space position x can be expressed as $n\Delta x$. Substitution of (4.18) into (4.16) and division by $Ae^{i(km\Delta t + \alpha n\Delta x)}$ yield

$$e^{ik\Delta t} - e^{-ik\Delta t} + \frac{c\Delta t}{\Delta x}\left(e^{i\alpha\Delta x} - e^{-i\alpha\Delta x}\right) = 0 \qquad (4.19)$$

Then by using Euler's formula in the form

$$\sin P \equiv \frac{e^{iP} - e^{-iP}}{2i} \qquad (4.20)$$

obtained from (4.3) by (1) replacing P with $-P$, (2) subtracting the resulting equation from (4.3), and (3) dividing by $2i$, (4.19) can be shortened to

$$\sin k\Delta t = -\frac{c\Delta t}{\Delta x}\sin \alpha\Delta x \qquad (4.21)$$

This trigonometric equation has a nonamplifying solution if $|c\Delta t/\Delta x| \leq 1$. If $|c\Delta t/\Delta x| > 1$, then the solution for k will be unstable, meaning that it will grow in absolute magnitude with time, and also will be imaginary. This is an inappropriate solution for the original wave equation (4.1), where c is constant. Hence $|c\Delta t/\Delta x| \leq 1$ becomes a formal numerical criterion for the numerical solution to the wave equation. This is the same as the numerical criterion found by Courant, Friedrichs, and Lewy in 1928 for the analytical solution, albeit in a slightly different form. This criterion often is referred to as the CFL condition in honor of the authors of the 1928 article.

In the primitive system of equations for the atmosphere, the wave speed c involves horizontally propagating external gravity-acoustic waves that have a speed of about 300 m s^{-1} and are the fastest horizontal waves in the atmosphere. The total wave speed in the more general set of equations includes an advection speed, u, of about 100 m s^{-1}, in addition to the 300 m s^{-1}, for a total of 400 m s^{-1}. Therefore Richardson, with his horizontal grid resolution of approximately $\Delta x = 200$ km, should have used about an 8-minute time step as a maximum to ensure that $\Delta t \leq \Delta x/c$. He instead used a 6-hour time step, which would have led to increasing difficulties if he had extended the calculation by repeated iterations.

Another aspect of the finite difference solutions to the wave equation is that the calculated finite difference waves will generally move more slowly than the waves of the exact analytical solution. This is an inherent shortcoming of many finite difference solutions and is called the phase error. In general, for a given time step, the higher the spatial resolution with respect to a wave moving through a grid, the closer the finite difference solution is likely to be to the true solution. A rule of thumb is that if a wave is represented by eight or more gridpoints it moves at a speed close to the true speed.

Although the centered method (sometimes called the leap-frog method) presented in (4.15) is the most common method of time differencing in climate models, it suffers from a spurious computational oscillation between even and odd time increments. Inspection of (4.16) reveals that for any m, either the odd increments are in the time change term and the even increments are in the advection term or vice versa. This eventually leads to separation of the solution for the even and odd time steps. This oscillation can be suppressed by occasionally using a forward difference, i.e.,

$$\left(\frac{\partial u}{\partial t}\right)_n \simeq \frac{u_n^{m+1} - u_n^m}{\Delta t} \tag{4.22}$$

which is of first-order rather than second-order accuracy. This has the advantage of eliminating the even-odd oscillation in time. Another technique widely used for suppressing the oscillation is the so-called Euler-backward scheme (see Kasahara, 1977), which occasionally uses a forward difference as a first approximation that is resubstituted into the numerical equation. Robert (1966) and Asselin (1972) have designed a time filter of the variables that also eliminates this time oscillation by strongly smoothing oscillations near or slightly longer than the time increment.

Another widely used time scheme is the implicit time approximation. This involves evaluating the space derivative terms at old and new times, often by using simple time averages. For example, if (4.1) is solved using forward differences in time from (4.22) and the following centered difference in space, applied to the average u's at times m and $m+1$,

$$\left(\frac{\partial u}{\partial x}\right)_{m+0.5} = \frac{\frac{1}{2}\left(u_{n+1}^{m+1} + u_{n+1}^m\right) - \frac{1}{2}\left(u_{n-1}^{m+1} + u_{n-1}^m\right)}{2\Delta x} \tag{4.23}$$

then the resulting finite difference solution to (4.1) becomes

$$u_n^{m+1} = u_n^m - \frac{c\Delta t}{4\Delta x}\left(u_{n+1}^{m+1} + u_{n+1}^m - u_{n-1}^{m+1} - u_{n-1}^m\right) \tag{4.24}$$

Note that u^{m+1} appears in two terms on the right-hand side of (4.24). The solution, obtained by solving a set of simultaneous equations, thus becomes implicit rather than explicit. Although this is a definite disadvantage, the time step limitation of the explicit method disappears. Hence the implicit finite difference method has the significant advantage of always being stable. Depending upon the problem being solved, for some sets of equations it can also be computationally efficient compared to explicit methods.

Many current climate models use partially (semi-) implicit methods to solve the prediction equations. In such models, terms that give rise to high frequency oscillations, such as those involving gravitational waves, are evaluated implicitly, while the remaining terms are evaluated explicitly. This method, developed by Robert (1969), permits a time step approximately six times larger than in a fully explicit approach. Tests of semi-implicit time differencing using the primitive equations have shown that truncation errors associated with the much larger time step remain an order of magnitude smaller than errors introduced by spatial truncation (Kwizak and Robert, 1971). Thus, because this time differencing scheme can save large amounts of computer expense (usually speeding the calculations by a factor of four), it enjoys great popularity for both long-term numerical integrations of climate and short-term operational forecasts.

Finite Differencing in Two Dimensions

The selection of appropriate spatial finite difference formulations to use in the horizontal and vertical directions varies widely depending on which properties are being modeled or are considered important. Most climate models are not intended primarily for predicting the exact location of daily weather events with great accuracy but are directed more toward the simulation of the long-term mean and variance states of the climate system. Since the models frequently run for long periods of time (often out to many years), the numerical schemes used should approximately maintain many of the integral constraints that are in the original differential equations, such as conservation of mass, energy, and momentum. Over recent years many climate models have employed horizontal finite difference methods that maintain these features. No attempt will be made to explain all the schemes that have been tried or all those presently in use, but some of the basic properties will be discussed. In order to do this efficiently it will be advantageous to introduce a simpler primitive equation system than the full system provided in Chapter 3, since many of the numerical properties still can be shown without the added complications of the more general set of equations.

Consider an incompressible fluid with a variable height, such as water in a container, and assume that the horizontal velocity components, u and v, are independent of height and that the vertical velocity, w, equals 0 at the bottom of the fluid. The incompressibility assumption allows the equation of continuity in Cartesian coordinates to be written as

$$\frac{\partial u}{\partial x} + \frac{\partial v}{\partial y} + \frac{\partial w}{\partial z} = 0 \qquad (4.25)$$

This can be further modified to:

$$\frac{dh}{dt} + h\left(\frac{\partial u}{\partial x} + \frac{\partial v}{\partial y}\right) = 0 \qquad (4.26)$$

by integrating (4.25) over height z from the bottom of the fluid at $z = 0$ to the top of the fluid at $z = h$, using the earlier assumption that u and v are independent of height. The height of the top of the fluid, h, sometimes is referred to as the height of the free surface. Since h is a function of x, y, and t, and $u = dx/dt, v = dy/dt$,

$$\frac{dh}{dt} \equiv \frac{\partial h}{\partial t} + u\frac{\partial h}{\partial x} + v\frac{\partial h}{\partial y} \tag{4.27}$$

Combining (4.26) and (4.27) yields

$$\frac{\partial h}{\partial t} + \frac{\partial(hu)}{\partial x} + \frac{\partial(hv)}{\partial y} = 0 \tag{4.28}$$

Similarly, the equations of motion can be integrated vertically over the depth of the fluid to yield the following set of momentum equations in Cartesian coordinates:

$$\frac{\partial u}{\partial t} + u\frac{\partial u}{\partial x} + v\frac{\partial u}{\partial y} - fv + g\frac{\partial h}{\partial x} = 0 \tag{4.29}$$

$$\frac{\partial v}{\partial t} + u\frac{\partial v}{\partial x} + v\frac{\partial v}{\partial y} + fu + g\frac{\partial h}{\partial y} = 0 \tag{4.30}$$

Note that the pressure gradient term has been simplified by use of the hydrostatic law (3.53) so that pressure is $g\rho h$ in the last term on the left of (4.29) and (4.30) where the density, ρ, cancels out of all terms.

Following the treatment of Arakawa and Lamb (1977), the finite difference approximations are applied to another form of (4.29) and (4.30), obtained by combining each with (4.28). Multiplying (4.29) and (4.30) by h and adding (4.28) times u and v, respectively, the adjusted momentum equations are

$$\frac{\partial(uh)}{\partial t} + \frac{\partial(huu)}{\partial x} + \frac{\partial(hvu)}{\partial y} - fhv + gh\frac{\partial h}{\partial x} = 0 \tag{4.31}$$

$$\frac{\partial(vh)}{\partial t} + \frac{\partial(hvu)}{\partial x} + \frac{\partial(hvv)}{\partial y} + fhu + gh\frac{\partial h}{\partial y} = 0 \tag{4.32}$$

The system of fluid equations consisting of (4.28)–(4.30) or of (4.28), (4.31), and (4.32), often referred to as the shallow water equations, conserves several important properties including momentum, mass, and energy, all of which are desirable to have conserved in climate modeling studies. The conservation of momentum follows immediately from (4.29) and (4.30) or from (4.31) and (4.32). The conservation of mass follows from (4.28) alone, by integration of that equation over the spatial domain, assuming the domain is cyclic (i.e., that it has the same end point on each end). Starting with (4.28) and integrating,

$$
\iint \frac{\partial h}{\partial t}\, dx\, dy = \iint \left(-\frac{\partial(hu)}{\partial x} - \frac{\partial(hv)}{\partial y} \right) dx\, dy
$$

$$
= -\iint \frac{\partial(hu)}{\partial x}\, dx\, dy - \iint \frac{\partial(hv)}{\partial y}\, dx\, dy \tag{4.33}
$$

$$
= -\int [0]\, dy - \int \left[\int \frac{\partial(hv)}{\partial y}\, dy \right] dx
$$

$$
= 0 - 0 = 0
$$

where the two 0's preceding the final equal sign result from the cyclic assumption. Because the integrated time change of the height is zero, the mean height over the domain of the fluid must be constant. Since the fluid is assumed to be incompressible, this forces the total mass to be constant also, therefore proving the conservation of mass.

To establish that the system of equations (4.28)–(4.30) conserves energy, we first seek an appropriate expression for the time change of the total energy of the system, that is, for the time change of the sum of the kinetic energy

$$
KE = \frac{1}{2}h(u^2 + v^2) \tag{4.34}
$$

and the potential energy

$$
PE = g\frac{h^2}{2} \tag{4.35}
$$

Multiplying (4.29) by hu and (4.28) by $u^2/2$ and adding, and multiplying (4.30) by hv and (4.28) by $v^2/2$ and adding, we obtain

$$
\frac{\partial}{\partial t}\left(h\frac{1}{2}u^2 \right) + \frac{\partial\left(hu\frac{1}{2}u^2 \right)}{\partial x} + \frac{\partial\left(hv\frac{1}{2}u^2 \right)}{\partial y} - fhvu
$$

$$
+ ghu\frac{\partial h}{\partial x} = 0 \tag{4.36}
$$

$$\frac{\partial}{\partial t}\left(h\frac{1}{2}v^2\right) + \frac{\partial\left(hu\frac{1}{2}v^2\right)}{\partial x} + \frac{\partial\left(hv\frac{1}{2}v^2\right)}{\partial y} + fhuv$$

$$+ ghv\frac{\partial h}{\partial y} = 0 \qquad (4.37)$$

Multiplying the continuity equation (4.28) by gh and rearranging we obtain

$$\frac{\partial}{\partial t}\left(g\frac{h^2}{2}\right) + \frac{\partial}{\partial x}\left(gh^2 u\right) + \frac{\partial}{\partial y}\left(gh^2 v\right)$$

$$- gh\left[u\frac{\partial h}{\partial x} + v\frac{\partial h}{\partial y}\right] = 0 \qquad (4.38)$$

Adding (4.35), (4.36), and (4.37) produces

$$\frac{\partial}{\partial t}\left(\frac{1}{2}h(u^2 + v^2) + g\frac{h^2}{2}\right)$$

$$= -\frac{\partial}{\partial x}\left(hu\frac{1}{2}(u^2 + v^2) + gh^2 u\right)$$

$$- \frac{\partial}{\partial y}\left(hv\frac{1}{2}(u^2 + v^2) + gh^2 v\right) \qquad (4.39)$$

This provides us with the desired expression for the time change of the total energy. Substituting (4.34) and (4.35) into (4.39) and integrating over the spatial domain,

$$\iint \frac{\partial}{\partial t}(KE + PE)\,dx\,dy$$

$$= -\int\left[\int \frac{\partial}{\partial x}\left(hu\frac{1}{2}(u^2 + v^2) + gh^2 u\right)dx\right]dy$$

$$- \int\left[\int \frac{\partial}{\partial y}\left(hv\frac{1}{2}(u^2 + v^2) + gh^2 v\right)dy\right]dx$$

$$= -\int 0\,dy - \int 0\,dx = 0 \qquad (4.40)$$

using, as before, the fact that the domain is cyclic (having identical end points). Equation (4.40) establishes the conservation of total energy. Hence this particular system of equations, (4.38)–(4.30), conserves momentum, mass, and energy. It is important to understand that many straightforward finite difference approximations, such as the use of centered finite differences, do not conserve all of these properties.

In (4.36)–(4.37), note that the Coriolis terms cancel each other so that there is no net energy gain or loss due to these terms. Note also that the sum of the final terms in (4.36)–(4.38) is zero:

$$gh \left[u \frac{\partial h}{\partial x} + v \frac{\partial h}{\partial y} - u \frac{\partial h}{\partial x} - v \frac{\partial h}{\partial y} \right] = 0 \qquad (4.41)$$

These terms convert between kinetic and potential energy, and it is important that the finite difference form used for (4.36)–(4.38) maintains this property of having these four terms cancel.

In Phillips's (1956) early general circulation experiment with the nonlinear quasi-geostrophic equation, the model was unstable even though the CFL criterion on the time step was fully satisfied. Phillips found that noise near the size of the grid spacing could be controlled by use of a large diffusion term in the equations, which smoothed the small-scale features. Three years later Phillips (1959) found this type of instability to be caused by nonlinear interaction between scales of motion that were too small to be resolved by the grid. These unresolved scales were interacting with the larger scales, resulting in a process called *aliasing*. In this process, the interactions of scales not resolved by the grid with the resolved scales can lead to a spurious growth of energy.

An extensive body of literature has been developed regarding aliasing in finite difference schemes (e.g., see Gary, 1979, and Mesinger and Arakawa, 1976, for extensive reviews). To give the reader some insight into the process, the following simple example from Mesinger and Arakawa is provided. Assume a sinusoidal function:

$$u = \sin \alpha x \qquad (4.42)$$

If (4.42) is substituted into the following advection equation:

$$\frac{\partial u}{\partial t} = -u \frac{\partial u}{\partial x} \qquad (4.43)$$

then

$$\frac{\partial u}{\partial t} = -\alpha \sin \alpha x \cos \alpha x = -\frac{1}{2} \alpha \sin 2\alpha x \qquad (4.44)$$

The shortest wavelength resolvable on a finite difference grid is $L_{min} = 2\Delta x$, which corresponds to a maximum wave number of

$$\alpha_{max} = \frac{2\pi}{L_{min}} = \frac{\pi}{\Delta x} \tag{4.45}$$

However, there appears in (4.44) a sine wave with twice the wave number of the original sine wave (4.42). A wave number greater than α_{max} (or, equivalently, a wavelength less than L_{min}) cannot be distinguished from a wave at the limiting wave number, and the error thereby introduced is called an aliasing error. Arakawa (1966) and Arakawa and Lamb (1977) devised a novel finite difference technique that restricts the interaction between the resolved and unresolved scales and thus *tends* to prevent the nonlinear instability. Strictly speaking, the instability still occurs, because the inclusion of time derivatives prevents perfect energy conservation with the finite difference approximations.

Following Kasahara (1977), the advection equation of (4.43) is used for demonstrating the basic property of the Arakawa scheme. First, (4.43) can be extended to

$$\frac{\partial u}{\partial t} = -u \frac{\partial u}{\partial x} = -\frac{\partial}{\partial x}\left(\frac{u^2}{2}\right) \tag{4.46}$$

and hence can be expressed in centered finite difference form as

$$\frac{\partial u}{\partial t} \simeq -\frac{(u_{n+1})^2 - (u_{n-1})^2}{4\Delta x} \tag{4.47}$$

at point $x = x_n$. Let N be the maximum n. Assuming boundary conditions of $u_0 = u_1 = u_{N-1} = u_N = 0$, or simply $u_0 = u_{N-1}$ and $u_1 = u_N$, the right-hand side of (4.47) can be summed over all n, approximating the integral of the left side with respect to x, to yield

$$\int \frac{\partial u}{\partial t} \, dx \simeq \sum_{n=1}^{N-1} \frac{(u_{n-1})^2 - (u_{n+1})^2}{4\Delta x}$$

$$= \frac{1}{4\Delta x}\left(u_0^2 + u_1^2 - u_{N-1}^2 - u_N^2\right) = 0 \tag{4.48}$$

due to cancellation of intermediate terms. This implies that the areal mean of u is constant with time, and furthermore, if density is constant then the momentum is constant with time and hence is conserved. However, energy is not conserved using the finite difference form of (4.47). Both momentum and energy can be conserved by using an alternate finite difference approximation to (4.46). Specifically, rewrite the differential equation (4.46) in the form

$$\frac{\partial u}{\partial t} = -\frac{1}{3}\left[u\frac{\partial u}{\partial x} + \frac{\partial}{\partial x}(u^2)\right] \tag{4.49}$$

and convert to the finite difference version

$$\frac{\partial u}{\partial t} \simeq -\frac{1}{3}\left[u_n\left(\frac{u_{n+1}-u_{n-1}}{2\Delta x}\right) + \frac{(u_{n+1})^2 - (u_{n-1})^2}{2\Delta x}\right] \tag{4.50}$$

at point $x = x_n$. As with (4.47), the finite difference form given in (4.50) maintains conservation of momentum, if density is constant, because

$$\int \frac{\partial u}{\partial t}\,dx \simeq -\frac{1}{6\Delta x}\sum_n(u_n u_{n+1} - u_n u_{n-1}$$

$$+ u_{n+1}^2 - u_{n-1}^2)$$

$$= -\frac{1}{6\Delta x}(-u_1 u_0 + u_{N-1}u_N - u_0^2 - u_1^2$$

$$+ u_{N-1}^2 + u_N^2)$$

$$= -\frac{1}{6\Delta x}(0) = 0 \tag{4.51}$$

However, the form given in (4.50) also maintains conservation of energy, because

$$\int \frac{\partial(u^2/2)}{\partial t}\,dx = \int u\frac{\partial u}{\partial t}\,dx$$

$$\simeq -\frac{1}{6\Delta x}\sum_n(u_n^2 u_{n+1} - u_n^2 u_{n-1}$$

$$+ u_n u_{n+1}^2 - u_n u_{n-1}^2)$$

$$= -\frac{1}{6\Delta x}\left[\sum_n\left(u_n u_{n+1}^2 - u_{n-1}u_n^2\right)\right.$$

$$+ \left.\sum_n\left(u_n^2 u_{n+1} - u_{n-1}^2 u_n\right)\right]$$

$$= -\frac{1}{6\Delta x}\left[\left(-u_0 u_1^2 + u_{N-1}u_N^2\right)\right.$$

$$+ \left.\left(-u_0^2 u_1 + u_{N-1}^2 u_N\right)\right] = 0 \tag{4.52}$$

which implies that the average value of $u^2/2$ over the grid remains constant with time. Hence the kinetic energy is conserved, and since in this system there is no change in potential energy, this implies a conservation of total energy. This prevents the spurious growth of energy that can occur with other finite difference schemes such as that in (4.47).

Lilly (1965) and Bryan (1966) used Arakawa's ideas to formulate variants on the above schemes for primitive equation models that conserve quadratic quantities such as energy as well as first-moment quantities such as mass and momentum. Later Grammeltvedt (1969) performed some systematic comparisons between the variants with several types of grid structures. This has been expanded by Arakawa and Lamb (1977) to conserve not only energy but also the square vorticity (enstrophy). This latter quantity should be conserved exactly in strictly two-dimensional flow. Its importance for three-dimensional climate models is not established, except in that the atmosphere and oceans, being quasi-geostrophic, can in some contexts be treated as approximately two-dimensional fluids (see Charney, 1971).

Since a considerable amount of climate modeling research uses conservative spatial finite difference methods for the two horizontal dimensions, some of the typical schemes will be shown. To do this, it will be convenient to introduce some special notation that will simplify the description of the numerical methods. The most widely used methods employ derivative and averaging operators introduced by Shuman (1962):

$$\delta_x A = \frac{A(x_j + (\Delta x/2), y_\ell) - A(x_j - (\Delta x/2), y_\ell)}{\Delta x} \tag{4.53}$$

$$\delta_y A = \frac{A(x_j, y_\ell + (\Delta y/2)) - A(x_j, y_\ell - (\Delta y/2))}{\Delta y} \tag{4.54}$$

$$\overline{A}^x = \frac{1}{2}\left[A\left(x_j + \left(\frac{\Delta x}{2}\right), y_\ell\right) + A\left(x_j - \left(\frac{\Delta x}{2}\right), y_\ell\right)\right] \tag{4.55}$$

$$\overline{A}^y = \frac{1}{2}\left[A\left(x_j, y_\ell + \left(\frac{\Delta y}{2}\right)\right) + A\left(x_j, y_\ell - \left(\frac{\Delta y}{2}\right)\right)\right] \tag{4.56}$$

Figure 4.2 shows the two-dimensional grid structure being considered. The variables such as u and v are defined explicitly at points with coordinates $x_j = j\Delta x$ and $y_\ell = \ell\Delta y$, with both j and ℓ integers. These points are indicated in Fig. 4.2 by heavy black dots. The intermediate cross points are taken midway between the gridpoints. Note that both $\delta_x A$ and $\delta_y A$ are centered differences although they use the adjacent intermediate points rather

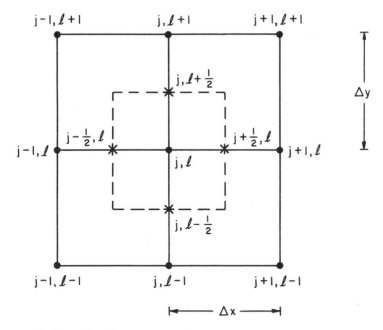

Fig. 4.2   Horizontal grid structure where $x_j = j\Delta x$ and $y_\ell = \ell\Delta y$ for conservative finite difference schemes. [From Kasahara (1977).]

than the adjacent gridpoints as before. The choice of which finite difference scheme to employ will depend upon which variables the modeler desires to conserve (such as total energy, momentum, mass, or squared vorticity). Kasahara (1977) mentions two schemes for solving (4.28), (4.31), and (4.32) that conserve mass, momentum, and total energy, namely,

$$\frac{\partial h}{\partial t} + \delta_x(\overline{hu}^x) + \delta_y(\overline{hv}^y) = 0 \qquad (4.57)$$

$$\frac{\partial hu}{\partial t} + \delta_x(\overline{hu}^x\overline{u}^x) + \delta_y(\overline{hu}^y\overline{v}^y)$$
$$-fhv + gh\delta_x(\overline{h}^x) = 0 \qquad (4.58)$$

$$\frac{\partial hv}{\partial t} + \delta_x(\overline{hu}^x\overline{v}^x) + \delta_y(\overline{hv}^y\overline{v}^y)$$
$$+fhu + gh\delta_y(\overline{h}^y) = 0 \qquad (4.59)$$

and

$$\frac{\partial h}{\partial t} + \delta_x(\overline{h}^x\overline{u}^x) + \delta_y(\overline{h}^y\overline{v}^y) = 0 \qquad (4.60)$$

$$\frac{\partial hu}{\partial t} + \delta_x(\overline{h}^x\overline{u}^x\overline{u}^x) + \delta_y(\overline{h}^y\overline{u}^y\overline{v}^y)$$

$$-fhv + g\overline{\delta_x(h^2/2)}^x = 0 \qquad (4.61)$$

$$\frac{\partial hv}{\partial t} + \delta_x(\overline{h}^x\overline{u}^x\overline{v}^x) + \delta_y(\overline{h}^y\overline{v}^y\overline{v}^y)$$

$$+fhu + g\overline{\delta_y(h^2/2)}^y = 0 \qquad (4.62)$$

Note that operators δ_x and δ_y are outside the parentheses in many of the above terms. Thus when a summation is done in x and y (which is the numerical equivalent of integrating over the area), the nearest neighbors in the grid cancel in the summation in a manner similar to that shown earlier for advection equation finite difference schemes (4.47) and (4.50).

Zeng and Ji (1981), Zeng and Zhang (1982), and Zeng, Yuan, and Zhang (1984) provide several examples of energy-conserving schemes in space and time for the shallow water equations and for the full primitive equation system. They obtain this property by substituting time averages of the variables at $n+1$ and n into the advection operators of the equations. This makes the equation implicit, thus requiring an iterative solution, but it does have the property of being essentially energy conserving.

The first spherical coordinate scheme (4.57)–(4.59) was devised by Grimmer and Shaw (1967), and Tiedtke later devised and tested the second scheme (4.60)–(4.62). Apparently both methods give similar results for short tests. Arakawa and Lamb (1977) have presented many alternative schemes with different placement of variables than shown in Fig. 4.2, all of which have advantages or disadvantages depending upon the problem being studied, making it impossible to define a single *ideal* scheme for all purposes. As an additional note of caution, the use of quadratic conserving schemes does not ensure the accuracy of the solution to the original differential equation. However, such schemes are less prone to nonlinear instability because they prevent a flow of energy to the scales smaller than the grid size. If the problem physically contains strong interactions on scales on the order of the grid size, then the answer will be incorrect even with quadratic conserving schemes. Fortunately for most large-scale atmospheric problems, the bulk of the energy is in the large scales; thus, detailed treatment of scales near the grid size is thought to be less crucial than the treatment of the larger scales. Without this property of having most energy in the larger

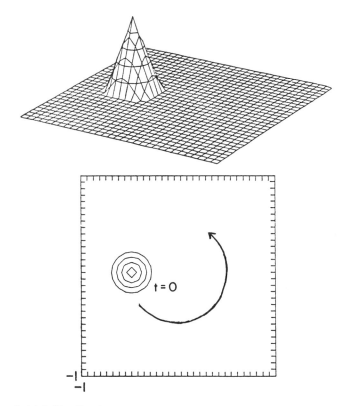

Fig. 4.3 Initial distribution of a cone that is made to move analytically with a constant solid-body rotation on a plane. [From Orszag (1971).]

scales for geophysical fluids, the simulation and prediction of the climate system would likely be virtually impossible.

To give some indication of how the accuracy of the finite-difference solution to a differential equation degrades with time, we present a simple example generated by Orszag (1971) concerning the advection of a cone with constant solid body rotation on a plane. Figure 4.3 shows the cone at time zero ($t = 0$). The differential equation should result in the cone's rotating without change of shape in a counterclockwise fashion. The extent to which the finite difference approximations fail to preserve the cone's shape is illustrated in Fig. 4.4, which shows the mountain after rotating one revolution around the plane using a second-order Arakawa scheme in one instance and a fourth-order Arakawa scheme in the other. Note how much better the shape and positioning are preserved with the fourth-order scheme. Moreover, note that the shorter waves trail the cone more in the second-order scheme, indicating larger phase (speed)

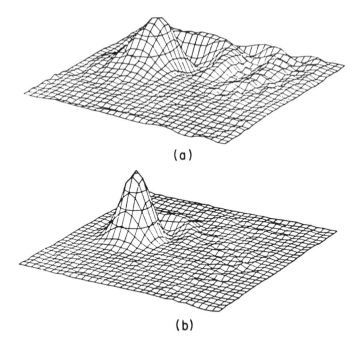

(a)

(b)

Fig. 4.4 Finite difference distribution of the cone in Fig. 4.3 after one rotation around the plane, using (a) a second-order Arakawa scheme and (b) a fourth-order Arakawa scheme.

errors. With time, as more revolutions are simulated, both schemes will become highly inaccurate.

A simple programming example of the finite difference method, kindly provided by John E. Kutzbach of the University of Wisconsin and further modified by Thomas W. Bettge of NCAR, is presented in Appendix D. This program computes atmosphere or ocean circulation from a nondivergent vorticity model. As shown in Appendix D, this example can be solved on an IBM PC or similar personal computer using the programming language BASIC. The model is very close to that originally solved on the ENIAC computer at Princeton University by Charney, Fjørtoft, and von Neumann (1950) in their numerical simulations of the atmosphere.

Another programming example is presented, in Appendix F, for the shallow water equations (4.28)–(4.30), although without the Coriolis term. In this example, described by Sadourny (1975), the Arakawa method is used and the potential enstrophy is conserved rather than energy. Potential enstrophy is defined here as

$$\eta^2 P \qquad (4.63)$$

where P is pressure and η is the potential vorticity, defined as the vorticity divided by pressure, i.e.,

$$\eta = \frac{\zeta}{P} \qquad (4.64)$$

The pressure is related to the height of the fluid by $P = g\rho h$. The reason that potential enstrophy may be a more useful conservative quantity than energy is that this quantity is very important in describing the energy transfer between various scales of motion for approximately two-dimensional fluid flow such as is found in the large scales of the atmosphere and ocean (Charney, 1971).

Spectral Method

The preceding sections have dealt with the use of finite differences to approximate horizontal derivatives in the model equations. Since finite differences are simple, straightforward approximations to derivatives, their use is a relatively easy concept to understand. There are, however, different methods of describing fields that are a function of both space and time. For many physical problems, the use of trigonometric functions such as sines or cosines has definite advantages over finite differences, because such functions can be differentiated quite easily and exactly. Usually the type of basis function selected for a particular problem is determined by the relevant geometry. Because the atmosphere flows over the entire globe, the types of functions that best represent global atmospheric fields should be those natural to the spherical coordinate system, thus allowing for variations in latitude, longitude, and distance from the earth's center. In the climate modeling literature, mathematical functions of this type are called spherical basis functions.

As an example of how basis functions can be used to solve a physical problem, let us look at the example of a simple vibrating string. Consider a string stretched along an x-axis and fixed at both ends, with the left end at $x = 0$ and the right end at $x = \ell$. The speed, c, of the waves traveling on the string can be calculated from the mass, m, per unit length of the string and the tension, T, on the string:

$$c = \sqrt{T/m} \qquad (4.65)$$

Formula (4.65) can be found in many elementary physics texts. The vertical displacement of the string is governed by the so-called classical wave equation

$$\frac{\partial^2 \Phi}{\partial t^2} = c^2 \frac{\partial^2 \Phi}{\partial x^2} \qquad (4.66)$$

where Φ is the displacement of the string perpendicular to the x-axis. The equation contains a common characteristic of wave equations, namely the inclusion of second derivative terms in both space and time. If consideration is first given only to a single wave with frequency ν and angular frequency $\omega = 2\pi\nu$, then

$$\Phi = \Psi(x)e^{i\omega t} \qquad (4.67)$$

The exponential term can also be expressed in terms of sines and cosines using Euler's formula

$$e^{i\omega t} = \cos\omega t + i\sin\omega t \qquad (4.68)$$

Substituting (4.67) into (4.66) yields

$$\frac{d^2\Psi}{dx^2} + k^2\Psi = 0 \qquad (4.69)$$

where k equals the angular frequency (ω) divided by the speed (c) of the waves. Note that (4.69) also is satisfied for either $\Phi = \Psi(x)\cos\omega t$ or $\Phi = \Psi(x)\sin\omega t$. Since the speed of the waves is the wavelength λ divided by the period τ, i.e.,

$$c = \frac{\lambda}{\tau} \qquad (4.70)$$

and frequency $\nu = 1/\tau$, then $c = \lambda\nu$. Hence $k = \omega/c = 2\pi/\lambda$. Note that k, termed the wave number in classical physics, is the inverse of the wavelength with a 2π factor. If $k = 1$ there is one complete wave between the two ends of the string, if $k = 2$ there are two waves, and so forth. A general solution of (4.69) is

$$\Psi(x) = A\sin(kx + \delta) \qquad (4.71)$$

where A and δ are arbitrary constants. That (4.71) is a solution of (4.69) can easily be verified by direct substitution. The boundary conditions on the string are that both ends are fixed; thus

$$\Psi(0) = \Psi(\ell) = 0 \qquad (4.72)$$

There are an infinite number of solutions to (4.69) using (4.71) with $\delta = 0$ in that $k = \pi/\ell,\ 2\pi/\ell,\ 3\pi/\ell, \ldots, n\pi/\ell, \ldots$,

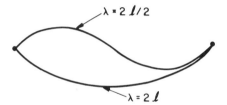

Fig. 4.5 The two lowest modes of a vibrating string with fixed end points.

for integral n, all allow (4.71) to satisfy the boundary conditions (4.72). The general solution is expressed as $k_n = n\pi/\ell$. As an aside, $\Psi = B\cos(kx)$ is also a general solution to (4.69), but with this solution the boundary conditions cannot be satisfied except with $B = 0$. Returning to the solution of (4.71) and rewriting it with the above-specified values of k and δ:

$$\Psi_n(x) = A_n \sin \frac{n\pi x}{\ell} \tag{4.73}$$

where A_n is an arbitrary amplitude associated with each n. In the mathematical literature k is called an eigenvalue and each Ψ_n is called an eigenfunction. Eigenvalues arise from imposing special conditions on equations, which in this case are the boundary conditions at the ends of the string.

Returning to the physical problem, since $\lambda = 2\pi/k$ and $k = n\pi/\ell$, then $\lambda = 2\ell/n$. Hence the wavelength λ can have only certain discrete values that are integral divisors of double the length of the string. Figure 4.5 shows the lowest possible modes. The first, or fundamental (harmonic) mode $\lambda = 2\ell$ has nodes, or points where there is no displacement of the string, only at the ends. The second, $\lambda = 2\ell/2$, has a node at the center as well as at the two ends. The third, $\lambda = 2\ell/3$, has yet another node, and so forth. The number of nodes is consistently $n + 1$.

Before leaving this example of the vibrating string, there are two additional concepts to be introduced: orthogonality of eigenfunctions and completeness. Two functions are said to be *orthogonal* if the integral of their product is zero. Recalling that the eigenvalues are $k_n = n\pi/\ell$ and the eigenfunctions are $\Psi_n = A_n \sin n\pi x/\ell$, then the full set of eigenfunctions is mutually orthogonal if

$$\int_0^\ell \Psi_n(x)\Psi_m(x)\,dx = C_n\delta_{nm} \tag{4.74}$$

where

$$\delta_{nm} = \begin{cases} 0 & \text{if } n \neq m \\ 1 & \text{if } n = m \end{cases} \tag{4.75}$$

and C_n is a constant dependent upon n and often forced to be unity by normalizing the function Ψ. To show that (4.74) is satisfied by the eigenfunctions of (4.73), substitute (4.73) into (4.74) to obtain

$$A_n^2 \int_0^\ell \sin^2\left(\frac{n\pi}{\ell}x\right) dx = \frac{A_n^2}{2} \int_0^\ell \left[1 - \cos\left(\frac{2n\pi}{\ell}x\right)\right] dx$$

$$= \frac{A_n^2 x}{2}\Big|_0^\ell - \frac{\ell A_n^2}{4\pi n} \sin\left(\frac{2n\pi}{\ell}x\right)\Big|_0^\ell \tag{4.76}$$

$$= \frac{\ell}{2}A_n^2$$

when $n = m$, and

$$A_n A_m \int_0^\ell \sin\left(\frac{n\pi}{\ell}x\right)\sin\left(\frac{m\pi}{\ell}x\right) dx$$

$$= A_n A_m \int_0^\ell \left[\frac{1}{2}\cos\frac{n-m}{\ell}\pi x - \frac{1}{2}\cos\frac{n+m}{\ell}\pi x\right] dx$$

$$= \frac{1}{2}A_n A_m \frac{\ell}{(n-m)\pi}\frac{1}{\pi}\sin\frac{n-m}{\ell}\pi x\Big|_0^\ell$$

$$- \frac{1}{2}A_n A_m \frac{\ell}{(n+m)\pi}\frac{1}{\pi}\sin\frac{n+m}{\ell}\pi x\Big|_0^\ell = 0 \tag{4.77}$$

when $n \neq m$.

Often it is convenient to normalize (4.76) to unity by defining the arbitrary constant A_n in (4.76) to be $\sqrt{2/\ell}$; thus, from (4.73),

$$\Psi_n = \sqrt{\frac{2}{\ell}}\sin\frac{n\pi x}{\ell} \tag{4.78}$$

The completeness concept says that the actual shape of the vibrating string can always be represented as an infinite series of eigenvector basis functions:

$$f(x) = \sum_{n=1}^\infty \Psi_n(x) \tag{4.79}$$

This infinite series is termed the Fourier series after the French algebraist and engineer Jean Baptiste Fourier (1768–1830).

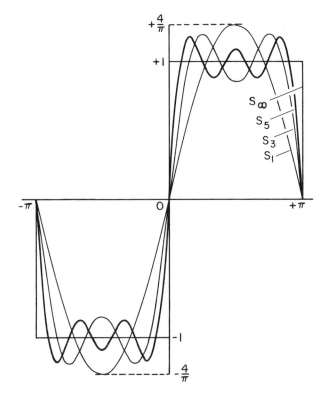

Fig. 4.6 Example of the Gibbs phenomenon for a step function. [From Sommerfeld (1949).]

In spectral representations of atmospheric and oceanic phenomena it is standard to use a limited or truncated series instead of an infinite series. For a simplified example of the effects of truncation, consider a step function as shown in Fig. 4.6. This function is defined as -1 from $-\pi$ to 0 and as $+1$ from 0 to $+\pi$. The function is discontinuous at 0 and can be represented by an infinite sum of sine functions:

$$f(x) = \frac{4}{\pi}\left(\sin x + \frac{1}{3}\sin 3x + \frac{1}{5}\sin 5x + \cdots\right) \qquad (4.80)$$

If the step function $f(x)$ is approximated only by the first term in (4.80), labeled S_1 in Fig. 4.6, then it overshoots the functional value at $\pi/2$ by equaling $4/\pi$ instead of 1. Including successively more terms in the series approximation improves the representation; however, even with the first three terms, labeled S_5 in Fig. 4.6, there is overshooting and undershooting of the true answer (Fig. 4.6). This misrepresentation is called the Gibbs phenomenon and exists with any truncated basis function expansion.

It decreases as more terms are added to the approximation and it is less if the function contains no discontinuities.

We note that the function of (4.80) satisfies the equation $f(-x) = -f(x)$, and for that reason it is termed an *odd* function. Sums of cosines, by contrast, are *even* functions since they satisfy $f(x) = f(-x)$. Sums of cosines are not discontinuous at $x = 0$.

More general considerations of Fourier series and integrals

Much of the mathematics involved in spectral models makes use of Fourier or spectral techniques, and therefore some of the basic spectral techniques will be elaborated.

Any arbitrary periodic function $f(x)$ on the interval $-\pi \leq x \leq +\pi$ can be approximated by the sum of $2n + 1$ elementary trigonometric terms:

$$f(x) \simeq A_0 + \sum_{k=1}^{n} A_k \cos kx + \sum_{k=1}^{n} B_k \sin kx \qquad (4.81)$$

where the $2n+1$ coefficients (A_0, A_k, B_k) have to be determined. Following Sommerfeld (1949), the general orthogonality relationships for this series are

$$\int \cos kx \sin \ell x \, dx = 0 \qquad (4.82)$$

$$\int \cos kx \cos \ell x \, dx = 0 \qquad \text{if } k \neq \ell \qquad (4.83)$$

$$\int \sin kx \sin \ell x \, dx = 0 \qquad \text{if } k \neq \ell \qquad (4.84)$$

If $\ell = k > 0$, then, with a derivation similar to that in (4.76),

$$\frac{1}{\pi} \int_{-\pi}^{+\pi} \cos kx \cos \ell x \, dx = \frac{1}{\pi} \int_{-\pi}^{+\pi} \cos^2 kx \, dx = 1 \qquad (4.85)$$

The $2n + 1$ unknown coefficients can be obtained by using the orthogonality relations (4.82)–(4.85) and by requiring the mean square difference between the function $f(x)$ and the approximating series in (4.81) to be a minimum. This yields the following set of equations for the coefficients:

$$A_0 = \frac{1}{2\pi} \int_{-\pi}^{+\pi} f(x) \, dx \qquad (4.86)$$

$$A_k = \frac{1}{\pi} \int_{-\pi}^{+\pi} f(x) \cos kx \, dx \tag{4.87}$$

$$B_k = \frac{1}{\pi} \int_{-\pi}^{+\pi} f(x) \sin kx \, dx \tag{4.88}$$

where $k > 0$. A_0 is the mean of the function over the interval $-\pi \le x \le +\pi$.

Since the spectral representation for atmospheric quantities is usually given in the exponential rather than the trigonometric form, a conversion from trigonometric to exponential will be shown. Using the variable ξ as a dummy variable of integration, (4.86)–(4.88) can be substituted into the untruncated version of $f(x)$, expanded from (4.81), to yield

$$f(x) = \frac{1}{2\pi} \int_{-\pi}^{+\pi} f(\xi) \, d\xi$$

$$+ \frac{1}{\pi} \sum_{k=1}^{\infty} \left[\int_{-\pi}^{+\pi} f(\xi) \cos k\xi \, d\xi \right] \cos kx$$

$$+ \frac{1}{\pi} \sum_{k=1}^{\infty} \left[\int_{-\pi}^{+\pi} f(\xi) \sin k\xi \, d\xi \right] \sin kx \tag{4.89}$$

Invoking the trigonometric identity

$$\cos k(x - \xi) = \cos k\xi \cos kx + \sin k\xi \sin kx \tag{4.90}$$

$f(x)$ becomes

$$f(x) = \frac{1}{2\pi} \int f(\xi) \, d\xi + \frac{1}{\pi} \sum_{k=1}^{\infty} \int f(\xi) \cos k(x - \xi) \, d\xi \tag{4.91}$$

Then by using Euler's formula in the form

$$\cos P \equiv \frac{e^{iP} + e^{-iP}}{2} \tag{4.92}$$

obtained by substituting (4.20) into (4.3),

$$f(x) = \frac{1}{2\pi} \left\{ \int f(\xi) \, d\xi + \sum_{k=1}^{\infty} \left(\int f(\xi) e^{ik(x-\xi)} d\xi \right. \right.$$

$$\left. \left. + \int f(\xi) e^{-ik(x-\xi)} d\xi \right) \right\} \tag{4.93}$$

If k is allowed to be both positive and negative then the last term in (4.93) is equivalent to a summation from $-\infty$ to -1 if $-k$ is replaced by $+k$. Recognizing also that the first term of (4.93) is the $k = 0$ case of the second term, (4.93) becomes

$$f(x) = \frac{1}{2\pi} \sum_{k=-\infty}^{+\infty} \int f(\xi) e^{ik(x-\xi)} d\xi \qquad (4.94)$$

Equation (4.94) can be written instead as

$$f(x) = \sum_{k=-\infty}^{+\infty} c_k e^{ikx} \qquad (4.95)$$

where

$$c_k = \frac{1}{2\pi} \int f(\xi) e^{-ik\xi} d\xi \qquad (4.96)$$

Here c_k, the complex Fourier coefficient, is introduced in place of A_k and B_k in (4.81). This coefficient is related to A_k and B_k by

$$c_k = \begin{cases} \frac{1}{2}(A_k - iB_k) & \text{for } k > 0 \\ \frac{1}{2}(A_{|k|} + iB_{|k|}) & \text{for } k < 0 \end{cases} \qquad (4.97)$$

and

$$c_0 = A_0 \qquad (4.98)$$

Equations (4.95)–(4.96) are often more convenient to use than the trigonometric form, which contains sines and cosines.

Similar orthogonality properties hold for the complex representation (4.94)–(4.98) as for the trigonometric representation (4.81)–(4.88). Namely, products of functions of wave number $-k$ and k integrated over $-\infty$ to $+\infty$ are nonzero while those at $|-k| \neq |+k|$ are zero. Although this book will not delve into Fourier transform theory, (4.95) is a beginning point of that branch of mathematics.

Spherical Representation

The representation of atmospheric fields in terms of spherical functions, based on latitude, longitude, and height, has long been an attractive method for numerical modelers. Haurwitz (1940) used spherical methods to investigate the movement of Rossby waves of planetary scales with a linear form of the Rossby wave equation (two-dimensional nondivergent vorticity equation referred to in Chapter 3). Silberman (1954) extended the treatment to the nonlinear Rossby wave equation. Later Platzman (1960), Baer and Platzman (1961), and Kubota et al. (1961) studied many of the mathematical characteristics of the spherical technique, such as that it conserves kinetic energy and squared vorticity and that it is not prone to the nonlinear instability usually found in gridpoint schemes. However, a major shortcoming of spherical techniques was the need for very cumbersome formulae to define the interactions between various waves. For years this was a distracting aspect of the spherical method. In recent years this handicap has been overcome by not solving explicitly for the interactions between various waves. Some fundamental concepts of this new method will be introduced here with a detailed simplified set of equations.

Following Merilees (1976), we begin an examination of the equations for spherical models with a simplified form of the shallow water wave equations (4.29) and (4.30),

$$\frac{\partial u}{\partial t} = -g\frac{\partial h}{\partial x} \tag{4.99}$$

$$\frac{\partial v}{\partial t} = -g\frac{\partial h}{\partial y} \tag{4.100}$$

and a modified form of (4.26) adjusted to a spherical coordinate system (see Chapter 3),

$$\frac{\partial h}{\partial t} = -H\left(\frac{\partial u}{\partial x} + \frac{\partial v}{\partial y} - \frac{v}{a}\tan\phi\right) \tag{4.101}$$

H is the mean depth and $h = H + h'$ with $h' \ll H$. The $(v\tan\phi)/a$ term arises from the spherical geometry and can be obtained by differentiating the second term on the right-hand side of (3.25). Since H is constant, derivatives in space and time of h are the same as those of h'. Therefore the prime notation can be ignored in the subsequent discussion. This set of equations describes only the movement of the external gravity wave mode, ignoring for the time being the other wave types.

It is assumed that the above equations can be put into a simple spherical system using the following relationships:

$$\partial x = a\cos\phi\partial\lambda \tag{4.102}$$

$$\partial y = a\partial\phi \tag{4.103}$$

where ϕ and λ are latitude and longitude, a is the radius of the earth, and ∂x and ∂y are the geographical east-west and north-south distances corresponding to latitude and longitude changes of $\partial\phi$ and $\partial\lambda$. Clearly for a given $\partial\lambda$, ∂x becomes smaller as the poles are approached since $\cos\phi$ ranges from 1 at the equator to 0 at the poles. Using (4.102) and (4.103), (4.99)–(4.101) can be written in the following spherical form:

$$\frac{\partial u}{\partial t} = -\frac{g}{a\cos\phi}\frac{\partial h}{\partial\lambda} \tag{4.104}$$

$$\frac{\partial v}{\partial t} = -\frac{g}{a}\frac{\partial h}{\partial\phi} \tag{4.105}$$

$$\frac{\partial h}{\partial t} = -\frac{H}{a\cos\phi}\left(\frac{\partial u}{\partial\lambda} + \frac{\partial}{\partial\phi}(v\cos\phi)\right) \tag{4.106}$$

If (4.106) is partially differentiated with respect to t (regarding u, v, and h as functions of λ, ϕ, and t) and then (4.104) and (4.105) are substituted into the equation, the result is

$$\frac{\partial^2 h}{\partial t^2} = gH\nabla^2 h \tag{4.107}$$

where

$$\nabla^2 \equiv \frac{1}{a^2\cos^2\phi}\left[\cos\phi\frac{\partial}{\partial\phi}\left(\cos\phi\frac{\partial}{\partial\phi}\right) + \frac{\partial^2}{\partial\lambda^2}\right] \tag{4.108}$$

Equation (4.107) is similar in form to the wave equation for the vibrating string, (4.66), except gH replaces c^2 and the two-dimensional Laplacian operator ∇^2 defined in (4.108) replaces the second partial derivative with respect to x. Similarly to the situation in (4.66), the speed of the waves represented in (4.107) is the square root of the coefficient on the right-hand side,

$$c = \sqrt{gH} \tag{4.109}$$

If the relevant mean depth of the atmosphere H is assumed to be about 10 km, then the wave speed of the external gravity wave is $c \simeq 300 \text{ m s}^{-1}$. Since g often appears in conjunction with h, the depth of the fluid, another variable can be defined. This variable is termed the geopotential, Φ, and is defined as $\Phi = gh$. Φ can be expanded into a set of spherical basis functions or harmonics

$$\Phi = \sum_m\sum_\ell \Phi_\ell^m P_\ell^m e^{im\lambda} \tag{4.110}$$

where P_ℓ^m is an associated Legendre function or polynomial of the first kind normalized to unity with the restriction that $|m| \le \ell$

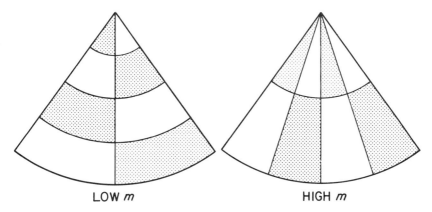

LOW *m*   HIGH *m*

Fig. 4.7 Alternating patterns of positive and negative deviations of the geopotential from the mean geopotential, gH, for low and high values of the planetary wave number m.

(see Appendix B for definitions and mathematical properties of Legendre functions). The east-west planetary wave number, meaning the number of waves around a latitude circle, is m, and $\ell - m$ denotes a so-called meridional wave number, which is defined as the number of nodal zeroes between but not including the two poles. To illustrate this, let $m = 1$, indicating one east-west wave around the sphere. If ℓ is also 1 then no north-south wave can exist. Figure 4.7 shows two different patterns (from Merilees, 1976) for given ℓ of high and low m. Note that as m changes from low to high the number of waves in the zonal direction increases, while the number of waves in the meridional direction decreases. These modes are similar to the example shown for the vibrating string in Fig. 4.5. Substitution of (4.110) into the wave equation (4.107), after some algebraic manipulation, yields

$$\frac{d^2 \Phi_\ell^m}{dt^2} = -\frac{\ell(\ell + 1)}{a^2} \overline{\Phi} \, \Phi_\ell^m \qquad (4.111)$$

The coefficient on the right-hand side of the above equation is the frequency of the waves, i.e.,

$$\omega^2 = \frac{\ell(\ell + 1)}{a^2} \overline{\Phi} \qquad (4.112)$$

which is a function only of ℓ, the radius of the earth, a, and the mean depth geopotential gH. As pointed out by Merilees and others, m is not involved directly since ℓ defines a scale size in the two horizontal dimensions of longitude and latitude. The

Φ_ℓ^m are eigenfunctions of the equation, and ω is the eigenvalue for specified m and ℓ.

Spectral Transform Technique

The use of truncated expansions of analytical spherical functions for describing atmospheric phenomena has certain definite advantages over the gridpoint method. For example, spectral methods of solving barotropic nondivergent models conserve area-averaged mean square kinetic energy and mean square vorticity, two quantities that fail to be conserved with some finite difference methods. Also, the mapping of the sphere in spectral models automatically makes a more uniform grid spacing than is common with finite difference models. One of the problems with the finite difference approach is that near the poles, constant longitudinal increments, $\Delta\lambda$, yield small geographical east-west distances between gridpoints, and these very small distances limit the time step due to the Courant, Friedrichs, and Lewy criterion.

The first use of spectral techniques for atmospheric prediction was by Silberman (1954), who obtained interaction coefficients arising from the nonlinear horizontal advection terms by substituting the spherical basis function (4.110) into the nondivergent vorticity equation. These coefficients represent the interactions of one wave with another. Unfortunately, their calculation is very time consuming, making the technique computationally unattractive for models containing more than a few spherical waves. Platzman (1960), Kubota et al. (1961), Baer and Platzman (1961), and Ellsaesser (1966) explored the conservation and numerical properties of spectral techniques but did not overcome the handicap of the cumbersome calculations involved in the interaction coefficients. Thus the method was not viewed in the 1950s and 1960s as a practical alternative to the finite difference method for atmospheric models. However, the course of atmospheric modeling quickly changed when Orszag (1970) and Eliasen et al. (1970) devised a new technique for solving the spectral equations, called the transform approach. In this section the basic elements of the transform technique will be explained following the development of Bourke (1972). Once again, for the purpose of simplicity, the more general primitive equations will be replaced by the simpler shallow water wave equations (4.26), (4.29), and (4.30). The extension to the more general set of equations is explained by Machenhauer and Daley (1972), Bourke et al. (1977), Baede et al. (1979), Machenhauer

(1979), Gordon and Stern (1982), and Williamson and Swarztrauber (1984).

The momentum equation and equation of continuity in vector form are, respectively,

$$\frac{dV}{dt} = -fk \times V - \nabla \Phi \qquad (4.113)$$

$$\frac{d\Phi}{dt} = -\Phi \nabla \cdot V \qquad (4.114)$$

where V is the horizontal wind $V = iu + jv$, and $\Phi = gh$ is the geopotential. Equations (4.113)–(4.114) follow directly from (4.26) and (4.29)–(4.30) with

$$\nabla = i \frac{\partial}{\partial x} + j \frac{\partial}{\partial y} \qquad (4.115)$$

$$\frac{dV}{dt} = \frac{\partial V}{\partial t} + (V \cdot \nabla)V \qquad (4.116)$$

and

$$fk \times V = f \begin{vmatrix} i & j & k \\ 0 & 0 & 1 \\ u & v & 0 \end{vmatrix} = -ifv + jfu \qquad (4.117)$$

Using a spherical rather than an x-y grid, u and v become longitudinal and latitudinal components of wind and the del operator is adjusted to

$$\nabla = \frac{i}{a \cos \phi} \frac{\partial}{\partial \lambda} + \frac{j}{a} \frac{\partial}{\partial \phi} \qquad (4.118)$$

Robert (1966) found that a redefinition of u and v was more convenient for the representation of scalar fields on the globe. Specifically,

$$U = u \cos \phi \qquad (4.119)$$

$$V = v \cos \phi \qquad (4.120)$$

where ϕ is latitude. Redefinition of the horizontal wind components from u and v to the U and V of (4.119)–(4.120) makes these components zero at the poles. Using (4.116) and the vector identity

$$(V \cdot \nabla)V = \nabla \left(\frac{V \cdot V}{2} \right) + \varsigma k \times V \qquad (4.121)$$

(4.113) can be rewritten in the form

$$\frac{\partial V}{\partial t} = -(\varsigma + f)\mathbf{k} \times V - \nabla \left(\Phi + \frac{V \cdot V}{2} \right) \quad (4.122)$$

Equation (4.121) can be readily verified by expanding each term of the equation and recognizing that ς, the vertical component of relative vorticity, is

$$\varsigma = \mathbf{k} \cdot (\nabla \times V) \quad (4.123)$$

Another critical concept is that of divergence, D:

$$D = \nabla \cdot V \quad (4.124)$$

By performing the curl, $\mathbf{k} \cdot \nabla \times$, and divergence, $\nabla \cdot$, operations on (4.122), equations for the time rate of change of the vorticity and divergence can be obtained:

$$\frac{\partial \varsigma}{\partial t} = -\nabla \cdot (\varsigma + f)V \quad (4.125)$$

$$\frac{\partial D}{\partial t} = \mathbf{k} \cdot \nabla \times (\varsigma + f)V - \nabla \cdot \nabla \left(\Phi + \frac{V \cdot V}{2} \right) \quad (4.126)$$

Derivation of (4.125) and (4.126) from (4.122) requires the identities

$$\mathbf{k} \cdot \left[\nabla \times \left(\frac{\partial V}{\partial t} \right) \right] = \frac{\partial}{\partial t} (\mathbf{k} \cdot [\nabla \times V]) \quad (4.127)$$

and

$$\nabla \cdot \left(\frac{\partial V}{\partial t} \right) = \frac{\partial}{\partial t} (\nabla \cdot V) \quad (4.128)$$

both of which follow from the interchangeability of the order of differentiation [e.g., $\partial/\partial\phi(\partial v/\partial t) = \partial/\partial t(\partial v/\partial\phi)$] and can be obtained by expanding all terms in the equations. The continuity equation (4.114) becomes

$$\frac{\partial(\Phi')}{\partial t} = \frac{\partial(\Phi)}{\partial t} = -(V \cdot \nabla)(\overline{\Phi} + \Phi') - (\overline{\Phi} + \Phi')\,D$$

$$= -[V \cdot \nabla\Phi' + \Phi'D] - \overline{\Phi}D \quad (4.129)$$

$$= -\nabla \cdot \Phi'V - \overline{\Phi}\,D$$

The geopotential term, Φ, is divided into two terms, the time-independent areal mean, $\overline{\Phi}$, and the deviation from the areal mean, $\Phi' = \Phi - \overline{\Phi}$. This separation into mean and deviation allows the introduction of a semi-implicit time integration scheme if desired. Also, $\nabla \cdot \nabla$ is abbreviated notationally to ∇^2, i.e.,

$$\nabla^2 = \nabla \cdot \nabla = \frac{1}{a^2 \cos^2 \phi} \left[\frac{\partial^2}{\partial \lambda^2} + \cos \phi \frac{\partial}{\partial \phi} \left(\cos \phi \frac{\partial}{\partial \phi} \right) \right] \quad (4.130)$$

The Helmholtz theorem separates the vector V into two terms, one containing a scalar streamfunction, ψ, and the other containing a scalar velocity potential, χ:

$$V = k \times \nabla \psi + \nabla \chi \quad (4.131)$$

This theorem is named after H. von Helmholtz, a nineteenth-century hydrodynamicist who found that any vector flow field could be expressed in terms of a rotational component, vorticity, and a nonrotational component, divergence. By performing the curl and dot operations respectively on (4.126), the vorticity and divergence, defined in (4.123) and (4.124), can readily be expressed by the Laplacian of the streamfunction and velocity potential, respectively:

$$\zeta = k \cdot (\nabla \times V) = \nabla^2 \psi \quad (4.132)$$

$$D = \nabla \cdot V = \nabla^2 \chi \quad (4.133)$$

By using (4.132) and (4.133) the vorticity and divergence equations (4.125) and (4.126) can be written, respectively, as

$$\frac{\partial}{\partial t}(\nabla^2 \psi) = - \frac{1}{a \cos^2 \phi} \left[\frac{\partial}{\partial \lambda}(U \nabla^2 \psi) + \cos \phi \frac{\partial}{\partial \phi}(V \nabla^2 \psi) \right]$$
$$- 2\Omega \left(\sin \phi \, \nabla^2 \chi + \frac{V}{a} \right) \quad (4.134)$$

and

$$\frac{\partial}{\partial t}(\nabla^2 \chi) = + \frac{1}{a \cos^2 \phi} \left[\frac{\partial}{\partial \lambda}(V \nabla^2 \psi) - \cos \phi \frac{\partial}{\partial \phi}(U \nabla^2 \psi) \right]$$
$$+ 2\Omega \left(\sin \phi \, \nabla^2 \psi - \frac{U}{a} \right)$$
$$- \nabla^2 \left(\frac{U^2 + V^2}{2 \cos^2 \phi} + \Phi' \right) \quad (4.135)$$

The continuity equation can be written as:

$$\frac{\partial \Phi'}{\partial t} = -\frac{1}{a \cos^2 \phi}\left[\frac{\partial}{\partial \lambda}(U\Phi') + \cos\phi\frac{\partial}{\partial \phi}(V\Phi')\right] - \bar{\Phi}D \quad (4.136)$$

Because of the Helmholtz theorem (4.131), U and V in the above equations can be written in terms of ψ and χ, i.e.,

$$U = -\frac{\cos\phi}{a}\frac{\partial\psi}{\partial\phi} + \frac{1}{a}\frac{\partial\chi}{\partial\lambda} \quad (4.137)$$

$$V = \frac{1}{a}\frac{\partial\psi}{\partial\lambda} + \frac{\cos\phi}{a}\frac{\partial\chi}{\partial\phi} \quad (4.138)$$

At this point the system is complete in that (4.134)–(4.136) are three prediction equations for the three unknown variables ψ, χ, and Φ. Note from (4.137) and (4.138) that U and V are known functions of ψ and χ. This system of equations obviously is more complex than the straightforward use of u, v, and h discussed for finite difference approaches. In principle it is essentially the same in terms of the number of equations and unknowns, but ψ and χ replace U and V. The ψ and χ fields are scalar functions that can be expressed easily as spherical harmonics.

Since nonlinear interaction of waves leads to complex algebraic computation, the transform technique allows for a relatively simple way of solving the equations. The essence of this technique is to transform from spectral space quantities to gridpoint space quantities, then to perform nonlinear products at each gridpoint, such as those in the square bracketed terms of (4.134)–(4.136), and to reverse transform the product quantities from gridpoint quantities back to spectral quantities.

In order to do this, the fields of ψ, χ, and Φ' are represented as truncated series of spherical harmonics. These rhomboidal truncated series are defined for streamfunction, velocity potential, and geopotential:

$$\psi = a^2 \sum_{m=-J}^{+J}\sum_{\ell=|m|}^{|m|+J} \psi_\ell^m Y_\ell^m \quad (4.139)$$

$$\chi = a^2 \sum_{m=-J}^{+J}\sum_{\ell=|m|}^{|m|+J} \chi_\ell^m Y_\ell^m \quad (4.140)$$

$$\Phi = a^2 \sum_{m=-J}^{+J}\sum_{\ell=|m|}^{|m|+J} \Phi_\ell^m Y_\ell^m \quad (4.141)$$

where J is the truncation wave number. Similarly, the horizontal wind components can be represented as

$$U = a \sum_{m=-J}^{+J} \sum_{\ell=|m|}^{|m|+J+1} U_\ell^m Y_\ell^m \qquad (4.142)$$

$$V = a \sum_{m=-J}^{+J} \sum_{\ell=|m|}^{|m|+J+1} V_\ell^m Y_\ell^m \qquad (4.143)$$

The quantities ψ_ℓ^m, χ_ℓ^m, and Φ_ℓ^m are time-dependent expansion coefficients that are complex. We denote by $*$ the complex conjugate $x - iy$ of the complex number $x + iy$. In order for the streamfunction to be real and not imaginary, the following condition must be satisfied:

$$(\psi_\ell^m)^* = (-1)^m \psi_\ell^{-m} \qquad (4.144)$$

where m, the longitudinal wave number, is an integer.

The function Y_ℓ^m is defined as $P_\ell^m(\sin\phi)e^{im\lambda}$, where P_ℓ^m is an associated Legendre polynomial (see Appendix B). If (4.139)–(4.141) are substituted into (4.137) and (4.138), the following equations can be obtained after suitable grouping of terms:

$$U_\ell^m = +(\ell-1)\epsilon_\ell^m \psi_{\ell-1}^m - (\ell+2)\epsilon_{\ell+1}^m \psi_{\ell+1}^m + im\chi_\ell^m \quad (4.145)$$

$$V_\ell^m = -(\ell-1)\epsilon_\ell^m \chi_{\ell-1}^m + (\ell+2)\epsilon_{\ell+1}^m \chi_{\ell+1}^m + im\psi_\ell^m \quad (4.146)$$

where

$$\epsilon_\ell^m = \sqrt{(\ell^2 - m^2)/(4\ell^2 - 1)} \qquad (4.147)$$

Since U and V are obtained by differentiating streamfunction and velocity potential, the expansion has one degree more than that of ψ and χ, i.e., $J+1$.

As shown by Bourke (1972) for the shallow water equations, the nonlinear products $U\nabla^2\psi, V\nabla^2\psi, U\phi', V\phi'$, and $(U^2+V^2)/2$ must be evaluated. This is done by transforming each term of the product onto a two-dimensional spatial grid and then multiplying the gridpoint values to form the products in gridpoint space. For example, U and $\nabla^2\psi$ are transformed to grid space, after which U and $\nabla^2\psi$ are multiplied at the grid location. Products in gridpoint space then are transformed back into spectral space.

For the vorticity equation (4.134), this reverse transform requires a representation of the gridpoint products as truncated Fourier series at each latitude circle as follows:

$$U\nabla^2\psi = a \sum_{m=-J}^{+J} A_m e^{im\lambda} \tag{4.148}$$

$$V\nabla^2\psi = a \sum_{m=-J}^{+J} B_m e^{im\lambda} \tag{4.149}$$

A_m and B_m can be found by using a technique called the fast Fourier transform (FFT). It is beyond the scope of this book to delve into details of the fast discrete Fourier transform. The integral form of the Fourier transform was discussed in (4.94); the integral is made discrete by replacing it with an explicit summation. The *fast* aspect was invented by Cooley and Tukey (1965) when they constructed certain efficient computational algorithms for the cases when the number of values to be transformed is a power of 2. Somewhat less efficient methods also can be constructed for situations where the points to be transformed are not powers of 2. Further gains in computational efficiency can be obtained in atmospheric models by doing the transforms in a *vector* fashion, with many of the calculations being performed at the same time. Several explicit formulations for FFTs in the computer language FORTRAN can be found in Swarztrauber (1984).

Returning to the model equations, substituting (4.148)–(4.149) into (4.134) yields, after some algebraic manipulation and transformation from gridpoint space to wave number space, the following:

$$\ell(\ell+1)\frac{\partial\psi_\ell^m}{\partial t} = -\int_{-\pi/2}^{+\pi/2}\frac{1}{\cos^2\phi}\left[imA_m P_\ell^m(\sin\phi)\right.$$

$$\left. - B_m\cos\phi\frac{\partial P_\ell^m(\sin\phi)}{\partial\phi}\right]\cos\phi\,d\phi$$

$$+ 2\Omega[\ell(\ell-1)\epsilon_\ell^m\,\chi_{\ell-1}^m$$

$$+ (\ell+1)(\ell+2)\epsilon_{\ell+1}^m\,\chi_{\ell+1}^m - V_\ell^m] \tag{4.150}$$

where

$$\varsigma = \sum_\ell\sum_m \varsigma_\ell^m\,Y_\ell^m = \sum_\ell\sum_m\left(\frac{-\ell(\ell+1)}{a^2}\right)\psi_\ell^m\,Y_\ell^m \tag{4.151}$$

and

$$D = \sum_\ell \sum_m D_\ell^m Y_\ell^m = \sum_\ell \sum_m \left(\frac{-\ell(\ell+1)}{a^2} \right) \chi_\ell^m Y_\ell^m \quad (4.152)$$

The first square bracket of (4.150) arises because the second term of the following equation can be integrated by parts:

$$\int_{-\pi/2}^{+\pi/2} \frac{1}{\cos^2 \phi} \left(im A_m + \cos \phi \frac{\partial B_m}{\partial \phi} \right) P_\ell^m (\sin \phi) \cos \phi \, d\phi \quad (4.153)$$

with the boundary conditions $B_m(\pm \pi/2) = 0$ arising because V is zero at the poles. If the latitudes are chosen at the nodal points of the Legendre polynomials, then the integral in (4.153) can be determined exactly. The usual technique for solving this integral is a technique referred to as the Gaussian quadrature, which is discussed in Appendix B. Furthermore, the nodal latitudes are called the Gaussian latitudes.

Similarly to the treatment of the products in the vorticity equation, the products in the divergence and continuity equations can be written as

$$U\Phi' = a^3 \sum_{m=-J}^{+J} C_m e^{im\lambda} \quad (4.154)$$

$$V\Phi' = a^3 \sum_{m=-J}^{+J} D_m e^{im\lambda} \quad (4.155)$$

$$\frac{U^2 + V^2}{2} = a^2 \sum_{m=-J}^{+J} E_m e^{im\lambda} \quad (4.156)$$

The prediction equations for velocity potential and geopotential become

$$-\ell(\ell+1)\frac{\partial \chi_\ell^m}{\partial t} = \int_{-\pi/2}^{+\pi/2} \frac{1}{\cos^2 \phi} \left[im B_m P_\ell^m (\sin \phi) \right.$$
$$+ A_m \cos \phi \frac{\partial P_\ell^m (\sin \phi)}{\partial \phi} \left. \right] \cos \phi \, d\phi$$
$$- 2\Omega[\ell(\ell-1)\epsilon_\ell^m \psi_{\ell-1}^m$$
$$+ (\ell+1)(\ell+2)\epsilon_{\ell+1}^m \psi_{\ell+1}^m$$
$$+ U_\ell^m] + \ell(\ell+1)(E_\ell^m + \Phi_\ell^m) \quad (4.157)$$

and

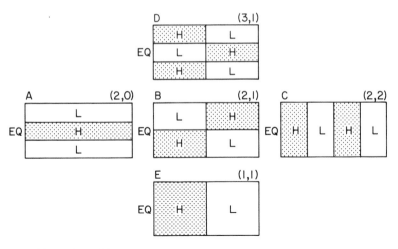

Fig. 4.8 Examples of low-resolution ℓ and m patterns. Note that high m can have more nodal structure in the north due to the $\ell - m$ restriction.

$$\frac{\partial \Phi_\ell^m}{\partial t} = - \int_{-\pi/2}^{+\pi/2} \frac{1}{\cos^2 \phi} \left[imC_m P_\ell^m (\sin \phi) \right.$$

$$\left. - D_m \cos \phi \frac{\partial P_\ell^m (\sin \phi)}{\partial \phi} \right] \cos \phi \, d\phi$$

$$+ \overline{\Phi} \ell(\ell + 1) \chi_\ell^m \tag{4.158}$$

where

$$E_\ell^m = \int_{-\pi/2}^{+\pi/2} \frac{E_m}{\cos^2 \phi} P_\ell^m (\sin \phi) \cos \phi \, d\phi \tag{4.159}$$

Equations (4.150), (4.157), and (4.158) are the prediction equations for ψ, χ, and Φ. Their extension to the full primitive equation model is shown in Bourke et al. (1977) and Haltiner and Williams (1980). The incorporation of semi-implicit time integration means that Φ_ℓ^m in (4.157) and χ_ℓ^m in (4.158) are considered implicitly.

Figure 4.8 is a schematic diagram of truncation at low wave numbers. Recall that m represents the number of waves in the east-west direction and $(\ell - m)$ represents the meridional wave number or nodal zeroes between but not including the two poles. The five subdiagrams of Fig. 4.8 show the patterns for several resolutions. A, representing $\ell = 2$ and $m = 0$, has two lows, L, and one high, H, in the north-south direction and is constant in longitude. The number of nodal zeroes is two. B has a high-low pattern in the north-south direction, one nodal zero, and

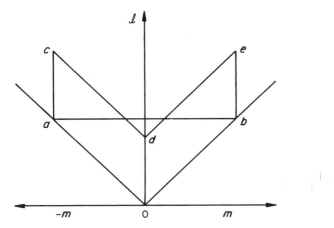

Fig. 4.9 ℓ and m space showing two commonly used spectral truncations, triangular $(0, a, b)$ and parallelogramic $(a, c, d, e, b, 0)$.

one wave in the east-west direction. Subdiagram C has no nodal zero due to the $\ell - m$ constraint and has two east-west waves. D has one east-west wave and two nodal zeroes; finally, E has no nodal zeroes and one east-west wave. Figure 4.9 shows types of commonly used truncation. In the figure, the region $(0, a, b)$ depicts a triangle in the $m - \ell$ space and the region $(a, c, d, e, b, 0)$ depicts a parallelogramic truncation, in which the truncation is to

$$\sum_{m=-M}^{M} \sum_{\ell=|m|}^{|m|+J} \psi_\ell^m Y_\ell^m \tag{4.160}$$

with $M \neq J$, in contrast to a rhomboidal truncation, in which $M = J$. Baer (1972) has displayed the spherical functions on a globe for $\ell = 5$ and $m = 0, 1, 2, \ldots, 5$ (Fig. 4.10). The number of east-west waves, m, is easily seen in the figure by counting the shaded areas. Also, since ℓ is held fixed, the number of north-south waves decreases to zero as $m \to \ell$.

To illustrate the effect on 500 mb patterns of different truncations, Fig. 4.11 shows data on a 2.5° latitude/longitude grid with different truncations. Many climate studies have used a rhomboidal truncation, R, of 15 wave numbers, i.e., $J = 15$. As can be seen in Fig. 4.11, R15 gives a good approximation to the large-scale motions of the atmosphere (ALL WAVES in Fig. 4.11) but misses some of the smaller-scale features. Increasing the resolution to R30 improves the representation to the original 2.5° data, while coarsening the resolution to R5 makes the pattern much too smooth. It is not clear which type of spectral truncation is

$m = 0$ $m = 1$

$m = 2$ $m = 3$

$m = 4$ $m = 5$

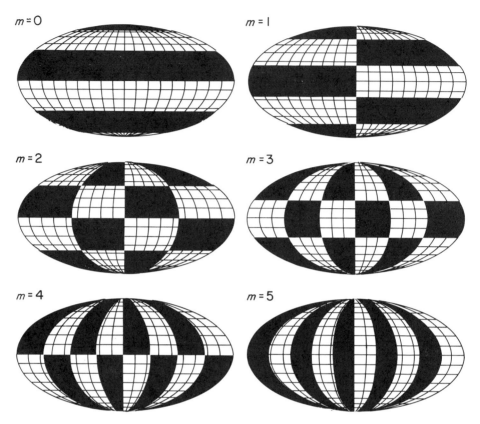

Fig. 4.10 Alternating patterns of positives and negatives for spherical functions with $\ell = 5$ and $m = 0, 1, 2, \ldots, 5$. [From Baer (1972).]

best. If triangular truncation (T) is used, then certain horizontal scales are neglected, namely those where ℓ and m are relatively large (see Fig. 4.9, where for large ℓ and m the regions of the two truncations do not overlap). If resolution is sufficiently high, it may not make much difference which form of truncation is used. Figure 4.12 shows the Gaussian grid on the globe for various resolutions: rhomboidal R15 and R21, and triangular T42 and T95. R15 is coarse; however, it is commonly used in climate simulation studies. Higher resolutions require substantially more computer time, but are needed for realistic short to medium range forecasts (up to 2 weeks).

Appendix F includes a programming example of the transform method, kindly provided by Joseph Tribbia of NCAR. The example solves the same nondivergent vorticity equation (see Chapter 3) used by Charney et al. (1950), except it uses the

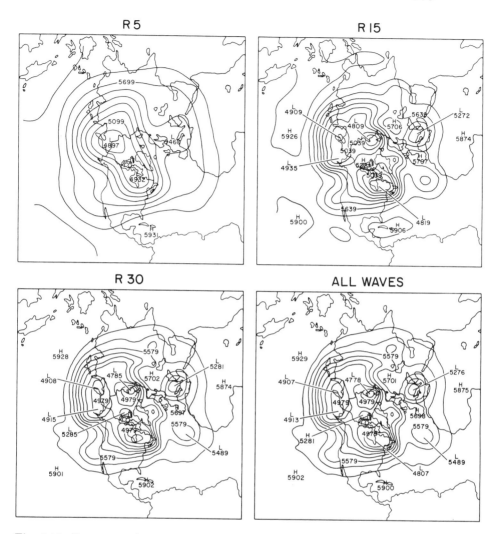

Fig. 4.11 Demonstration of the effects of various horizontal truncations of 500 mb patterns of geopotential height (m) of 2.5° latitude/longitude data: rhomboidal, R5, R15, and R30, and the original 2.5° data (ALL WAVES). [David Baumhefner, personal communication.]

204

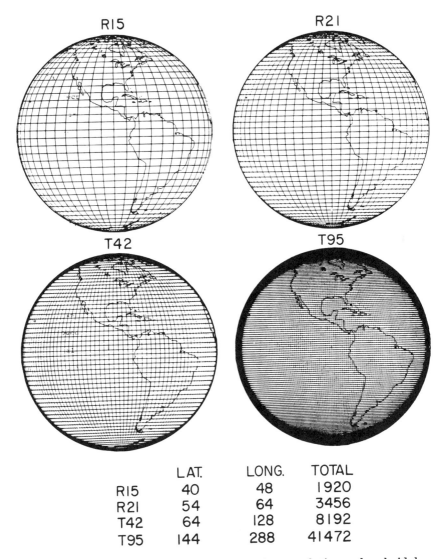

	LAT.	LONG.	TOTAL
R15	40	48	1920
R21	54	64	3456
T42	64	128	8192
T95	144	288	41472

Fig. 4.12 Gaussian grid on the globe for various resolutions: rhomboidal, R15 and R21, and triangular, T42 and T95. [David Williamson, personal communication.]

transform method. The formalism follows that of Machenhauer (1979).

Vertical Representation

There are many possible representations of the vertical structure of the atmosphere and oceans. As mentioned in Chapter 3, most of the current atmospheric models employ either sigma, $\sigma \equiv p/p_s$ (the ratio of pressure to surface pressure), or height, z, as the vertical coordinate. One of the many other possible choices would be potential temperature (see Kasahara, 1974, for a general derivation of vertical coordinate systems). Most climate models use finite difference and quadrature approximations for vertical derivatives and integrals, even if they treat the horizontal structure with spectral methods.

The transformed σ-coordinate system takes mountains into account but has problems in that the horizontal pressure gradient in the equations of motion contains two large terms (Chapter 3). One term represents the gradient of pressure along the σ surface and the other represents the vertical hydrostatic correction. The difference between the two terms determines the horizontal pressure gradient, so that if computational errors occur in either term, the horizontal pressure gradient and resulting winds will be in error also. Experience has shown that this error will be minimized by forcing the gradient of mountain heights to be small locally. Smagorinsky et al. (1965), Corby et al. (1977), Gary (1973), and Sundqvist (1979) show several alternative techniques of dealing with this problem in large-scale models, including converting the pressure gradient term from σ surfaces to constant pressure surfaces before taking horizontal pressure gradients. This avoids the computation of the pressure gradient on a sloping σ surface. Another method used by Kasahara and Washington (1967, 1971) in atmospheric models and Bryan (1969a) in an ocean model is to use the z-coordinate system but allow the air or water flow to be obstructed or blocked by the mountains or coastlines. This method has the advantage of making the errors more local. The disadvantage is that special treatment is required in the vicinity of mountains. The same problem occurs in ocean models where continents, shelves, and submarine mountains are encountered.

Even though the placement of variables in the vertical also differs from one model to another, there are some important general characteristics. The Australian spectral model by Bourke, McAvaney, Puri, and Thurling (1977), which has a vertical

VERTICAL
INDEX   VARIABLES

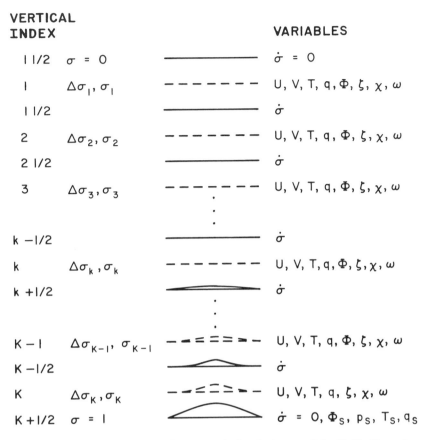

1 1/2	$\sigma = 0$	───────	$\dot{\sigma} = 0$
1	$\Delta\sigma_1, \sigma_1$	– – – – – – –	$U, V, T, q, \Phi, \zeta, \chi, \omega$
1 1/2		───────	$\dot{\sigma}$
2	$\Delta\sigma_2, \sigma_2$	– – – – – – –	$U, V, T, q, \Phi, \zeta, \chi, \omega$
2 1/2		───────	$\dot{\sigma}$
3	$\Delta\sigma_3, \sigma_3$	– – – – – – –	$U, V, T, q, \Phi, \zeta, \chi, \omega$
k −1/2		───────	$\dot{\sigma}$
k	$\Delta\sigma_k, \sigma_k$	– – – – – – –	$U, V, T, q, \Phi, \zeta, \chi, \omega$
k +1/2		───────	$\dot{\sigma}$
K − 1	$\Delta\sigma_{K-1}, \sigma_{K-1}$	– – ≈ ≈ ≈ – –	$U, V, T, q, \Phi, \zeta, \chi, \omega$
K −1/2		───────	$\dot{\sigma}$
K	$\Delta\sigma_K, \sigma_K$	– – ≈ ≈ ≈ – –	$U, V, T, q, \zeta, \chi, \omega$
K +1/2	$\sigma = 1$	───────	$\dot{\sigma} = 0, \Phi_S, p_S, T_S, q_S$

Fig. 4.13 Vertical structure of a general circulation model. U, V, T, and q, Φ, ζ, D, ω are computed at the dashed-line locations, and $\dot{\sigma}$ is computed at the solid-line locations. [From Williamson (1983).]

structure similar to that devised by Manabe, Smagorinsky, and Strickler (1965), will be used to illustrate how vertical discretization is done. Figure 4.13 shows the vertical structure of the Australian model, which is the same vertical structure as in the NCAR Community Climate Model (CCM). In the σ-system the forecast variables U, V, T, q (defined as mixing ratio), Φ, ζ, D, ω (defined as dp/dt) are computed at the dashed-line locations, and the vertical velocity $\dot{\sigma}$ is computed at the solid-line locations. At the earth's surface, Φ_s, the surface pressure, p_s, is computed, while the surface temperature is either computed or specified. The subscript k is the layer index. The vertical advections of momentum, temperature, and moisture are computed at the half levels; then the mass weighted averages are computed at the interfaces between the layers. At the top and bottom levels uncentered finite difference operators are used to approximate the vertical advection terms. As an example, the vertical advection of momentum shown in Williamson (1983) is

$$
\left(\dot{\sigma} \frac{\partial U}{\partial \sigma} \right)_k = \alpha_{k+} \dot{\sigma}_{k+1/2} \left(\frac{U_{k+1} - U_k}{\sigma_{k+1} - \sigma_k} \right)_{k+1/2}
$$

$$
+ \alpha_{k-} \dot{\sigma}_{k-1/2} \left(\frac{U_k - U_{k-1}}{\sigma_k - \sigma_{k-1}} \right)_{k-1/2} \qquad (4.161)
$$

where $1 \leq k \leq K$ and the α_{k+} and α_{k-} are interpolation coefficients. The terms in the parentheses on the right-hand side represent the vertical approximations to derivatives $\partial u / \partial \sigma$ at the $\dot{\sigma}$ levels. Then the weighted average of $\dot{\sigma} \, \partial u / \partial \sigma$ is obtained at the dashed-line locations. This average makes use of the thicknesses of each of the layers, e.g., $\sigma_k - \sigma_{k-1}$ and $\sigma_{k+1} - \sigma_k$. Regarding (4.161), Williamson (1983) shows generalized notation for the approximations to the vertical derivatives and the interpolation formulae at the top and the bottom as well as the intermediate levels in the models. Phillips (1974) and Arakawa and Lamb (1977) point out that the vertical difference method and the computation of vertically integrated quantities, such as vertical velocity and the hydrostatic equation, should be done in an energy-conserving manner so that total energy is conserved for an adiabatic atmosphere. The above approximation, (4.161), does not have this property. The relationship between accuracy in finite difference approximations and globally averaged energy constraints is not clear, particularly in the vertical representation, since gradients are often very large. It is possible to devise a scheme that conserves energy very well but gives a relatively poor estimate of vertical derivatives. The converse is also true. Ideally, both properties should be computed accurately.

Returning to (4.161), the interpolation coefficients are

$$
\alpha_{1+} = \frac{\sigma_1}{\sigma_{1+1/2}} \qquad\qquad (4.162)
$$

$$
\alpha_{1-} = 0 \qquad\qquad (4.163)
$$

at the top,

$$
\alpha_{k+} = \frac{\sigma_k - \sigma_{k-1}}{\sigma_{k+1} - \sigma_{k-1}} \qquad 2 \leq k \leq K - 1 \quad (4.164)
$$

$$
\alpha_{k-} = \frac{\sigma_{k+1} - \sigma_k}{\sigma_{k+1} - \sigma_{k-1}} \qquad 2 \leq k \leq K - 1 \quad (4.165)
$$

at intermediate layers, and

$$\alpha_{K+} = 0 \tag{4.166}$$

$$\alpha_{K-} = \frac{1 - \sigma_K}{1 - \sigma_{K-1/2}} \tag{4.167}$$

at the bottom. The coefficients can be different for other variables such as temperature and moisture since they have large variations near the ground. In this case logarithmic variation is assumed, which gives a better fit to the observed vertical profiles.

The treatment of vertical differences in large-scale ocean models (e.g., Bryan, 1969a; Semtner, 1974) is somewhat similar to the treatment in large-scale atmospheric models. The vertical advection equation for some quantity A is written in flux form so that the vertical advection operator is

$$\frac{\partial(wA)}{\partial z} \doteq \delta_z(w\overline{A}^z) \tag{4.168}$$

where w is defined at the top and bottom of a grid volume. The $\overline{A}^z$ can be momentum components (u, v), temperature (T), or salinity (S). The averaging overbar indicates a simple average of quantities on either side of an interface. Since w is zero at the top and bottom of the ocean model, the vertical sum of (4.168) is zero, and thus the quantity A is conserved in the vertical.

CHAPTER 5

Examples of Simulations
of Present-Day Climate

In this chapter we complement the theoretical description of computer models presented in Chapter 3 with sample results from various numerical simulations, first from stand-alone models of the atmosphere, oceans, and sea ice, respectively, and then from coupled models combining at least two of these three major climatic components.

Before proceeding to state-of-the-art simulations we illustrate how a realistic simulation of the atmosphere evolves from a very artificial initial state of no motion. Figure 5.1 shows a time sequence of flow patterns of the sea level pressure and 3 km temperature fields for a January simulation using the NCAR 1967-vintage two-layer model (Kasahara and Washington, 1967; Washington and Kasahara, 1970). At day 0 the model atmosphere is at rest (i.e., no winds) and is isothermal at 240 K. The temperature at the earth's surface is specified as the observed climatological January distribution over the oceans, although it is computed from a local surface energy balance over the continents. Because the atmosphere is cold initially and the oceans are relatively warm, there is rapid heating of the atmosphere over the oceans by direct sensible heat from the surface and by the release of latent heat. Even at day 1 the ocean areas generally show relatively low sea level pressure and warmer low-level air temperatures than the continental areas (Fig. 5.1). Within 5 days this land/sea pressure contrast has intensified to produce

210

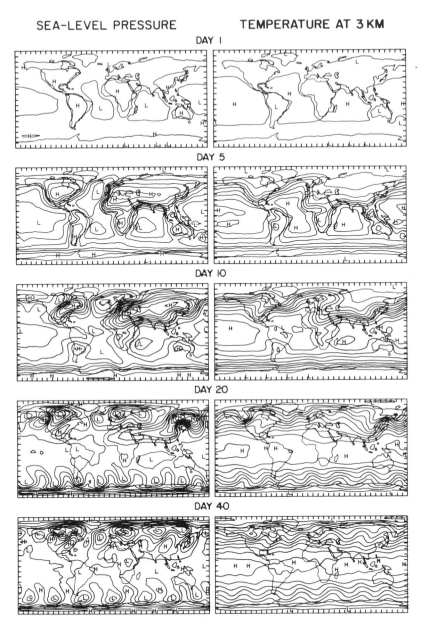

SEA-LEVEL PRESSURE   TEMPERATURE AT 3 KM

DAY 1

DAY 5

DAY 10

DAY 20

DAY 40

Fig. 5.1 Time sequences of sea level pressures and 3 km temperatures from a perpetual January simulation of a two-layer atmospheric GCM starting at rest with a 240 K isothermal atmosphere. The pressure contours are drawn at intervals of 4 mb and the temperature contours at intervals of 5 K. The contour lines on day 1 are at 1000 mb and 245 K. [From Washington (1968).]

large sea-breeze-type circulation patterns with generally sinking motions over the continents (where the sea level pressure is relatively high), low-level outflow to the oceans, rising motions over the oceans (where the sea level pressure is relatively low), and corresponding high-level flow from over the oceans to over the land regions. This cellular type of motion is most prominent in the winter hemisphere, which at day 5 in this simulation is in the Northern Hemisphere. At day 10 the simple flow pattern becomes unstable and breaks down into large-scale horizontal eddies that are weather storm systems (often referred to as baroclinic waves). These eddies intensify and move slowly downstream, which in mid-latitudes is from west to east. Each of these storms has a low pressure pattern, with counterclockwise flow in the Northern Hemisphere and clockwise flow in the Southern Hemisphere. In either hemisphere the air on the western side of the low pressure is moving equatorward, bringing colder polar air toward the tropics in a pattern referred to as a cold front. The warmer air moves toward the pole on the eastward side of the low pressure and forms a warm front. Although these cold and warm intrusions, or fronts of air, can be seen in the temperature fields of Fig. 5.1, in nature they usually have much sharper boundaries, which are limited in the model simulation by the model grid resolution of 5° of latitude and longitude.

Baroclinic eddies have the capacity to transport heat and moisture both horizontally and vertically in the atmosphere. To see how this is possible it is useful to introduce the concept of the zonal average and deviation from it:

$$[A] = \frac{1}{2\pi} \int_0^{2\pi} A \, d\lambda \tag{5.1}$$

and

$$A' = [A] - A \tag{5.2}$$

where A is a quantity such as temperature, T, or moisture, q, $[A]$ is the longitudinal (or zonal) average of A at some latitude, ϕ, and A' is the difference between an individual local value, $A = A(\lambda, \phi)$ and $[A] = [A(\phi)]$. If A shows no deviation from the zonal average then $A' = 0$. Obviously $[A'] = 0$. For the average of the product of quantities A and B,

$$[AB] \equiv [([A] + A')([B] + B')] = [A][B] + [A'B'] \tag{5.3}$$

Returning to Fig. 5.1, if one examines the low pressure regions along a given Northern Hemisphere latitude, there tends to be cold ($T' < 0$), dry ($q' < 0$; not shown) air on the western side of the low, where $v' < 0$, and warm $(T' > 0)$, moist $(q' > 0$; not shown) air on the eastern side, where $v' > 0$. As a result of these correlations, with cold, dry air moving equatorward and warm, moist air moving poleward, the horizontal eddy transports $[v'T']$ and $[v'q']$ of sensible heat and moisture result in net heat and moisture transports toward the poles. The horizontal transports by the mean flow are represented as $[T][v]$ and $[q][v]$. In regions of strong baroclinic activity the poleward eddy transports can be much larger than the mean transports. In an analogous manner the vertical transports of sensible heat and moisture can be expressed as $[w'T']$ and $[w'q']$, and can be determined, on average, to transport heat and moisture upward. These eddies must be accounted for by general circulation models of the atmosphere and oceans. In the atmospheric case, baroclinic theory suggests 6–8 eddies or waves around the globe at mid-latitudes. In the oceans the relevant scale is smaller by at least an order of magnitude, with many more eddies of smaller scale being important in ocean circulations (see Gill, 1982; important length scales in the ocean are on the order of 10–100 km). An atmospheric model should have sufficient horizontal resolution to resolve the major atmospheric waves. By contrast, coarse resolution ocean models used in climate studies usually do not resolve the baroclinic ocean eddies. Semtner and Mintz (1977), Cox (1985), and Bryan (1985) have found that small-scale eddies explicitly simulated by fine-resolution ocean models (with grid intervals of 1° latitude/1° longitude or less) can be reasonably approximated by using coarser resolution ocean models (with grid intervals of 2° latitude/2° longitude) if the effects of such eddies are included through the use of an eddy diffusion parameterization.

Simulations of the Atmosphere

A typical model simulation of the present atmosphere will be shown using results from the NCAR Community Climate Model (CCM). There are many other model results that could be used that give comparable, and in some cases better, comparisons with observed quantities, but the NCAR results are representative of the type of simulation possible with the current generation of atmospheric GCMs. The NCAR model has evolved from contributions by many general circulation modelers around the world. It uses the σ vertical coordinate system devised by N. Phillips (1957) and parameterizations of surface fluxes, precipitation, and cumulus convection following those used at GFDL (see Manabe et al., 1965). The spectral transform technique, introduced independently by Orszag (1970) and Eliasen et al. (1970) and incorporated by Bourke et al. (1977) and McAvaney et al. (1978) into the Australian Forecast Model, was modified by K. Puri, E. Pitcher, and R. Malone for the NCAR model, along with a new radiation scheme (Ramanathan et al., 1983). This version of the NCAR model has 9 σ levels and a rhomboidal truncation of 15 waves, and was first used to run perpetual (more than 400 days) January and July simulations with specified ocean temperatures (see Pitcher et al., 1983; Ramanathan et al., 1983). Several examples of fields of variables averaged over the last 120 days of the perpetual January and July simulations will be shown and compared with observed data. The NCAR model has since been run with a seasonal cycle (Chervin, 1986) and has been coupled to a variety of ocean formulations by Washington and Meehl (1983, 1984) and Washington and VerPlank (1986).

Zonal mean temperature

January and July simulated zonally averaged temperatures are shown in Fig. 5.2, along with the observed fields. The model has reproduced the principal observed features. For example, the tropopause temperatures in the two seasons are near 195–200 K and the lower stratosphere patterns are also similar to the observations, although a close examination at particular locations shows discrepancies of the order of 5–10 K. The major deficiency of the simulation is that the troposphere and stratosphere are too cold by 3–7 K, which is perhaps linked to the cirrus cloud treatment (see Ramanathan et al., 1983). The Northern Hemisphere polar surface temperatures agree qualitatively with the observations, including the surface temperature inversion, although the magnitude of the modeled inversion is somewhat greater than that of the observed.

Zonal mean wind

Figure 5.3 shows the simulated January and July zonal winds and compares them with observations from Newell et al. (1972). The four simulated mid-latitude tropospheric jets (the strongest being at roughly 35°N and an altitude of 12 km in January) are roughly in the correct locations in both latitude and height. The strength of the jets compares closely with the observations, especially in the winter hemispheres. The summertime jets tend to be weaker than the observed by 5 m s^{-1}. One of the successes of

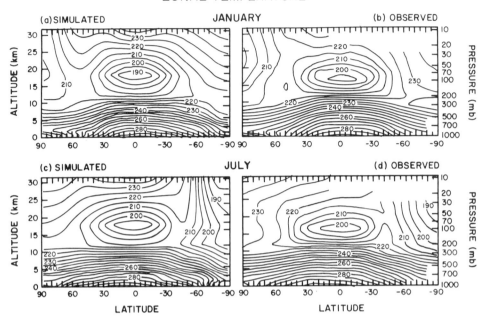

Fig. 5.2 Zonally averaged temperatures in K computed for perpetual January and July simulations and calculated from observations averaged over December-February and June-August. [Simulated from Pitcher et al. (1983); observed from Newell et al. (1972).]

this particular model simulation is that it produces a separation between the stratospheric polar jet and the tropospheric mid-latitude jet in the winter hemisphere. The wind speeds near the top of the model in the Northern Hemisphere January are within a few meters per second of the observed. Separation between upper and lower jets in the Southern Hemisphere July is not as distinct as in the Northern Hemisphere January in either the observations or the model. Ramanathan et al. (1983) explore in different numerical experiments some of the possible reasons for this separation of jets from the point of view of cloud-radiation theory and find that changes in the cirrus-radiation parameterization can cause large changes in the upper troposphere wind and temperature structures.

ZONAL WIND

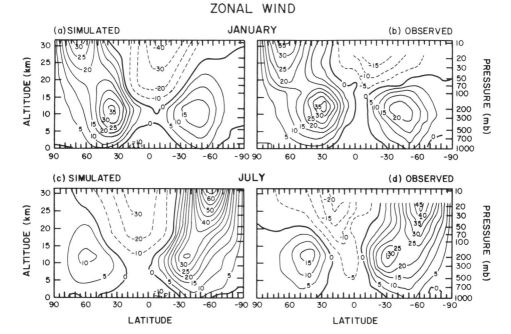

Fig. 5.3 Zonally averaged *u* component of the wind in m s^{-1} computed for perpetual January and July simulations and calculated from observations averaged over December-February and June-August. [Simulated from Pitcher et al. (1983); observed from Newell et al. (1972).]

Meridional mean wind

The zonally averaged meridional wind distributions are shown in Fig. 5.4. The upper-level branch of the winter hemisphere Hadley cell is well simulated in the model in terms of the location of the maximum in the poleward flow: In January at 10°N the intensity is smaller than the observed 2 m s^{-1}, whereas in July at 5°S the intensity is close to the observed. The mean meridional motions are asymmetric with respect to seasons, with the highest values being in the winter hemisphere. The model-generated wintertime equatorward flow at low latitudes near the ground—the low-level branch of the Hadley cell—is stronger than observed. The remaining meridional flow patterns generally are in agreement with the observed features except in the Northern Hemisphere polar regions in both January and July. A detailed comparison of simulated meridional winds with observations is unwarranted due to large observational and analysis uncertainties (see Peixóto and Oort, 1984), which limit how far an intercomparison with observations should be taken. The problem

MERIDIONAL WIND

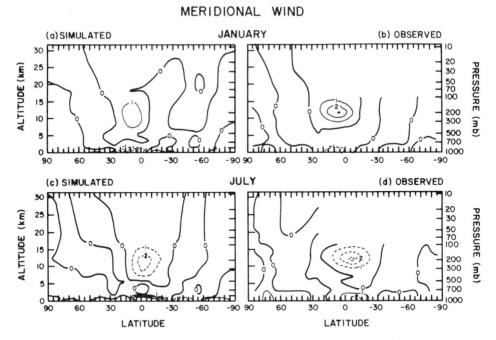

Fig. 5.4 Zonally averaged v component of the wind in m s^{-1} computed for perpetual January and July simulations and calculated from observations averaged over December-February and June-August. [Simulated from Pitcher et al. (1983); observed from Newell et al. (1972).]

is more acute with meridional than zonal winds because of the generally much smaller magnitude of the former.

Zonal mean vertical velocity

Figure 5.5 shows a comparison of the zonal average of the simulated vertical velocities and the vertical velocities as calculated from observed horizontal wind fields via the continuity equation. In both cases the vertical velocities are given as $\omega = dp/dt$ rather than as $w = dz/dt$ because the model uses a σ-coordinate system rather than a z-coordinate system. The ascending and descending branches of the wintertime Hadley circulation are well simulated in both hemispheres, with rising tropospheric vertical motions near the equator and descending vertical motions at about 20° latitude. By comparing Figs. 5.4 and 5.5 and examining regions of upward and north-south motions, zonally averaged cellular patterns can be seen as described in Chapter 2. As with the

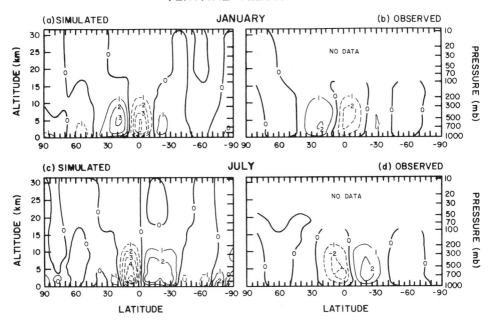

Fig. 5.5 Zonally averaged dp/dt in 10^{-4} mb s^{-1} computed for perpetual January and July simulations and calculated from observations averaged over December-February and June-August. Positive values (solid lines) signify descending motions, and negative values (dashed lines) signify ascending motions. [Simulated from Pitcher et al. (1983); observed from Newell et al. (1972).]

observations, the meridional overturning simulated by the model is weaker in the summertime hemisphere, and the strongest cell is the winter Hadley cell. A broad region of descending motion in the stratosphere, centered about the Northern Hemisphere mid-latitudes in January, as well as regions of ascent to the north and south, are present in the simulation, giving evidence of thermally indirect circulations. In July the northern edge of ascent in the model troposphere simulation near 20°N is poleward of its position in January, which is in agreement with observations.

Geographical distribution of surface air temperature

The simulated and observed January and July surface air temperature distributions are shown in Fig. 5.6. Because the temperature of the ocean surface is prescribed, it is not surprising that there is close agreement between simulated and observed

SURFACE TEMPERATURE

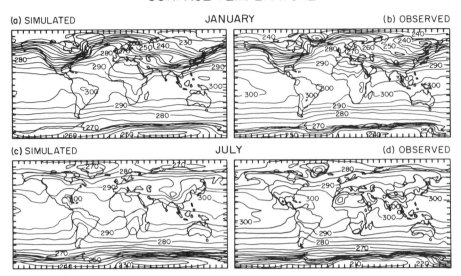

Fig. 5.6 Global distributions of surface air temperatures in K simulated and observed for January and July. [Simulated from Pitcher et al. (1983); observed from Schutz and Gates (1971, 1972b).]

surface air temperatures over oceanic regions. Over land and sea ice regions the surface temperature is calculated from the surface energy balance with zero surface heat capacity as discussed in Chapter 3, so that there is a greater possibility of large deviations between modeled and observed surface air temperatures over the land than over the ocean.

Generally the simulated surface temperature field agrees well with the observed field (Fig. 5.6). In January the temperature minima in Siberia and northern Canada are well reproduced by the model, although the model temperatures over continents are systematically colder than the observed temperatures by about 5–10 K. By contrast, other modeling groups, using similar models, instead find temperatures warmer than observed in these continental regions (e.g., Manabe and Hahn, 1981; McAvaney et al., 1978). Such differences among model results reflect some of the uncertainties in atmospheric modeling, where the surface temperature distribution is a result of a great deal of interactive dynamics and physical processes. For example, this version of the NCAR model does not use an explicit surface hydrological cycle but assumes evaporation is 1/4 the potential evaporation rate. This unrealistically simple parameterization of evaporation has significant effects on the resulting surface temperatures, as do

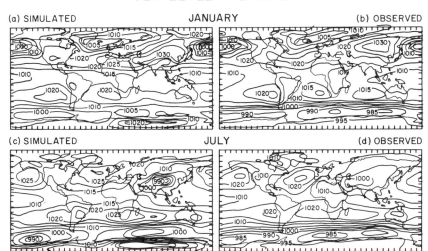

Fig. 5.7 Global distributions of sea level pressures in mb simulated and observed for January and July. [Simulated from Pitcher et al. (1983); observed from Schutz and Gates (1971, 1972b).]

other model inputs, such as the value of the drag coefficient, and other model variables, such as surface wind speed and clouds.

The July results show successful simulation of many of the principal observed features in the surface air temperature field, including the minima over Greenland, the Himalayas, western South America, and southeastern Africa (Fig. 5.6). With the exception of Antarctica, where the model temperatures are somewhat warmer than observed, the continental regions, as simulated by the model, tend to be a few degrees cooler than the observations.

Geographical distribution of sea level pressure

Figure 5.7 shows the simulated and observed global January and July distributions of sea level pressure. Model surface pressures have been extrapolated to sea level pressures by assuming a hydrostatic balance and a "standardized" temperature lapse rate of 6.5 K km^{-1} (see thickness equation in Holton, 1979). Thus in mountainous regions the pressures are artificial and may show spuriously high or low values.

The January simulation is successful in capturing most of the principal features of the observed distribution of sea level pressure. The strength and position of the Icelandic low compare favorably with the observations, although the simulated eastward extension of the low is somewhat south of the observed trough (low pressure) axis. The Aleutian low is also basically well simulated, appearing in the northern North Pacific near the International Date Line as a single center of low pressure, which, as in the case of the observations, is elongated in the east-west direction. The success of the simulation in the case of these two major low pressure systems is the more remarkable in that no individual year in the observations looks like the long-term mean, so that exact agreement between model and observations cannot be expected.

The Siberian high present in the January model results is fairly well positioned, although somewhat south of its observed location. A major shortcoming of the simulation is the westward extension of the high pressure ridge into the Middle East and North Africa. The model does capture the broad equatorial belt of low pressure, and positions the subtropical high pressure systems to the west of continents in the Southern Hemisphere. The model locates the Antarctic circumpolar trough of low pressure in approximately the correct position, but underestimates its intensity by some 10–15 mb. Also, the climatological low pressure troughs at lower latitudes over South America and Africa are not nearly as evident in the simulation as in the observations.

In the July simulation the centers of the North Atlantic and North Pacific high pressure systems are properly located west of the principal land masses, but the centers are 5°–10° too far north and the pressures are about 5 mb too high. The general Northern Hemisphere summer monsoon structure is quite evident, with high pressures over the oceans and relatively low pressures over the major land areas of North America and Asia. The Indian monsoon, as indicated by low pressure over southern Asia in July and relatively high pressure in January, is present in the model. The deep low pressure center to the northeast of India in the model's July results is positioned directly over the Himalayas and results basically from the method of reducing the model's surface pressures to sea level. The model generates an Arctic zone of high pressure that does not appear in the observed July pressure field; and the pressures simulated over South America and Africa are higher than those in the observations. Central pressures associated with the Antarctic circumpolar trough of low pressure are shifted too far north, although the trough itself is an important feature which the model succeeds in simulating. The reduction of surface pressure to sea level in mountainous regions with strong surface temperature inversions is responsible

300 mb ZONAL COMPONENT OF WIND

SIMULATED   OBSERVED

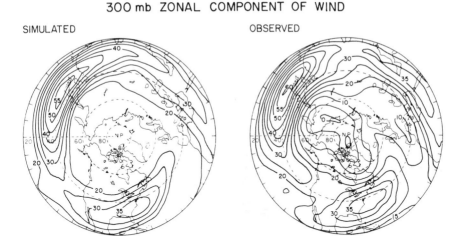

Fig. 5.8 Northern Hemisphere distributions of 300 mb zonal wind in m s^{-1} simulated for January and observed for mid-November to mid-March. [Simulated from Pitcher et al. (1983); observed from Lau et al. (1981).]

for the spuriously high sea level pressures obtained near the high-altitude Antarctic continent.

Geographical distribution of the
300 mb zonal component of the wind

The 300 mb level in the atmosphere is close to the jet stream level, making this level a particularly interesting and important one in simulation results. The simulated mean January winds in the Northern Hemisphere at the 300 mb level are shown in Fig. 5.8 and compared to the Northern Hemisphere winter mean from Lau et al. (1981). The model correctly simulates the positions of maxima over the east coasts of Asia and North America but fails to locate the smaller maximum near southern Europe or to obtain the prominent minima in the Arctic region. This indicates, through use of the thermal wind relationship, that the high latitude surface to 300 mb temperature field is not being simulated correctly.

PRECIPITATION

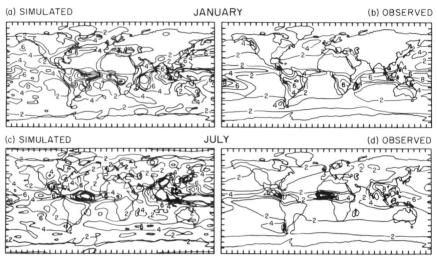

Fig. 5.9 Global distributions of precipitation in mm day^{-1} simulated for January and July and observed for December-February and June-August. [Simulated from Pitcher et al. (1983); observed from Schutz and Gates (1972a,b).]

Geographical distribution of precipitation

The model's mean daily precipitation rates during January and July appear in Fig. 5.9 along with observed values. There are clear differences in the details, with the modeled and observed fields appearing quite different on first glance, especially because of the much smoother plotted contours for the observations. Nonetheless, there still exist some notable successes, such as the simulation of the intense equatorial rainbelts in January, including the elongated twin maxima on each side of the equator in the eastern tropical Pacific. The precipitation maxima east of North America and Asia, associated with the storm tracks in the North Atlantic and Pacific, the minima over central North America and North Africa, and to a lesser extent the subtropical oceanic minima to the west of continents are also simulated in the model. These features are located in about the right positions, with approximately the correct relative strengths. The model also simulates the minima over the North African deserts in spite of the fact that the simple hydrology there is grossly in error. The positioning of the rainfall maximum in northeast Brazil is not well simulated, as it is too far from the heart of the continent. On the other hand, this quantity has very large

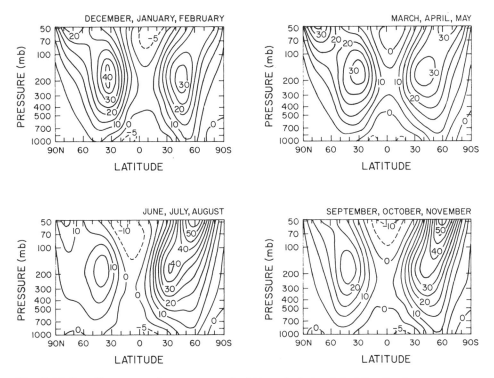

Fig. 5.10 Zonally averaged *u* component of the wind simulated for the four seasons from the Canadian Climate Centre (CCC) model. [From Boer et al. (1984).]

interannual variations both in the observations and in the model simulations, so that more sophisticated procedures for comparing model and observed climate statistics are desired. The reader is referred to Peixóto and Oort (1984) for an up-to-date review of observed atmospheric general circulation statistics of the angular momentum balance, water balance, and energy balance.

Intermodel comparisons

To indicate the atmospheric patterns simulated by models other than the NCAR model discussed above, sample results are presented here from several other GCM simulations. Figure 5.10 shows seasonal averages of zonal winds from a 5-year annual cycle experiment with the Canadian Climate Centre (CCC) general circulation model (Boer et al., 1984). These can be compared to the zonal wind patterns shown in Fig. 5.3 with perpetual

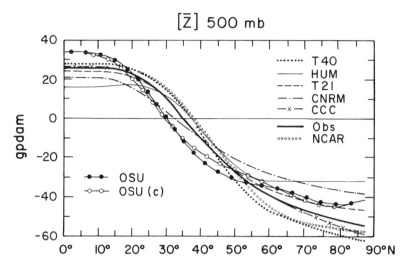

Fig. 5.11 Comparison of Northern Hemisphere January zonally averaged
normalized geopotential height (in geopotential decameters, gpdam) as
a function of latitude as simulated by various modeling groups: T40–
ECMWF—Baede et al. (1979); HUM—Storch and Roeckner (1983);
CNRM—Déqué and Royer (1983); T21–ECMWF—Baede et al. (1979);
CCC—Boer et al. (1984); OSU—Schlesinger and Gates (1980); OSU(c)—
Schlesinger et al. (1985); NCAR—Pitcher et al. (1983); Obs—observed.
[From Storch et al. (1985).]

single-month forcing. The overall patterns are quite similar in
the two simulations, with the biggest differences being in the
values of the wind maxima. Notice that the vertical axes in
Fig. 5.10 are in pressure units rather than altitude as in Fig. 5.3.

In Fig 5.11, modified from Storch (1984), a normalized height
for the January 500 mb surface is plotted as a function of latitude
for several model simulations and compared with observations.
Essentially all the major atmospheric models in use today are
capable of simulating the general shape of this function in terms
of decreasing from a maximum of about 20–30 gpdam (geopotential height decameters) at the equator to a minimum of about
-40 to -60 gpdam at the pole, with the 0 value coming somewhere
in the mid-latitudes at about 30–40°N. The model results that
appear to best match the observations for this particular variable
are those of the Canadian Climate Centre. Other atmospheric
variables could be displayed that would show better or worse
agreement with observations for any particular model, making it
impossible to assess fully the quality of a model simulation based
upon only a few variables. As Chervin (1981) has pointed out,
model intercomparisons and comparisons with observations can

be quite misleading if they fail to account for interannual variabilities and observational uncertainties, both of which can be substantial.

Simulations of the Ocean

The simulation of ocean circulation, like that of atmospheric circulation, has had a short but productive history of about three decades. The present stage of development is such that most of the large-scale oceanic features, including the major currents, primary regions of upwelling, areas of bottom water formation, and poleward heat transport, are being successfully simulated. However, many of the details are not well simulated or are included as parameterizations. In particular, most general circulation studies oriented toward climate do not include mesoscale eddies explicitly, but instead account for them through a diffusion parameterization. Resolving such eddies explicitly would require very high resolution models, which, if taken over the entire global oceans, would require excessive computer time, even with large, present-day computers. Some limited computer experiments have been performed to determine the effects of the mesoscale eddies, but these have been over restricted geographic regions.

A successful ocean simulation must account for turbulent mixing of the upper ocean by the wind, vertical mixing by locally driven shear and buoyancy forces (which depend upon the Richardson number), horizontal and vertical mixing due to mesoscale eddies, large-scale horizontal heat transport by ocean currents, deep- and intermediate-level convection in the polar regions, mixing along isopycnal (constant density) surfaces, and boundary changes with the atmosphere and sea ice. The reader is referred to Warren and Wunsch (1981) and Semtner (1984a) for recent reviews of the status of ocean modeling for climate studies.

Ocean circulation

To indicate the success of an ocean general circulation model in simulating the major ocean gyres, using prescribed atmospheric forcing, Fig. 5.12a presents a simulated field of the volume transport streamfunction (see Chapter 3, (3.226)), along with streamlines computed from the observed density field. The model results are from a simulation by Cox (1975) using the GFDL ocean

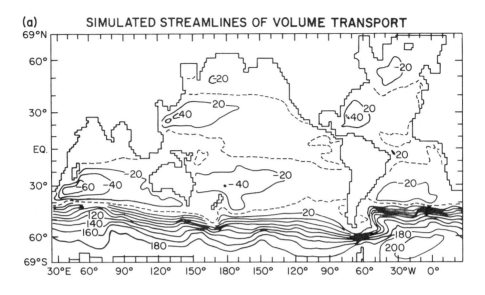

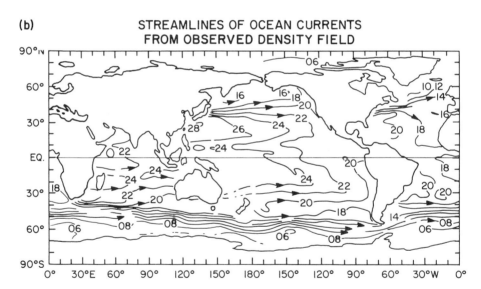

Fig. 5.12 (a) Global distribution of simulated streamlines of volume transport in the oceans. [From Cox (1975).] (b) Global distribution of streamlines of ocean currents computed from the observed density field. [From Levitus (1982).]

model with high vertical resolution and a 200 km horizontal resolution. The vertically integrated flow direction is parallel to the streamfunction and indicates clockwise motions around positive centers, counterclockwise motions around negative centers.

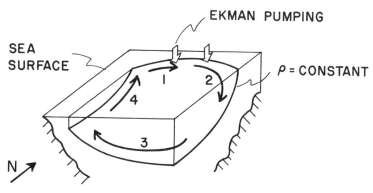

Fig. 5.13 Schematic of the injection of surface waters downward into a Northern Hemisphere wind-driven ocean gyre system. (1) Surface water is forced downward by Ekman pumping; (2) the water travels eastward and along a constant density, ρ, surface; (3) at lower latitudes, after traveling southward, the water turns and moves westward across the basin; (4) the water rises along the western boundary of the basin to "ventilate" with the surface. [Based on Holland et al. (1984).]

The intensity of the flow is proportional to the gradient between the streamfunction lines. Note that the major ocean gyres, the Gulf Stream off the east coast of North America, the Kuroshio Current off the east coast of Japan, the Antarctic Circumpolar Current flowing eastward around Antarctica, and the tropical currents flowing generally westward, are all visible in the simulation. Similar simulations have been obtained by Bryan et al. (1975), Han et al. (1985), and Meehl et al. (1982) employing variants of the Bryan-type model discussed in Chapter 3. There are no direct measurements of vertically integrated flow to use for comparison with Cox's simulation; however, the observed flow (shown in Fig. 5.12b) has been computed indirectly from the density field by Levitus (1982) using an assumption analogous to a thermal wind relationship whereby ocean currents flow parallel to density contours and with a speed proportional to the strength of the density gradient.

Several recent modeling papers (e.g., Cox and Bryan, 1984; Holland et al., 1984) have examined the general circulation of mid-latitude gyres. To clarify the discussion, we reproduce in Fig. 5.13 a schematic diagram of Holland et al. (1984) for a Northern Hemisphere gyre. At high latitudes, ocean vertical motions induced by wind stress produce an injection or downward motion, through Ekman pumping, of surface waters to the water underneath (1 in Fig. 5.13). This water then follows a path of constant potential density, flowing in a clockwise direction in Northern Hemisphere basins such as the North Pacific and North Atlantic (2). As the water reaches the eastern wall of the basin, the potential density surface descends and the water travels southward toward the equator before moving westward across the basin, still following the same potential density surface (3). It finally rises in the western boundary flow to come to the surface, or ventilate, in the northwestern part of the basin (4). The outcropping or surfacing location is a function of many factors, such as the amount of mixing that has taken place along the trajectory around the gyre. There is some evidence that a fluid parcel may make several circuits around the gyre before it resurfaces. The dynamics of this process are largely determined by potential vorticity, q, as defined in (3.238) and are governed dynamically by (3.237). As Holland et al. (1984) and Cox (1985) demonstrate, both q and ρ are approximately conserved during their circulation about the basin. These same quantities are approximately conserved in both quasi-geostrophic models ((3.228)–(3.238)) and primitive equation models ((3.202)–(3.227)).

Ocean heat transport

One of the most important aspects of ocean circulation for climate studies is the poleward transport of heat in both hemispheres. As discussed in Chapter 2, there is a net radiative heating in the subtropics and tropics, with solar heating exceeding outgoing infrared radiation, while the opposite is true poleward of the subtropics, where a net cooling occurs. For the climate system to be in the approximate equilibrium state in which it appears to be, the atmosphere and oceans must transport heat from the radiatively heating regions to the radiatively cooling regions. The oceanic portion of this transport can be a sizable fraction of the total, perhaps as large as 40%, with the remainder contributed by the atmosphere. Oort and Vonder Haar (1976) computed the ocean heat transport portion as a residual by first determining from satellite data the radiation imbalance at the top of the atmosphere (see Figure 3.14 in Chapter 3), which should be in approximate annual balance with the total heat transport, and then subtracting out the atmospheric heat transport and seasonal heat storages in the system. Figure 5.14a shows the resultant ocean heat flux for the Northern Hemisphere.

Note in Fig. 5.14a that there is a large seasonal variation in the ocean heat flux at 20°N from 5×10^{15} watts in April-May to zero in July-August. The seasonal changes can be caused in part by changes in the winds, which in turn cause stronger or weaker

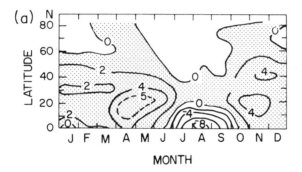

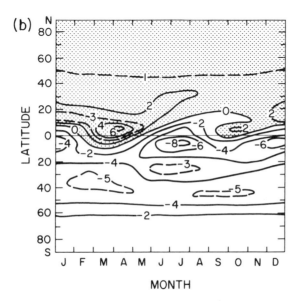

Fig. 5.14 (a) Northward flux of heat in the oceans (10^{15} W) calculated from observations. [From Oort and Vonder Haar (1976).] (b) Simulated northward flux of heat in the oceans (10^{15} W). [From Meehl et al. (1982).]

forced circulations in the upper oceans. Since there is upwelling at the equator and poleward surface circulations on both sides of the equator, heat can be transported from the tropics to the subtropics in these mostly wind-driven direct circulation cells.

Some of the features of the observed meridional heat flux have been simulated by Meehl et al. (1982) with a low resolution ocean general circulation model containing seasonally varying atmospheric forcing. The simulated meridional heat transport is shown in Fig. 5.14b. The predominant poleward transfer in both hemispheres is clear by the positive values in the Northern Hemisphere and the negative values in the Southern Hemisphere. The strong seasonal cycle in the tropical regions calculated by Oort and Vonder Haar for the Northern Hemisphere (Fig. 5.14a) appears also as a strong feature in the model results (Fig. 5.14b).

Formally, the horizontal flux of sensible heat is

$$\int_{-h}^{0} \int_{0}^{2\pi} \rho C_p T v \cos \phi \, d\lambda \, dz \tag{5.4}$$

In a manner similar to that shown in (5.2), variables can be separated into time mean and temporal (transient) fluctuations; i.e.,

$$A = \overline{A} + A^* \tag{5.5}$$

where $\overline{A}$ is the time mean, as opposed to the zonal mean $[A]$ of (5.2), and A^* is the difference $A - \overline{A}$. The flux $[\overline{Tv}]$ can then be separated into three parts: a mean meridional flux $[\overline{T}][\overline{v}]$; a large-scale transient eddy flux $[\overline{T^*v^*}]$, for instance from a mid-latitude gyre such as the Atlantic or Pacific Ocean gyre; and a stationary (time-mean) eddy flux $[\overline{T'}\,\overline{v'}]$. There is also a diffusion transport by subgrid-scale processes such as mesoscale eddies, defined as $A_H \partial T / \partial \phi$, where A_H is the horizontal subgrid-scale eddy diffusion coefficient defined in Chapter 3 (see (3.207)). The contributions of each component to the total transport are shown in Fig. 5.15a from the coarse resolution model computation of Meehl et al. (1982). These can be compared with observed indirect estimates of Oort and Vonder Haar (1976), Trenberth (1979), and Hastenrath (1980) in Fig. 5.15b. As shown by Bryan and Lewis (1979) and Meehl et al. (1982), the horizontal heat flux is dominated by the mean flux, $[\overline{T}][\overline{v}]$, in the tropics and by the diffusion and eddy fluxes in the mid- to high latitudes. If A_H is decreased, then the large-scale gyre flux will increase, as shown by Meehl et al. (1982).

Meridional ocean heat transports for three major individual ocean basins (the Pacific, Atlantic, and Indian oceans) are shown in Fig. 5.16. Estimates of observed fluxes are given by (1) Bryden and Hall (1980) for the Atlantic at 25°N, where the estimate is made by computing components of (5.4) indirectly from ocean data, and (2) Hastenrath (1980), where the implied horizontal heat flux is computed from an observed surface energy balance. Also indicated in Fig. 5.16 are results from a world ocean model with atmospheric forcing. Simulated values for the Atlantic are consistently low compared to the estimates of Hastenrath, but the model does simulate northward transport in both the North and South Atlantic, as obtained also in the Hastenrath computations. Simulated values in the Pacific are too low in the mid-latitudes and simulated values in the Indian Ocean shift from

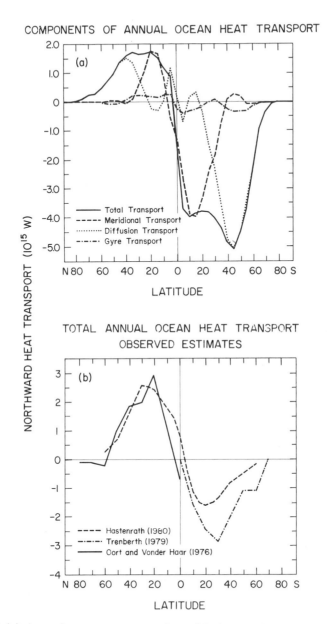

Fig. 5.15 (a) Annual ocean transport of sensible heat and its division into meridional flow, gyres (combining the transient and time-mean eddy fluxes), and diffusion, as simulated by an ocean model forced by specified atmospheric boundary conditions. (b) Observed estimates of total annual ocean transport of sensible heat. [From Meehl et al. (1982).]

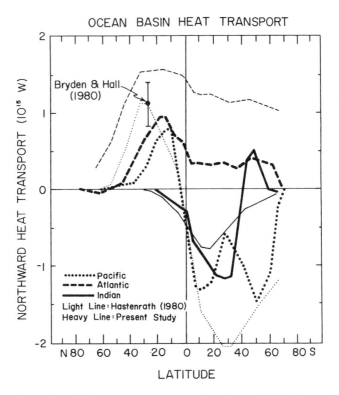

Fig. 5.16 Ocean basin heat transports as a function of latitude as simulated by an ocean model of Meehl et al. (1982) and as estimated from observations by Hastenrath (1980). Also plotted is an estimate for the Atlantic at 25°N from Bryden and Hall (1980). [From Meehl et al. (1982).]

southward to northward at about 40°S, a shift that does not occur in Hastenrath's observed estimates. Overall, the simulated meridional component of the ocean heat transport crudely reproduces the correct sense of the observed estimates in most locations, but does not reproduce many of the details. However, detailed comparisons are not warranted for this variable, due to the uncertainties involved in producing heat transport estimates from limited observational data.

Recently Cox (1985) carried out two experiments for an idealized ocean basin crudely approximating the North Atlantic, one with a horizontal resolution of 1/3° in latitude and 0.4° in longitude, and the other with a coarser resolution of 1° in latitude and 1.2° in longitude. The lower resolution model does not include time-varying mesoscale ocean eddies, while the higher resolution model does. Cox found that the net heat transport is not greatly affected by whether mesoscale ocean eddies are explicitly

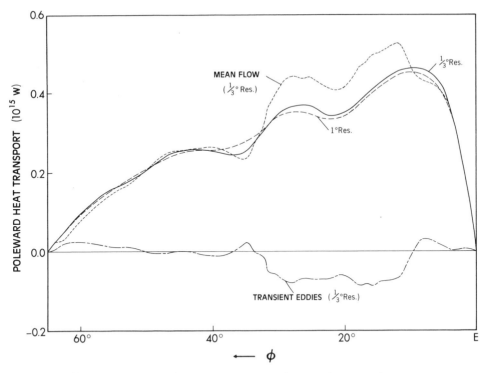

Fig. 5.17 Poleward transports of sensible heat simulated by coarse (1°
latitude/1.2° longitude) and fine (1/3° latitude/0.4° longitude) resolution
ocean models. The net poleward heat transports of the coarse- and fine-
resolution models are denoted by long dashes and a solid line, respectively,
while the time-mean transport from the fine resolution model (labeled
"mean flow") is denoted by short dashes, and the transient eddy transport from the fine-resolution model is denoted by alternating short and long
dashes. [From Cox (1985).]

included (as in his fine resolution model) or not (as in his coarse
resolution model) (Fig. 5.17). If time-varying mesoscale eddies
are included, then the increased poleward transport by the time
mean of the heat transport compensates for the equatorward
transport by the transient mesoscale eddies, $[\overline{T^*v^*}]$. Thus the
net flux $[\overline{Tv}]$ is about the same. This has important implications for climate-oriented ocean models, where explicit computation of mesoscale eddies is presently not feasible. The smaller
mesoscale eddies can be approximated with subgrid-scale parameterizations, as done by Cox, without significantly affecting
the net heat transport. To account for the eddies in his coarse
resolution computation, Cox simply increased the horizontal diffusion of momentum, sensible heat, and salinity.

SURFACE HEIGHT (cm)

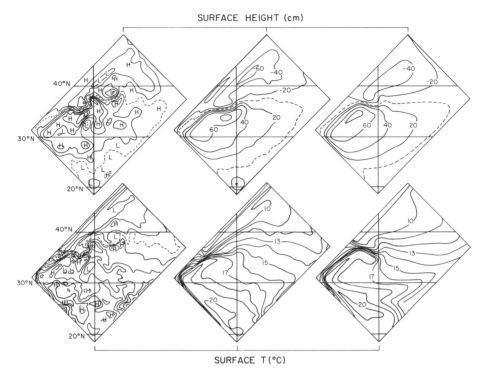

SURFACE T (°C)

Fig. 5.18 (Left) Atlantic Basin distributions of instantaneous surface height (as deviations from a mean) and surface temperature from an eddy resolving simulation. (Center) Time-averaged distributions of surface height (as deviations from the mean, in cm) and surface temperature in °C from an eddy resolving simulation. (Right) Steady-state distributions of surface height (as deviations from the mean, in cm) and surface temperature in °C from a coarse-grid simulation using horizontal diffusion coefficients large enough to compensate for the lack of mesoscale eddies. [From Semtner and Mintz (1977).]

Surface heights and temperatures

Figure 5.18 shows maps of instantaneous and time-averaged surface heights and surface temperatures for a schematic Atlantic Basin from Semtner and Mintz (1977). On the left are the instantaneous results using high horizontal resolution ($\simeq$ 37 km), in the center are time-averaged fields using high resolution, and on the right are time-averaged fields using lower resolution ($\simeq$ 75 km). In spite of differences in details between the time-averaged results of the coarse and high resolution experiments, the overall patterns are very similar. This, in conjunction with the results of Cox (1985) in Fig. 5.17, strongly implies that climate-oriented ocean models may not need explicit resolution of the mesoscale eddies in order to properly model time-averaged height and temperature fields and net heat transport.

Quasi-geostrophic results

As mentioned in Chapter 3, a major advantage of quasi-geostrophic models over primitive equation models is that they require much less computer time, thus allowing a higher resolution and a better simulation of mesoscale eddies. Figure 5.19 illustrates results from an eight-vertical-layer quasi-geostrophic model of Holland et al. (1984) for an idealized Northern Hemisphere basin (1400 km × 2800 km) with a 20 km resolution. The ocean is forced by an idealized wind stress that produces a western boundary current. Figures 5.19a and 5.19b show 5-year means of the streamfunction and potential vorticity at mean depths of 150, 850, and 1750 m, respectively. The two gyres at 150 m depth are fairly symmetrical, with the southern gyre circulating clockwise (indicated by solid lines) with a strong western boundary flow and the northern gyre circulating counterclockwise (dashed lines). The boundary between the gyres produces a strong eastward flowing "Gulf Stream" type current of 200 cm s^{-1} which is seen to meander significantly if one looks at the instantaneous flow. The meanders sometimes shed Gulf Stream type eddies as described in Chapter 2. The time mean potential vorticity field in Fig. 5.19b shows a sharp gradient across the eastward current and along the western boundary. At the 850 and 1750 m mean levels the two gyres shrink somewhat and the potential vorticity fields become more homogeneous in the interior of the region, with the largest gradients occurring near the boundaries.

Figure 5.19c shows the isopycnal thickness, $\partial\psi/\partial z$, of the 850 m layer in the same quasi-geostrophic simulation. This thickness, whose prediction equation is (3.236), is proportional to temperature. An interesting aspect of the figure is that the two gyres are much larger in the $\partial\psi/\partial z$ field than in the streamfunction field for the same 850 m layer. Finally, Fig. 5.19d shows the complex structure of an instantaneous potential vorticity field at mean depth 850 m, which can be compared with the time mean field shown in Fig. 5.19b for 850 m. Cox (1985) has found very similar flow structures with a primitive equation model with somewhat lower resolution.

236

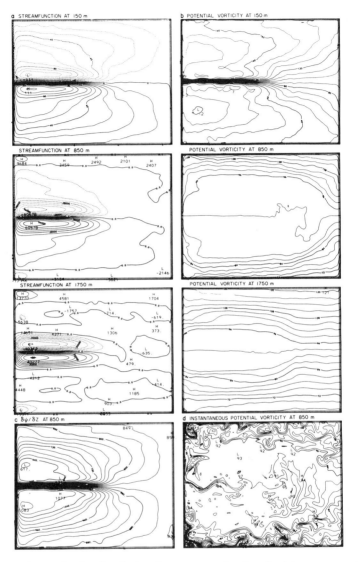

Fig. 5.19 Ocean basin circulation features as simulated with a mesoscale eddy resolving quasi-geostrophic ocean model: (a) 5-year time means of the streamfunction at depths of 150 m, 850 m, and 1750 m (the mean depths at the first, third, and fifth layers of the eight-layer model); (b) time means of the potential vorticity distributions at depths of 150 m, 850 m, and 1750 m; (c) thickness of the third layer, given as $\partial\psi/\partial z$, a variable proportional to temperature; (d) instantaneous distribution of potential vorticity of the third layer. [From Holland et al. (1984).]

The above examples of results from ocean simulations illustrate the basic understanding of ocean behavior emerging from numerical modeling studies. For climate purposes and efficiency in computation, one of the major issues addressed above regards the incorporation of mesoscale eddies. It would be desirable if explicit calculation of these eddies were not necessary, in light of the considerable expenditure of computer resources that such explicit resolution requires. The results of Cox (1985) and Semtner and Mintz (1977) shown in Figs. 5.17 and 5.18 are encouraging, but more research is required before it can be concluded that most of the important effects of the eddies can be included implicitly in coarse resolution models. If they can, then the savings in computer time would greatly facilitate long time integrations of ocean models and of coupled atmosphere/ocean/sea ice models.

Simulations of Sea Ice

Numerous efforts to model Arctic sea ice have been made over the past two decades, with the earlier models usually concentrating exclusively on simulating either the motions of the ice (ice dynamics) or the melting and growth of the ice (ice thermodynamics) and the later models usually presenting a combined dynamic/thermodynamic approach. Efforts at modeling the sea ice of the Southern Ocean have not been as numerous as those concerned with the Arctic and in general have been contained in studies that considered the Arctic and Antarctic jointly or that used models originally developed for the Arctic. However, since the mid-1970s, increased attention has been paid to modeling the Antarctic ice, and now models frequently are applied to both Northern and Southern Hemispheres. The most important variables simulated by the models are generally ice thicknesses, ice concentrations (or fractional ice coverage within a grid square), and ice velocities. Among the secondary outputs calculated in the course of determining the three primary variables are the ice temperatures.

Sea ice thickness and vertical temperature profiles

Figure 5.20 shows the annual cycle of sea ice thicknesses and temperatures for a point in the central Arctic as simulated by Maykut and Untersteiner (1969, 1971) using an elaborate one-dimensional thermodynamic model of the vertical growth and

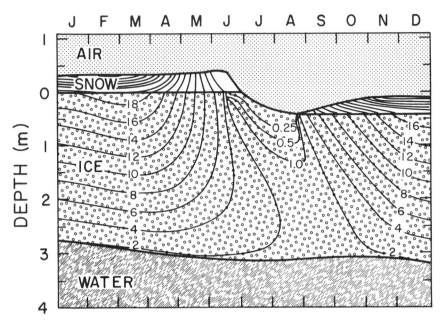

Fig. 5.20 Simulated annual cycles of sea ice thickness, snow thickness, and temperature profiles within the ice and snow covers in the central Arctic. The temperatures are given in units of −°C. [Redrawn from Maykut and Untersteiner (1971).]

decay of ice. Their equilibrium results for winter show ice thicknesses of approximately 3 m and temperatures within the ice ranging from a high of −1.8°C at the bottom of the ice to a low of approximately −19°C at the top of the ice. Maykut and Untersteiner included the effects of ice salinity, conductivity, and specific heat, plus brine pockets trapped within the ice, heating from penetrating shortwave radiation, and vertical variations in ice density. However, the model takes considerable computer time and 38 simulation years to reach equilibrium, inhibiting its extension to a three-dimensional framework.

Because of the computational expense of the Maykut and Untersteiner model, Semtner (1976) simplified aspects of the model and formulated a new one-dimensional model that was more appropriate for extending to three-dimensional simulations. He did this largely through a reduction in the number of vertical layers, a change in the time-differencing scheme to allow for a larger time step, the elimination of a heat source term in the diffusion equation, and the use of constants rather than variables for the specific heats and conductivities of ice and snow. Semtner inserted atmospheric and oceanic forcing identical to that

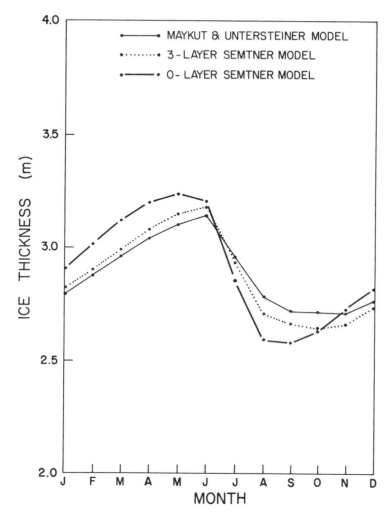

Fig. 5.21 Annual cycles of ice thicknesses simulated by the standard case of the model of Maykut and Untersteiner (1971) and the zero-layer and three-layer models of Semtner. [Based on Semtner (1976).]

of Maykut and Untersteiner and compared the results from two versions of his model with the results of the earlier, more complete work (Fig. 5.21). One of the two versions was a three-layer version, with two layers in the ice and one in the snow, and the other a so-called zero-layer version, with a single ice layer and a single snow layer. Semtner's comparisons with the Maykut and Untersteiner results, done for 25 separate cases with differing parameters, yielded an average deviation in the simulated ice thickness of only 0.22 m for the three-layer version of the model and 0.24 m for the zero-layer version. This encouraged later modelers basically to follow the Semtner zero-layer version, recognizing its considerable savings in computer resources over the Maykut and Untersteiner formulations. The Semtner three-layer model agrees with the Maykut and Untersteiner model in having the thickest ice in June and the thinnest ice in October–November (Fig. 5.21), reflecting the considerable lag of the ice thickness behind the atmospheric temperature cycle. With the zero-layer model, the thickness maximum and minimum occur 1 month earlier, which still reflects a significant lag behind the atmospheric temperatures.

Geographical distribution of sea ice thickness and concentration

Figures 5.22 and 5.23 present the annual cycles of sea ice thicknesses and 90% concentrations as simulated by a thermodynamic/dynamic sea ice model of Parkinson and Washington (1979). This model uses an 8-hour time step and a horizontal resolution of roughly 200 km. Ice dynamics are calculated with a momentum equation balancing air stress, water stress, and the stresses resulting from dynamic topography and the Coriolis effect, followed by an adjustment for internal ice resistance. Ice thermodynamics are calculated with equations for energy balances at the various interfaces between air, snow, ice, and water. A parameterization for open water within the ice pack (or "lead parameterization") is incorporated into both the dynamic and thermodynamic calculations.

The Parkinson and Washington (1979) model formulations result in a fairly realistic yearly cycle of sea ice growth and decay (Figs. 5.22 and 5.23 compared to the observed ice extents in Figs. 2.17 and 2.18 of Chapter 2). In the Arctic the simulated ice varies in extent from a minimum in September to a maximum in March, correctly matching the observed months of minimum and maximum Arctic ice coverage. In September, the simulated ice covers only a portion of the Arctic and has receded from most coastlines, with ice thicknesses reaching 3.0 m in the center of the pack. At the March maximum, the ice extent has greatly increased, reaching Iceland, well into the Bering Sea, and beyond the southern coast of Greenland. Central Arctic thicknesses in winter reach 3.6 m, and ice concentrations exceed 97% for the entire Arctic basin (Fig. 5.22).

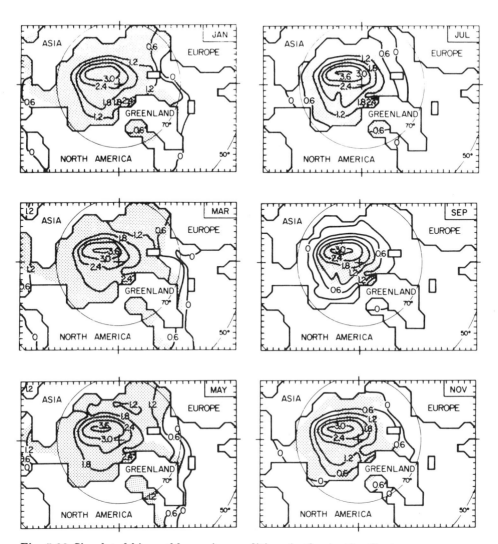

Fig. 5.22 Simulated bimonthly sea ice conditions in the Arctic. Contours show ice thicknesses in meters; shading indicates ice concentrations exceeding 90%. [Rearranged from Parkinson and Washington (1979).]

In the Antarctic simulation the simulated ice extent varies from a minimum in March to a maximum in late August/early September (Fig. 5.23). The timing of maximum ice extent corresponds well with observations, but the observed minimum extent is usually in late February rather than in March, so that the simulated minimum is late by approximately 2 weeks. In agreement with the observed sea ice covers, maximum ice thicknesses simulated for the Antarctic, about 1.4 m, are noticeably lower than

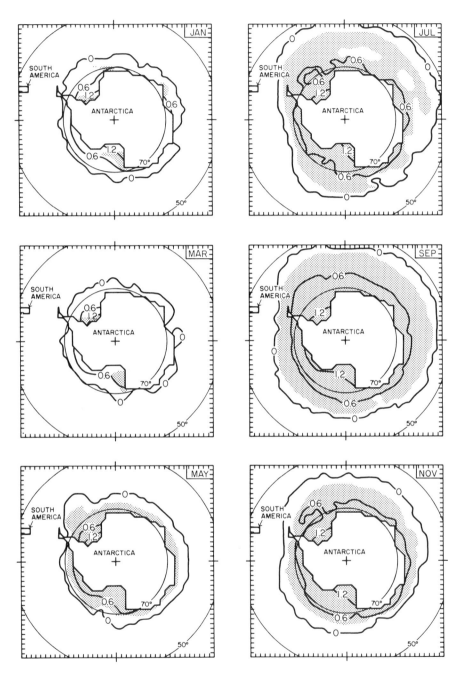

Fig. 5.23 Simulated bimonthly sea ice conditions in the Antarctic. Contours show ice thicknesses in meters; shading indicates ice concentrations exceeding 90%. [Rearranged from Parkinson and Washington (1979).]

those in the Arctic simulation, and ice concentrations overall are also less than in the Arctic. The modeled Antarctic ice remains close to the Antarctic continent at the March minimum, at which time almost all the ice of thickness greater than 0.6 m is in the Western Hemisphere; the ice at the August/September maximum extends equatorward to about 55°S, with the extent generally being farther equatorward in the Eastern Hemisphere than in the Western Hemisphere. The asymmetric nature of the ice cover distribution agrees with the Soviet *Atlas of Antarctica* (Tolstikov, 1966) and with satellite-derived observations (see Fig. 2.18).

When the simulated ice extents (Figs. 5.22 and 5.23) are compared to the observed ice extents (Figs. 2.17 and 2.18), there are some obvious contrasts, in particular in the North Atlantic region, where the simulated ice is more uniform latitudinally than the observed ice. This is true of simulations from other stand-alone sea ice models as well and results from the failure of these models to include the ocean currents that strongly affect the observed ice cover. This will be discussed further under "Sea ice modeling successes and failures."

Geographical distribution of sea ice velocities

Figure 5.24 shows simulated and observed average annual sea ice velocities in the Arctic. The simulated results are from a thermodynamic/dynamic model of Hibler (1979) that is more detailed in its formulations of ice dynamics than is the model of Parkinson and Washington (1979), but less detailed in its formulations of ice thermodynamics. Using a time step of 1 day and a horizontal resolution of 125 km, Hibler models the ice as a linear viscous fluid for very small deformation rates and as a rigid plastic for larger deformation rates, with the ice strength dependent on ice thickness and concentration. The momentum balance includes each of the five major stresses acting on the ice, while the thickness distribution includes an open water fraction plus an otherwise uniform ice thickness in any given grid square. The ice growth rates are prescribed as a function of season and ice thickness, rather than being calculated as they are in Parkinson and Washington (1979) and in later work by Hibler and Walsh (1982) with an extended version of the Hibler model.

The average annual ice velocity field simulated by Hibler (Fig. 5.24a) successfully reproduces the major observed long-term average features of Arctic ice drift (Fig. 5.24b; also see Dunbar and Wittman, 1963, Felzenbaum, 1958, and Fig. 2.19). In particular, a prominent clockwise gyre exists in the Beaufort

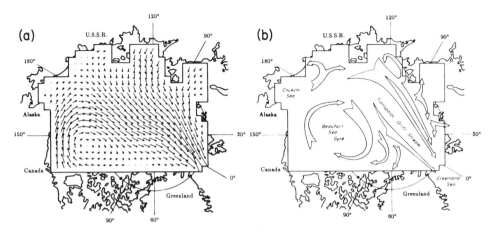

Fig. 5.24 (a) Simulated average annual ice velocities in the Arctic. The velocity vectors are scaled so that a vector with a length of one grid square represents a speed of 0.02 m s^{-1}. (b) Schematic of the observed average annual ice drift in the Arctic. [Partially redrawn and relabeled from Hibler (1979).]

Sea, and a significant Transpolar Drift Stream flows across the Arctic basin from off the Siberian coast to the strait between Greenland and Spitsbergen.

Weddell polynya

In the winters of 1974, 1975, and 1976, an unexpected, very large open water region, or polynya, appeared in the midst of the ice-covered waters of the Weddell Sea off Queen Maud Land, Antarctica. Since then, this Weddell polynya has been the subject of numerous investigations, largely because its existence, the irregular nature of its occurrence (it did not occur in 1973 or in 1977–1985), and its systematic westward movement over the 1974–1976 period have puzzled scientists and have suggested ice/ocean or ice/atmosphere or ice/ocean/atmosphere interconnections that have yet to be fully understood. It is the largest known recurring polynya in either hemisphere.

The simulation of the Weddell polynya with a stand-alone sea ice model provides some insight into modeling studies. Figure 5.25 shows a realistic simulation by Parkinson (1983) of the occurrence of the Weddell polynya and its formation through the encircling of an open water region by ice. This result was surprising initially, because the model used for the simulation does

(a) WEDDELL SEA GRID

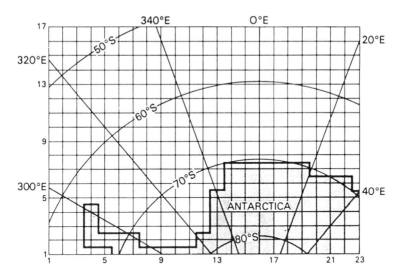

(b) SIMULATED ICE CONCENTRATION CONTOURS

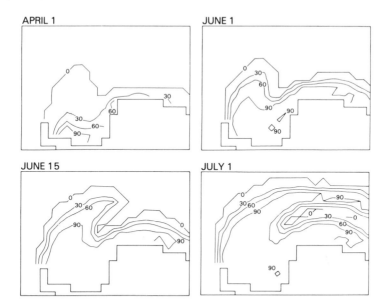

Fig. 5.25 (a) Grid used for studies of the Weddell polynya. (b) Simulation of the development of the Weddell polynya using mean monthly climatological atmospheric data. Contours show ice concentrations in %. The polynya is fully formed in the July 1 map. [Rearranged from Parkinson (1983).]

not incorporate spatial variability in the ocean heat flux and has no feature in any of its specified oceanographic fields or in its specified mean monthly climatological atmospheric temperature fields that would provide an explanation of the polynya. After several trials, the determining factor for the modeled results was found to lie in the prescribed mean monthly climatological wind fields: the modeled polynya forms beneath a low pressure system, the low speeds in the center of which restrict the sensible and latent heat fluxes from the ocean to the atmosphere, thereby also restricting oceanic cooling. Although the winds affect ice transport as well, the effect on the fluxes is shown to be the critical effect on the simulated polynya (Parkinson, 1983). This polynya formed realistically in the simulation, but it did freeze over by the end of winter, which is in contrast to the three observed Weddell polynyas of the 1974, 1975, and 1976 winter/spring seasons.

In a further sequence of clarifying test cases, Parkinson (1983) showed that artificial maintenance of the low pressure system in the atmospheric inputs throughout the year produced a polynya which, like the observed polynya, remains intact (although reduced in size and intensity) until its breakout with the general opening of the ice cover in spring. Hence the continued restriction of sensible and latent heat transfer to the atmosphere suffices to maintain the modeled polynya. Offering further confirmation of the decisive impact of wind on the simulated polynya, an artificial leftward shift in the position of the low pressure system in the prescribed wind fields resulted in an identical leftward shift in the position of the simulated polynya, and an elimination of the low pressure system resulted in a simulation with no polynya in any month (Fig. 5.26).

Although numerical modeling studies such as the above on the Weddell polynya shed light on modeled phenomena, extrapolations from model results to the corresponding geophysical phenomena should not be done without caution. As Parkinson (1983) mentions, an alternative sequence of experiments with the same sea ice model but with appropriate alterations in the oceanographic fields rather than in the wind fields might also produce similarly varying simulated polynyas. In fact, many geophysical analyses of the Weddell polynya have emphasized the probable overriding importance of the oceanography (e.g., Carsey, 1980; Martinson et al., 1981). The model results show only that the simulated polynya can be produced and/or moved through adjustments in the wind fields; they show nothing one way or the other regarding causative factors for the development of the actual Weddell polynya. Here, as elsewhere, the controlled nature of modeling experiments allows far more precise

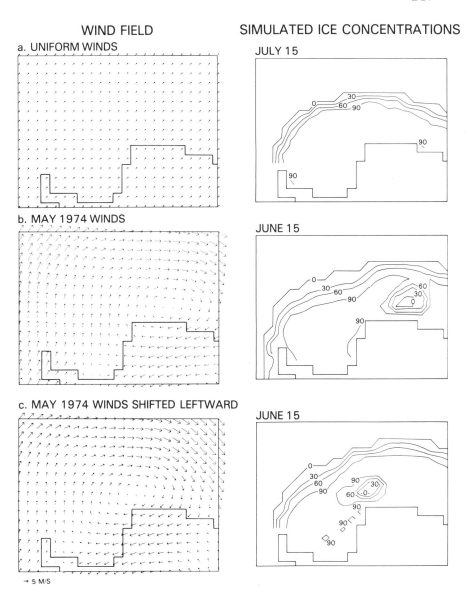

Fig. 5.26 Constant wind fields used in three 1-year experimental simulations of the Weddell Sea region, with resultant simulated mid-winter ice concentration contours in %: (a) uniform wind field with both u and v components set at 2 m s^{-1} throughout the grid, and the resultant July 15 ice concentrations, with no Weddell polynya; (b) May 1974 wind field, and the resultant June 15 ice concentrations, with a polynya formed beneath the prominent atmospheric low pressure system reflected in the wind field; (c) May 1974 wind field shifted four grid squares to the left, and the resultant June 15 ice concentrations, with a polynya formed beneath the leftward-shifted low pressure system. [Rearranged from Parkinson (1983).]

conclusions to be stated about modeled phenomena than are possible regarding the corresponding phenomena in the real world.

Impact of ice dynamics on sea ice simulations

Hibler and Walsh (1982) have extended the grid of the Hibler (1979) model to include the Bering Sea, Baffin Bay, Davis Strait, and the ice-covered region of the North Atlantic, and have added details to the thermodynamic calculations that bring them closer to the formulations of Parkinson and Washington (1979). By running the model with atmospheric forcing from three separate years, 1973–1975, and doing so both with the full model and with a thermodynamics-only simplification, Hibler and Walsh were able to examine the impact of interannual differences in the forcing and the effect of ice dynamics on the simulations. The resulting ice extents vary far more in the simulations for the three years than they do in the observations (Fig. 5.27), and although the inclusion of ice dynamics appears to improve the simulation in terms of creating a distribution of thicknesses such that the thickest ice is immediately north of Greenland rather than more centrally located in the Arctic basin, the ice-edge position is only slightly closer to the observations when the full model is used rather than the thermodynamics-only version. It is presumably the model's neglect of ocean currents that leads to the ice-edge results of the thermodynamics-only and full models being much closer to each other than either one is to the observations (Fig. 5.27).

Hibler and Ackley (1983) have applied the Hibler model to the Weddell Sea, and, as in the work of Hibler and Walsh (1982) for the Arctic, have examined the impact on the simulation results of including ice dynamics in the model calculations. Hibler and Ackley find the need for including dynamics to be strongest during the decay phase of the seasonal cycle, when the dynamics in the Weddell Sea area result in (1) the movement of the ice near the edge into warmer water and (2) an increase in the creation of open water within the ice pack. Both these processes enhance further ice melt, encouraging a large impact of ice dynamics on the spring and summer results (Fig. 5.28). The impact of dynamics on total sea ice coverage during the rest of the seasonal cycle is considerably less pronounced; for instance the winter simulated results are almost identical whether or not ice dynamics are included (Fig. 5.28). These winter ice cover amounts are about 1×10^6 km^2 higher than the observed values. When the Hibler and Ackley results are mapped (Fig. 5.29), they fortunately do not

(a) OBSERVED

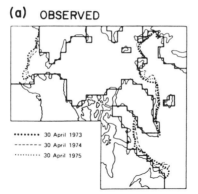

(b) SIMULATED WITH A
DYNAMIC/THERMODYNAMIC MODEL

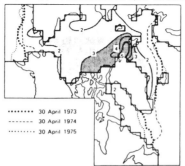

(c) SIMULATED WITH A
THERMODYNAMIC MODEL

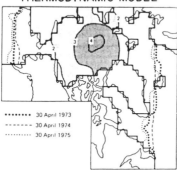

Fig. 5.27 (a) Observed Arctic sea ice extents on April 30 for the three years 1973–1975. (b) Arctic sea ice extents and thickness contours (m) on April 30 for the three years 1973–1975, as simulated with a dynamic/thermodynamic sea ice model. (c) Arctic sea ice extents and thickness contours (m) on April 30 for the three years 1973–1975, as simulated with a thermodynamics-only sea ice model. [Relabeled from Hibler and Walsh (1982).]

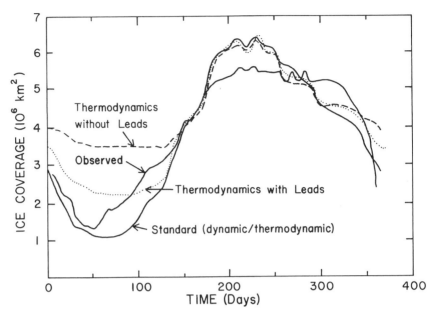

Fig. 5.28 Effect of ice dynamics on the simulated seasonal cycle of ice coverage in the Weddell Sea, along with the observed seasonal cycle. [Redrawn and relabeled from Hibler and Ackley (1983).]

include the full-scale Weddell polynya simulated by Parkinson (1983), since the atmospheric forcing used by Hibler and Ackley was for 1979, when no full-scale polynya was observed.

Sea ice modeling successes and failures

The above examples illustrate both successes and failures of current stand-alone sea ice models. Successes include reasonable timing of minimum and maximum ice extents in both the Southern and Northern Hemispheres (Figs. 5.22 and 5.23); asymmetric patterns of both summer and winter ice in the Antarctic, with summer ice predominantly in the Western Hemisphere and with the winter ice edge extending farthest equatorward at about the Greenwich meridian (Fig. 5.23); the autumnal development of a tongue of ice in the Weddell Sea, with the subsequent formation of a Weddell polynya (Fig. 5.25b); an asymmetric distribution of ice thicknesses in the Arctic, with the thickest ice north of Greenland (Fig. 5.27b); and a prominent Beaufort Gyre and Transpolar Drift Stream in the Arctic ice velocity field (Fig. 5.24).

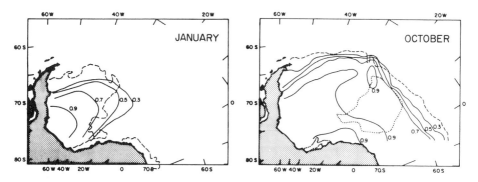

Fig. 5.29 January and October Weddell Sea sea ice concentrations computed with a dynamic/thermodynamic sea ice model (solid lines). The dashed lines indicate the locations of the observed 50% ice concentration contour and the dotted lines indicate the locations of observed polynyas. [Relabeled from Hibler and Ackley (1983).]

In spite of the successes, however, there remain significant contrasts between the observed and simulated sea ice distributions. The simulated ice tends to be too extensive in the North Atlantic; the gradient in ice concentrations tends to be stronger than the observed gradient in many regions; and the simulations do not reproduce several observed coastal polynyas. For many of the major contrasts, the explanation lies at least in part with inadequate formulations of the atmospheric and oceanic boundary conditions. For instance, the ice-edge results from both the Hibler and Walsh (1982) model and the Parkinson and Washington (1979) model deviate substantially from observations in the Davis Strait and Norwegian Sea regions, where the simulated ice is far more extensive than the observed ice. The failure of the models to simulate these ice-edge positions derives in large part from the lack of realistic detail in the prescribed ocean forcings. The warm north-flowing West Greenland and North Atlantic currents help maintain ice-free conditions throughout the year in much of the Davis Strait and almost all of the Norwegian Sea. The omission of these currents in the forcing fields of the aforementioned stand-alone sea ice models prevents them from properly simulating ice conditions in the affected regions. Hence the simulations could almost certainly be improved by improving the boundary conditions in the models, and they could probably be improved eventually by properly coupling the ice models with oceanic and/or atmospheric models. The proper coupling of the models has the added advantage of allowing a wide variety of further numerical experiments related to the fully interactive ice/ocean/atmosphere system. A major goal is thus the coupling of sea ice models to ocean models that include the major ocean currents influencing the polar regions. Initial progress in this direction has been made by Hibler and Bryan (1984), as discussed in the following section.

Coupled Atmosphere, Ocean, Sea Ice Simulations

Since the atmosphere, oceans, and sea ice interact in important ways (e.g., see Chapter 2), a precise simulation of the changes in the vicinity of the atmosphere/ocean/ice interface requires that all three media be incorporated in the calculations (or, in the case of ice-free waters, that both the atmosphere and oceans be incorporated). The problem is particularly acute for long time scale simulations, such as climate simulations addressing the issues of future impacts of increased carbon dioxide and trace gases in the atmosphere. Recognizing this, many researchers have worked at length on coupling ocean, ice, and atmosphere models. However, although such coupling is essential for a full numerical treatment of the processes at the earth's surface, the reader is forewarned that coupled models, especially in the early stages of development, will not necessarily produce more realistic results than stand-alone models. Errors in the calculations of one model can be expected to have an adverse effect on the calculations of the other models, and indeed this has turned out to be the case in some of the initial coupled simulations.

A major problem in coupling the ocean, ice, and atmosphere calculations is the difference in characteristic time scales for typical events in the various media. In particular, on the whole the atmosphere tends to respond far more rapidly to adjustments in external conditions than the oceans do, with the time scales of the atmospheric adjustments being on the order of days to months (Gates, 1981). Although the sea ice, snow cover, and surface waters in the oceans also can respond on time scales on the order of months, the response time of the deeper ocean typically is on the order of centuries. This book omits the modeling of mountain glaciers and ice sheets, but these typically have response times on the order of centuries and millenia, respectively.

Because of the contrast in the characteristic time scales in the atmosphere and oceans, several attempts at atmosphere/ocean coupling have been asynchronous. In such cases the atmosphere and ocean models are run separately and sequentially, with different time steps, and with the results of the atmospheric calculations being inserted into the ocean model as boundary conditions for the next ocean iteration, and the results of the ocean calculations then being inserted into the atmospheric model. This was done, for instance, by Manabe et al. (1975) and Bryan et al. (1975) with a 9-level atmosphere and a 12-level ocean in the first major global coupled general circulation model, and by Washington et al. (1980) with an 8-level atmosphere and a 4-level primitive equation ocean. The Manabe et al. and Bryan et al. studies, which expand upon earlier work by Manabe and Bryan with a coupled but geographically limited model (Bryan, 1969b; Manabe, 1969a,b; Manabe and Bryan, 1969), use mean annual insolation rather than seasonally varying forcing, while a later version of the model, described in Manabe et al. (1979), uses a realistic seasonal cycle of solar radiation.

By contrast, the calculations have been run synchronously in certain other coupled models, such as the model of Bryan et al. (1982), where a spectral atmospheric general circulation model is coupled synchronously to a deep ocean model. The Bryan et al. (1982) model builds upon an earlier model used by Manabe and Stouffer (1980) in which a 9-level spectral atmospheric model was coupled to a fixed-depth mixed layer ocean without advection or diffusion. Results obtained by Manabe and Stouffer with the earlier model in simulating the impact of increased atmospheric carbon dioxide, as well as results from the models of Washington and Meehl (1984), Hansen et al. (1984), and Schlesinger et al. (1985) in experiments concerning the same carbon dioxide impact, are discussed in Chapter 6.

Most of the above coupled models include sea ice in some fashion, although with parameterizations that are simplified over those of the stand-alone sea ice models described in the previous section. Bryan et al. (1975) and Manabe et al. (1975), for example, include ice thermodynamics based on energy balances and some ice transport, but leads in the ice pack are not included and all ice motion stops when the ice thickness exceeds 4 m. For thicknesses below 4 m, the ice moves strictly with the water of the upper ocean. The resulting calculations lead to an unreasonable build-up of Arctic ice with time, the average and maximum thicknesses continually increasing, with values of 5.32 m and 24.7 m, respectively, in year 200 of the simulation (Manabe et al., 1975). By contrast, the simulated Antarctic ice cover is far less than the observed Antarctic ice cover, as a result of excessively warm simulated atmospheric surface temperatures in the south polar region. This illustrates the expected initial difficulties with coupled models (predicted, for instance, by Parkinson and Herman, 1980) and specifically that errors in the calculated ocean and atmosphere fields can adversely affect the sea ice calculations.

Indeed, most of the studies done with coupled models (e.g., Manabe et al., 1975; Manabe et al., 1979; Manabe and Stouffer, 1980; Washington et al., 1980; Washington and Meehl, 1984; Bryan et al., 1982; Washington and VerPlank, 1986) have produced less realistic sea ice distributions than those produced by uncoupled models. As another example, in the simulations of Washington et al. (1980), where sea ice is treated strictly thermodynamically and with no leads, there is also far too little Southern Ocean ice, with ice restricted for most of the year mainly to the Ross and Weddell Seas (Fig. 5.30). The difficulty derives from simulated sea surface temperatures in the Southern Ocean that are warmer than the observed values by 1–5 K, a result in this case of the coarseness of the model grid and the excess subgrid-scale diffusion of heat (Washington et al., 1980).

In other cases coupled models have produced results improved over stand-alone models, in particular in aspects where the stand-alone model results were known to be poor because the models did not include coupled effects or detailed specified boundary conditions to account for them. For instance, the excess sea ice simulated in the northern North Atlantic by stand-alone sea ice models that neglect the warming produced by the North Atlantic and Norwegian currents is greatly reduced in a coupled ocean/ice simulation of Hibler and Bryan (1984). Hibler and Bryan couple the sea ice model of Hibler (1979) with an ocean model of Bryan (1969a) for a limited-area simulation including the Greenland and Norwegian seas. The resulting positioning of the wintertime sea ice edge is significantly improved over the stand-alone results of Hibler (1979) when compared to the observed ice-edge positioning in the Greenland and Norwegian seas (Fig. 5.31). The results can be expected to worsen when the specified atmospheric fields are replaced by calculated atmospheric fields, but the marked improvement in the positioning of the sea ice edge upon coupling with an ocean model is encouraging.

Modeling Groups

General circulation models of the atmosphere, oceans, and sea ice are in widespread use throughout the world. Several of the major models have standardized versions that are used by many researchers both within and outside the home institution. At the same time, at many of the centers conducting research with such models, more than one model is in use, and in some cases there are a dozen or more different models or at least different

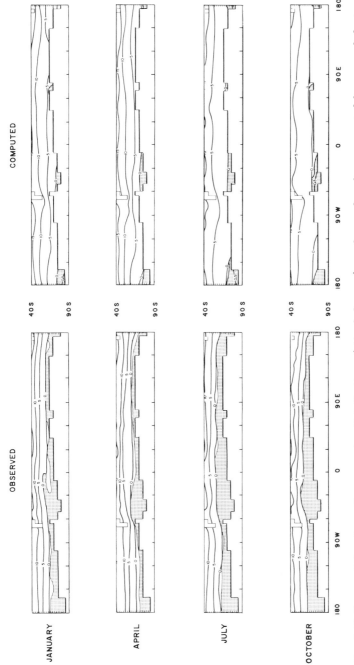

Fig. 5.30 Seasonal Southern Ocean sea ice distributions (stippled areas) as observed and as computed by a coupled atmosphere/ocean/ice model. Contours are ocean surface temperatures in °C. [Modified from Washington et al. (1980)].

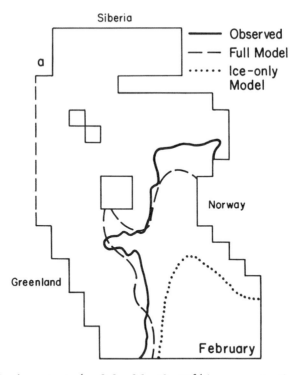

Fig. 5.31 Sea ice extents (as defined by the 50% ice concentration contour) in the Greenland and Norwegian seas in February as charted from observations by the U.S. Navy Fleet Weather Facility and as simulated by an ice-only model and by a coupled ocean/ice model. [Redrawn from Hibler and Bryan (1984).]

versions of models. This reflects the variety of applications these models are used for and the differing needs regarding such factors as horizontal resolution, vertical resolution, time resolution, and incorporation of various physical components and processes. Hence it is virtually impossible to identify a model simply by attaching a center's or researcher's name to it. Furthermore, this is a rapidly changing field, and new versions of the models are constantly being developed. Nonetheless, to inform the reader of some of the major centers where general circulation modeling of the climate system is currently being conducted, we include the following partial list of modeling centers, with one or more references providing details on at least one of the major models associated with each center.

Atmospheric models

Australian Numerical Meteorology Research Centre (ANMRC)
Bourke et al. (1977); McAvaney et al. (1978)
Canadian Climate Centre (CCC)
Boer et al. (1984)
European Centre for Medium-Range Weather Forecasts (ECMWF)
Baede et al. (1979); Simmons (1983)
French Meteorological Service (FMS)
Tourre et al. (1985)
Geophysical Fluid Dynamics Laboratory (GFDL/NOAA)
Gordon and Stern (1982); Manabe and Hahn (1981)
Goddard Institute for Space Studies (GISS/NASA)
Hansen et al. (1983)
Goddard Laboratory for Atmospheres (GLA/NASA)
Kalnay-Rivas et al. (1977); Kalnay et al. (1983)
Goddard Laboratory for Atmospheres and University of California
at Los Angeles (GLA/NASA and UCLA)
Randall et al. (1985); Suarez et al. (1983)
Hamburg University (HUM/Federal Republic of Germany)
Storch and Roeckner (1983)
Institute of Atmospheric Physics (IAP)
Zeng and Ji (1981)
Institute of Atmospheric Sciences (IAS)
Zeng et al. (1984)
Japan Meteorological Agency (JMA)
Kanamitsu et al. (1983)
Laboratoire de Météorologie Dynamique de Centre National
(LMD/CNRS)
Sadourny and Laval (1984)
National Center for Atmospheric Research (NCAR/NSF)
Pitcher et al. (1983); Williamson (1983)
National Meteorological Center (NMC/NOAA)
Sela (1980)
Oregon State University (OSU)
Gates and Schlesinger (1977); Gates et al. (1985); Schlesinger
et al. (1985)
United Kingdom Meteorological Office (UKMO)
Burridge and Gadd (1977); Gadd (1978); Gadd (1980)
University of California at Los Angeles (UCLA)
Arakawa and Lamb (1977)
University of Reading/United Kingdom (Reading/UK)
Hoskins and Simmons (1975)
USSR
Fux-Rabinovich (1974); Krichak and Fux-Rabinovich (1972)

Ocean models

Academy of Sciences, Shirshov Institute of Oceanography (USSR)
Sarkisyan et al. (1985)
Geophysical Fluid Dynamics Laboratory (GFDL/NOAA)
Bryan (1969a,b, 1975); Cox (1984)

National Center for Atmospheric Research (NCAR/NSF)
  Holland et al. (1984); Semtner (1986)
Oregon State University (OSU)
  Han (1984a,b); Han et al. (1985)
University of California at Los Angeles (UCLA)
  Semtner (1974)

Sea ice models

Cold Regions Research and Engineering Laboratory/US Army (CRREL)
  Hibler (1979)
Geophysical Fluid Dynamics Laboratory (GFDL/NOAA)
  Bryan (1974)
Goddard Space Flight Center (GSFC/NASA)
  Parkinson and Washington (1979)
National Center for Atmospheric Research (NCAR/NSF)
  Semtner (1976, 1987)
University of Puget Sound/USGS
  Ling et al. (1980)
University of Washington (UW)
  Coon et al. (1974); Coon (1980)

CHAPTER 6

Climate Sensitivity
Experiments

Once numerical models reached the stage where the simulations reproduced major features of the atmosphere, oceans, and sea ice, a logical next step was for researchers to begin using the models to simulate conditions other than those existent at the present time, such as paleoclimates or possible future climates affected by increased carbon dioxide amounts or nuclear warfare. In this chapter we examine several types of such numerical experiments, illustrating each with results from one or more numerical simulations. Such results reflect a variety of exciting applications of numerical climate models and hint at an even greater variety likely to emerge in the future. We caution, however, against overinterpreting the results, recognizing that the models are imperfect and will need considerable further improvement before a high level of confidence should be placed on the details of their simulations for past or future climate states.

Paleoclimate Simulations

In the first half of the nineteenth century, European geologists began to interpret erratic granite boulders scattered around the landscape, various markings on rocks, and certain moraines located far from currently glaciated regions as evidence that Europe at one time was far more heavily glaciated than at present. Similar evidence in North America examined in the mid-nineteenth century helped confirm that the Northern Hemisphere at some time in its past experienced widespread glaciation referred to as an Ice Age. Further studies over the past century and a half have identified and approximately dated not only ice ages but also intervening periods of much warmer climatic conditions than exist at present. In the early twentieth century evidence also began to accumulate indicating that the continents themselves have not remained stationary but instead slowly drift over time, so that their positioning in the distant past was far different from what it is today. By the 1960s this hypothesis of continental drift was being buttressed by an emerging theory of plate tectonics, providing an explanation of the geophysics behind the drifting of the continents.

Several attempts have been made in recent years to use general circulation models to simulate the likely atmospheric, oceanic, and sea ice conditions at various periods in the past, including ice ages, interglacials, and periods with far different continental alignments than exist at present. These attempts have been assisted by precise calculations from astronomers of the variations in the earth's orbital parameters over time and by approximate determinations from geologists, glaciologists, and oceanographers of various surface conditions in the past.

One of the primary requirements for paleoclimate simulations is the determination of appropriate boundary and initializing conditions. These are not obtainable from direct observations, as when simulating for present conditions, but must be determined by theoretical or geophysical evidence. In general, the more ancient the period being simulated, the more uncertain are the boundary conditions. If ice sheets and vegetation are not modeled, at a minimum the geography of the major ice sheets and the albedo characteristics of the land surface must be specified. Even if these features are modeled, they must be initialized in some realistic fashion. For simulations distant enough in the past, as indicated above, even the locations of the continents must be altered (e.g., Fig. 6.1), and generally to positions that are by no means firmly established. Recent massive efforts at paleoclimatic reconstruction, generally carried out for reasons other than numerical modeling, have provided modelers with scientifically based estimates of many of the needed fields of variables. Among these efforts is the Climate: Long-range Investigation, Mapping and Prediction (CLIMAP) program for the Wisconsin glacial period about 18,000 years ago and the Cooperative Holocene Mapping Project (COHMAP) for 3000, 6000, 9000, 12,000, and 15,000 years ago. The CLIMAP data set includes

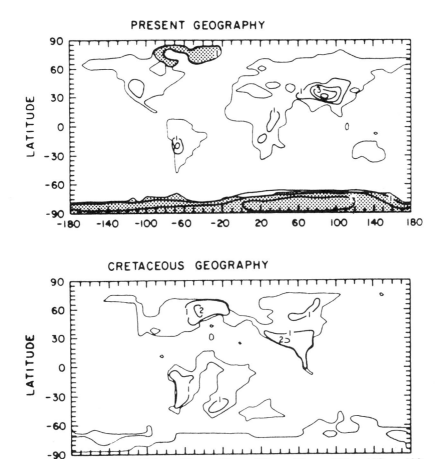

Fig. 6.1 Present-day and Cretaceous geography as used in experiments with the NCAR Community Climate Model. In addition to continental positions, mountain topography is shown, in 1 km contour intervals, and prescribed land snow cover is shown by stippling. [Relabeled from Barron and Washington (1984).]

global sea surface temperatures and sea ice, ice sheet extents and topography, land surface albedo, and sea level (see Peterson et al., 1979). Crowley (1983) reviews many of the principal results from a variety of paleoclimate studies.

Kutzbach (1981) and Kutzbach and Otto-Bliesner (1982) used a low resolution general circulation model with a full annual cycle to examine the effect on climate of changes in the cycle of solar radiation produced by changes in the earth's orbital parameters. By comparing the results of a control run containing present orbital parameters with a simulation using orbital parameters for the early Holocene 9000 years ago, they show that the orbital differences result in dramatic changes in the simulated climatic fields, including intensified summer and winter monsoon circulations over Africa and southern Asia in the paleoclimate simulation. The three orbital changes included were an increase in the tilt of the earth's axis from 23.44° to 24.24°, a change in the date of perihelion from January 3 to July 30, and a small increase in the earth's orbital eccentricity, from 0.016724 to 0.019264. The seasonal cycle of ocean surface temperatures was set at its present values, since theoretical estimates and observational evidence both indicate that the thermal response of the oceans would have been far less than that of the land.

Kutzbach and Guetter (1984b) repeated the basic experiments of Kutzbach (1981) and Kutzbach and Otto-Bliesner (1982), using a version of the NCAR Community Climate Model. The Community Climate Model has higher resolution and more realistic parameterizations than the low resolution model of the two earlier studies. On the other hand, the new experiments used perpetual January and July conditions rather than simulating the full annual cycle. In spite of the differences in the models, major features of the large-scale results remain qualitatively unchanged. For instance, Kutzbach and Guetter also found monsoonal circulations intensified, due in part to the increased land/sea temperature contrast resulting from increased continental temperatures in July and decreased continental temperatures in January. Kutzbach and Guetter, again following the earlier studies with the low resolution model, also simulated the ice sheet sensitivity by running the Community Climate Model with and without a 5,400,000 square kilometer ice sheet in North America, with a maximum thickness of 800 meters and a 0.5 albedo. The ice conditions were determined from geologic evidence documenting ice extent at the peak of the Wisconsin ice age 18,000 years ago, as summarized by CLIMAP. Results illustrating the effect of these changes on the temperature and pressure fields are shown in Figs. 6.2 and 6.3.

Kutzbach and Guetter (1984a) summarize results from simulations for 18,000 and 6000 years ago as well as for 9000 years ago. In all cases differences in inputs from the control run for the present include orbital changes and land surface changes. The results suggest that the climate was colder and drier 18,000 years ago than now, with a similar seasonal contrast between summer and winter. The simulation for 6000 years ago suggests that

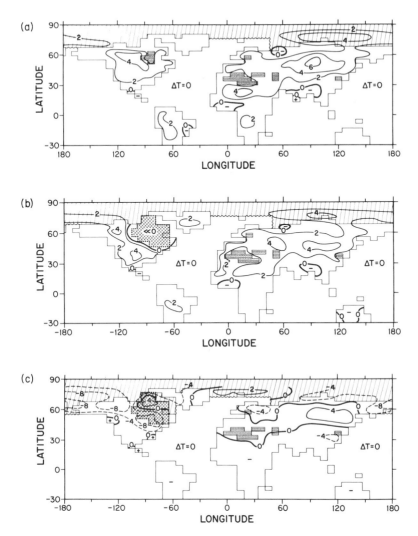

Fig. 6.2 Simulated differences from the present-day land surface temperatures (in K) for experiments with: (a) orbital parameters for July 9000 years ago but no North American ice sheet; (b) orbital parameters and a North American ice sheet for July 9000 years ago; (c) orbital parameters and a North American ice sheet for January 9000 years ago. Sea ice is hatched, ice sheets are cross-hatched, and inland seas are shown by wavy lines. Ocean surface temperatures are prescribed in each case at present-day values. [From Kutzbach and Guetter (1984b).]

the climate had warmed markedly from the conditions 18,000 years ago, and the range in seasonal temperature extremes was larger.

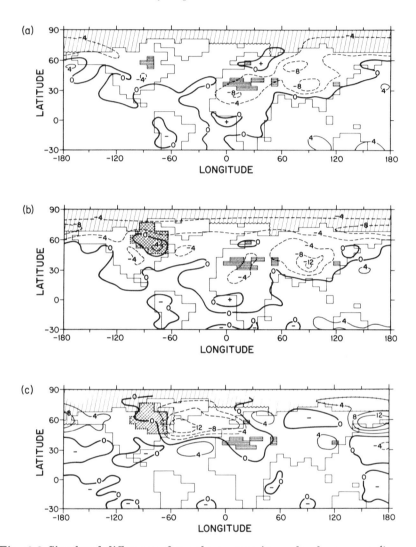

Fig. 6.3 Simulated differences from the present in sea level pressures (in mb) for the same set of three experiments as in Fig. 6.2. [From Kutzbach and Guetter (1984b).]

The dramatically different climatic conditions 18,000 years ago (during the Wisconsin ice age of the Pleistocene Epoch) have been modeled by several other numerical-modeling groups as well. Alyea (1972) was among the first, using a geostrophic atmospheric model for the Northern Hemisphere and idealized surface boundary conditions. He simulated generally cooler and drier atmospheric conditions. Williams et al. (1974) obtained similar results, using an early version of the NCAR global general circulation model. Gates (1976a,b) with an OSU global general circulation model and Manabe and Hahn (1977) with a GFDL global general circulation model used data from the CLIMAP program in their simulations of the July climate of the Wisconsin ice age. Gates (1976a) simulated surface air temperatures over large areas of the continents to be greater than 10 K cooler than at present, although the specified sea surface temperatures averaged only about 2 K cooler than present July temperatures.

Figure 6.4 shows July sea level pressure for the Wisconsin ice age and for the present, as simulated by Gates (1976b). Prominent high pressure systems located over the paleoclimate's continental ice sheets in northern North America and Europe are among the strongest deviations of the ice-age simulation from the present-day simulation. [The reader should recall, however, the cautions stated earlier regarding simulated sea level pressures in mountainous regions and in particular the somewhat arbitrary nature of the calculation of these values from surface pressures and an assumed vertical lapse rate for atmospheric temperatures.] The results of both Gates (1976b) and Manabe and Hahn (1977) show somewhat drier conditions in the ice-age simulations than in the present-day simulations, but this result should be viewed somewhat tentatively, since precipitation is even less certain than some of the other simulated variables.

In another attempt to simulate the influence of the glacial ice of 18,000 years ago, Manabe and Broccoli (1985) use a general circulation model of the atmosphere coupled with a simple mixed-layer ocean. They find that the Laurentide ice sheet greatly alters the tropospheric flow field of the Northern Hemisphere, which exhibits a split flow straddling the ice sheet, and that the Laurentide and Eurasian ice sheets produce regions of statistically significant decreases in soil moisture just south of the ice sheets (Fig. 6.5).

Although this book centers on general circulation models, less complicated energy balance models, without explicit prediction of winds, have also been used to advantage in examining ice-age issues. These simpler models often can help determine specific experiments to be done in more detail with GCMs. Furthermore, at times the conclusions from the less complicated models can be more definitive than those from GCMs in terms of elucidating the processes or influences involved in producing the model results. In particular, North et al. (1983) use a seasonal energy balance model with a realistic land/sea geography to simulate present and ice-age conditions, done with appropriate changes in the earth's orbital parameters. They conclude from their various numerical experiments that the continentality produced by the

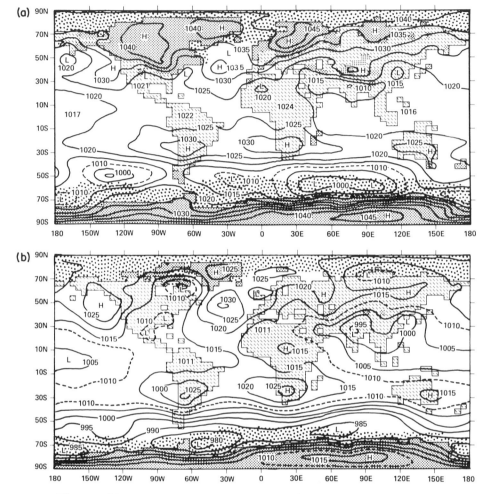

Fig. 6.4 Simulated July sea level pressures (in mb) using a two-level atmospheric general circulation model for (a) Wisconsin ice-age conditions and (b) present-day conditions. Ice sheets are indicated by dense stippling, sea ice by less-dense stippling. [From Gates (1976b).]

Eurasian landmass results in summers that are too hot even under Pleistocene conditions to maintain a perennial snow cover over Siberia. The more pervasive water influence due to the narrower, broken land cover in North America moderates the summer heat, allowing the North American Pleistocene ice sheets to grow, since the winter North American snowfall is not completely melted in summer. North et al. thereby explain the geophysical evidence that a major continental ice sheet formed in North America during the Pleistocene period but that none formed in

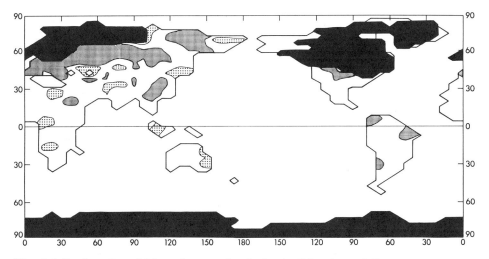

Fig. 6.5 Regions for which an ice-age simulation by Manabe and Broccoli (1985) shows statistically significant changes in soil moisture from a control run simulating present-day conditions. Dense stippling indicates regions of lower soil moisture for the ice-age simulation than for the control simulation, whereas less-dense stippling indicates regions of higher soil moisture. Black areas signify ice sheets. [Redrawn from Manabe and Broccoli (1985).]

Siberia. For a review of other results of ice-age simulations with energy balance models, the reader is referred to Held (1982).

As for simulations of even earlier periods, Barron and Washington (1984) used the NCAR Community Climate Model, with no ocean circulation and no changing seasons, to perform an ordered sequence of six simulations, changing one variable at a time, to investigate the impact of geographical factors on the global climate of the Cretaceous, 100 million years ago. They found that the simulated globally averaged surface temperature is 4.8 K higher with mid-Cretaceous geography specified (see Fig. 6.1) than with present-day geography and present ice sheets. The warming was only slight in the tropical regions but was 20–30 K in the polar regions. Through the sequence of experiments they showed that the changed continental positions were largely responsible for the warming in the Northern Hemisphere, whereas the absence of an Antarctic ice sheet was largely responsible for the warming in the Southern Hemisphere. Crowley et al. (1986), using a simpler climate model but one with an annual cycle, examined the changing climatic conditions of the past 100 million years and particularly the effects of seasonality and the land/sea distribution. Their results confirm the importance of both these factors.

In an earlier study, Barron and Washington (1982) performed another atmospheric simulation with an assumed Cretaceous geography. The results are of interest both for the contrasts that result between the atmospheric circulations simulated for the Cretaceous and those simulated for the present and for the contrasts between the simulated Cretaceous circulation and the classical hypotheses of what that circulation should have been. Major differences between the Cretaceous and present-day circulations include a weak mid-latitude rainbelt in the Cretaceous (Fig. 6.6), a well-developed polar high and associated polar easterlies in the Cretaceous, and a more latitudinally restricted region of westerlies in the Cretaceous than at present.

The Barron and Washington (1982) study calls into question at least three major classical hypotheses regarding the atmospheric circulation during the Cretaceous period: that the circulation was "sluggish"; that the surface circulation in the polar regions was predominantly westerly; and that the subtropical high and surface westerlies were displaced poleward. Importantly, the simulation does obtain a reduced equator-to-pole surface temperature gradient, which is the central basis for the classical expectation of a sluggish circulation. Hence the result of a more intense circulation than anticipated suggests that knowledge of the surface temperature gradient is insufficient to form firm conclusions on the intensity of the circulation. Barron and Washington suggest that the geologic record be examined for indications of stronger or weaker ocean and atmospheric circulations, for instance in sediment size distributions along continental margins and in desert regions. Similarly, they suggest analysis of paleobiogeographic data to decide between the classical hypothesis of polar westerlies and the model-simulated polar easterlies. They believe that the geologic data on the distribution of evaporites confirm the model simulation regarding the fact that the simulated subtropical high was not displaced poleward.

For a review of more of the issues and results regarding paleoclimate simulations, the reader is referred to Barron (1984), and for a review of pre-Pleistocene simulations from general circulation models and a variety of less comprehensive models, the reader is referred to Barron (1985).

PRECIPITATION

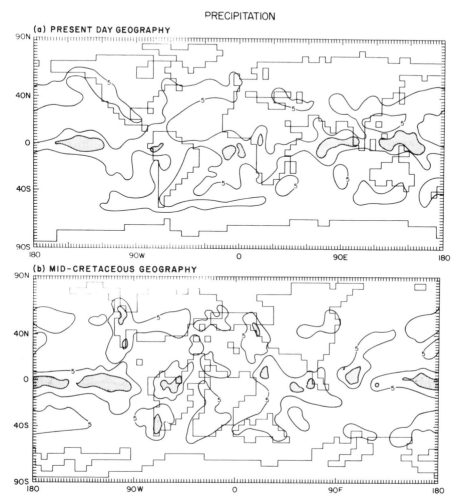

Fig. 6.6 Simulated average daily March precipitation rates (in mm day^{-1}) for (a) present-day conditions and (b) mid-Cretaceous conditions. [From Barron and Washington (1982).]

Simulations of El Niño/Southern Oscillation

As mentioned in Chapter 2, the El Niño/Southern Oscillation (ENSO) has dramatic, well-documented effects on the tropical atmosphere and ocean circulations (Rasmusson and Wallace, 1983). A series of atmospheric general circulation modeling studies has shown that an increase in sea surface temperatures between 180°W and the west coast of South America in the equatorial Pacific causes local increases in cloudiness and precipitation, which in turn result in release of latent heat. The rising motion induced by the heating drives direct circulations in both the north-south and east-west directions. In addition to the local effects and the modulations of the east-west circulations in the tropics, there is interaction with the Hadley circulation, which can lead to extra-tropical effects (see Bjerknes, 1966, 1969). One of the dominant modes of Northern Hemisphere circulation thought to be linked to ENSO events is a Pacific-North American (PNA) pattern described in the observational study of Wallace and Gutzler (1981). In the positive phase of the PNA pattern, there are anomalously northerly winds in eastern North America and anomalously southerly winds in western North America (see Fig. 2.21 in Chapter 2). Many atmospheric general circulation modeling groups have attempted to simulate this phenomenon by adding a sea surface temperature anomaly in the central tropical Pacific or in the eastern tropical Pacific near the west coast of South America (e.g., Rowntree, 1972; Julian and Chervin, 1978; Keshavamurty, 1982; Blackmon et al., 1983; Shukla and Wallace, 1983; Blackmon, 1985; Boer, 1985; Cubasch, 1985; Esbensen, 1985; Fennessy et al., 1985; Geisler et al., 1985; Lau and Oort, 1985; Palmer, 1985; Suarez, 1985; Tourre et al., 1985).

The Southern Oscillation pattern in the tropics has been simulated reasonably well by several different groups. As described in Chapter 2, when a warm sea surface temperature (SST) anomaly pattern, such as that shown in Fig. 6.7 from Cubasch (1985), exists in the central tropical Pacific, this pattern causes a negative anomaly in sea level pressure in the eastern Pacific and a positive sea level pressure anomaly in the western Pacific and Indo-Asian regions. Figure 6.8 shows results of two December/January/February simulations of the Southern Oscillation using the ECMWF model from Cubasch (1985). Figure 6.8a shows the difference in sea level pressure between an experiment with a positive SST anomaly as in Fig. 6.7 and a control experiment. The results show that in the tropical eastern Pacific the sea level pressure anomaly is negative, while in the tropical Indian Ocean and western Pacific it is positive (Fig. 6.8a). For comparison, the observed difference in sea level pressure for years with a Warm Event (positive SST anomaly in the central Pacific) versus an undisturbed mean is shown in Fig. 6.8c from van Loon (1986). Figure 6.8b shows the results of a model experiment in which a negative SST anomaly, rather than a positive anomaly, was prescribed in the model. The simulated sea level pressure in the equatorial region in the latter simulation almost has a phase reversed from that when the sea surface temperature anomaly was warm. Figure 6.8d shows the sea level pressure change for the years before Warm Events that do not have warm

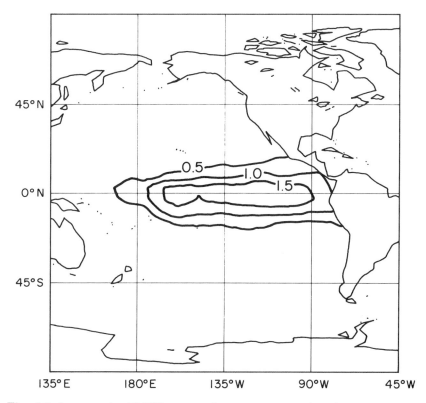

Fig. 6.7 A composite El Niño sea surface temperature (SST) anomaly pattern, with temperature anomalies in K. [Redrawn from Cubasch (1985), after Rasmusson and Carpenter (1982).]

SST anomalies in the Pacific. The patterns in Fig. 6.8d show similarities to those in Fig. 6.8b, although a more exact comparison is inappropriate, since the year before a Warm Event may not always have a negative SST anomaly.

Certain mid-latitude circulation anomalies resembling the Pacific-North American (PNA) pattern have been generated in several model studies with anomalous tropical Pacific sea surface temperatures. These include studies of Blackmon (1985), Geisler et al. (1985) with the NCAR CCM, Boer (1985) with the CCC model, Palmer (1985) with the UKBMO model, Cubasch (1985) with the ECMWF model, Esbensen (1985) with the OSU model, Fennessy et al. (1985) with the GLAS (later GLA) model, Keshavamurty (1982) with the GFDL model, Tourre et al. (1985) with the FMS model, and Suarez (1985) with the UCLA model. All of the models produce features that resemble the PNA pattern shown in Fig. 2.21 of Chapter 2. However, the position

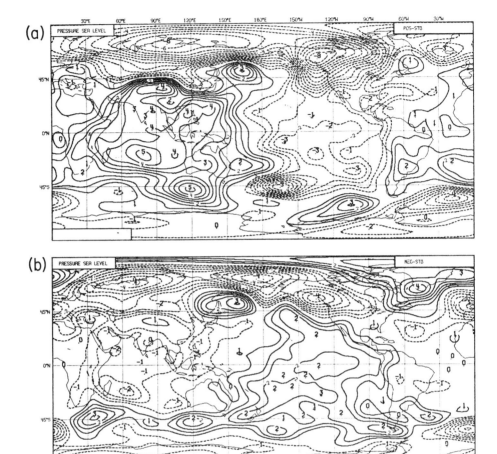

Fig. 6.8 December/January/February sea level pressure differences (in mb) for the following: (a) simulated difference between an experiment with a positive composite SST anomaly and a control experiment; (b) simulated difference between an experiment with a negative composite SST anomaly and a control experiment. [From Cubasch (1985).]

and amplitude of the PNA pattern differ greatly from model to model. Although there is no consensus on what causes the differences in the model studies, it is clear that the variability of both observed and model-simulated mid-latitude circulation systems is large (Chervin, 1986), so that there is a great deal of variability concerning the mid-latitude response to individual ENSO events, and the signal-to-noise ratio may be small. Therefore, using only the tropical Pacific sea surface temperature as a predictor of whether a PNA pattern will appear in the

(c)

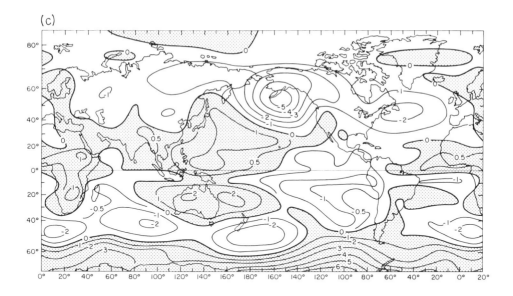

(d)

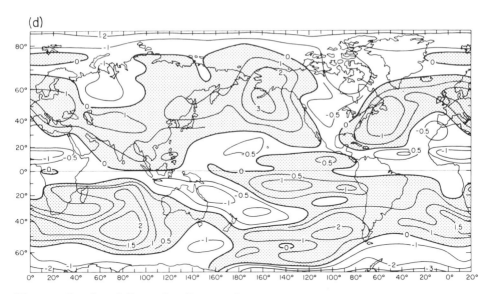

Fig. 6.8 Continued, December/January/February sea level pressure differences (in mb) for the following: (c) observed difference between years with a Warm Event and an undisturbed mean; and (d) observed difference between years immediately preceding Warm Events and an undisturbed mean. [From van Loon (1986).]

mid-latitudes may not be very useful as a long-range forecasting tool. Apparently other factors may govern the position of the long-term middle to high latitude planetary wave pattern, so

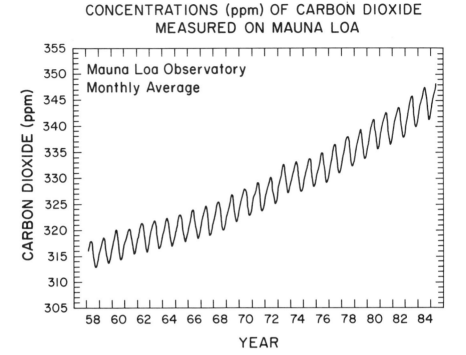

Fig. 6.9 Time sequence of atmospheric carbon dioxide concentration (in parts per million) measured on Mauna Loa. [From Schmidt (1985), after Keeling et al. (1982).]

that a large amount of inherent and/or unpredictable variability is not correlated with the ENSO event. The search for the limits of predictability in the tropics and higher latitudes is clearly an important subject of research because of its possible economic and societal importance.

Climatic Effects of Carbon Dioxide

In the past decade considerable speculation has occurred regarding the probable effects on climate of the dramatic increase in atmospheric carbon dioxide (CO_2) over the past century (Keeling et al., 1976a,b; Lacis et al., 1981). This increase, due in part to the burning of fossil fuels, is illustrated in Fig. 6.9. Since carbon dioxide absorbs infrared radiation, some of which would otherwise escape to space, an expected effect of adding CO_2 to the lower atmosphere is to raise the average surface temperature of the earth. As mentioned in the radiation section of Chapter 3, the addition of CO_2 and other trace gases will "dirty" the 10 μm window of the infrared radiation, which in turn will trap a higher percentage of outgoing radiation. The first-order response of the atmosphere system to increasing CO_2 is expected to be warmer temperatures in the troposphere due to this trapping of outgoing radiation and cooler temperatures in the stratosphere due to a more effective infrared radiation to space by the increased stratospheric CO_2.

Several different types of numerical models have been used to estimate the impact on the atmosphere of a CO_2 increase (e.g., Manabe and Wetherald, 1967, 1975, 1980; Sellers, 1974; Augustsson and Ramanathan, 1977; Hunt and Wells, 1979; Manabe and Stouffer, 1980; Ramanathan et al., 1979; Hansen et al., 1981; Washington and Meehl, 1984; Schlesinger and Mitchell, 1985). Although there is still some uncertainty, modeling results consistently predict an increase in tropospheric temperatures. The magnitude of the increase varies markedly among the various models, but the predicted global-average warming generally lies between 1 and 4 K for a CO_2 doubling. Most of the models show the largest temperature increase occurring in the polar or subpolar regions, the polar warming sometimes being several times the global average (Manabe and Wetherald, 1975; Ramanathan et al., 1979; Manabe and Stouffer, 1980; Washington and Meehl, 1984; Hansen et al., 1984; Schlesinger and Mitchell, 1985; Schlesinger et al., 1985).

Among the reasons that a greater temperature response to CO_2 increases is expected in the polar regions than elsewhere are specific positive feedbacks associated with the variable snow and sea ice covers. For example, increases in air temperatures cause greater sea ice melt and this in turn results in a lower overall surface shortwave albedo since ice reflects a far greater percentage of incoming solar radiation than does open water. The lower albedo produces greater absorption of solar radiation at the surface, with a consequent storage of heat in the ocean and therefore further ocean warming. The lessened ice extent also reduces the insulation between ocean and atmosphere, so that greater sensible and latent heat exchanges occur between these two media. This is true especially in winter, when the air is much colder than the water, so that the heat transfer from water to air can be considerable in the absence of a sea ice cover. The albedo and heat exchange factors thereby provide short-term positive feedbacks for further increasing surface atmospheric temperatures, although the reduced insulation of the ocean from the atmosphere could provide a long-term negative feedback since more heat becomes lost to space.

If cloud interactions are allowed in the modeling experiments, these can contribute either positive or negative feedbacks to the CO_2 question. The method for incorporating clouds in present models is still relatively crude, so that cloud parameterizations represent a major source of uncertainty in model CO_2 experiments (see Somerville and Remer, 1984).

Although using a highly idealized land/sea geography and not including a realistic ice climate feedback mechanism, Manabe and Wetherald (1975, 1980) were able to calculate at least an initial estimate of the surface temperature increases to be expected with a doubling of atmospheric CO_2. The results of their two studies show 10 K and 7–8 K polar surface temperature increases, respectively, with far lesser increases at lower latitudes.

Several more recent studies of CO_2 effects have been performed with general circulation models that are more realistic than the Manabe and Wetherald model in several respects: they have a more realistic geography, a seasonal cycle in solar forcing, and interactive atmosphere/ocean/sea ice calculations (e.g., Manabe and Stouffer (1980) with the GFDL model; Washington and Meehl (1984) with the NCAR model; Hansen et al. (1984) with the GISS model; Schlesinger et al. (1985) with the OSU model). The more realistic model configurations in these later studies allow better comparisons of the model results with observations for the present climate, although considerable uncertainty remains regarding the realism of the response of the models to presumed climate changes.

The GFDL, GISS, and NCAR models all show similar general patterns of response to prescribed CO_2 increases, even though the models differ in terms of their physical parameterizations, resolutions, and the type of ocean to which they are coupled. The models show an atmospheric cooling above 20 km and a warming below, with the largest warming occurring at the surface in the polar regions, where all three models show considerable warmings, particularly in the winter season (Fig. 6.10). The largest warming in the free atmosphere occurs in the tropical upper troposphere and is caused by cloud-convective interactions. This warming is substantially larger in the GISS experiment than in the GFDL and NCAR experiments.

Even though the temperature differences at the surface are great in the model results depicted in Fig. 6.10, it is difficult to judge the statistical significance of this fact since the interannual variations in the real atmosphere (Chervin, 1986) and in model studies with prescribed boundary conditions can also be quite

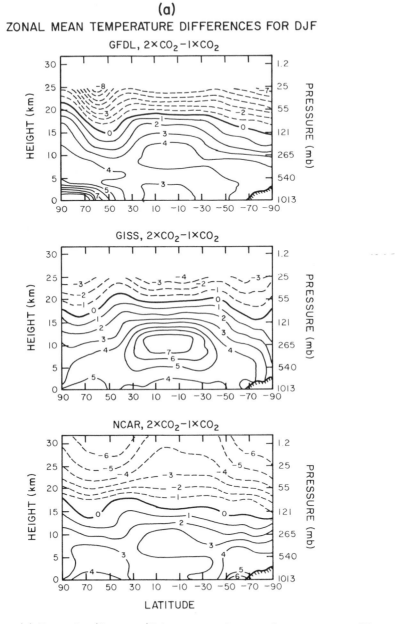

(a)

ZONAL MEAN TEMPERATURE DIFFERENCES FOR DJF

Fig. 6.10 (a) December/January/February zonal mean air temperature differences (in K) between $2 \times CO_2$ experiments and control experiments ($1 \times CO_2$) done with the GFDL, GISS, and NCAR models. Results are plotted as functions of latitude and height. [From Schlesinger and Mitchell (1985).]

278

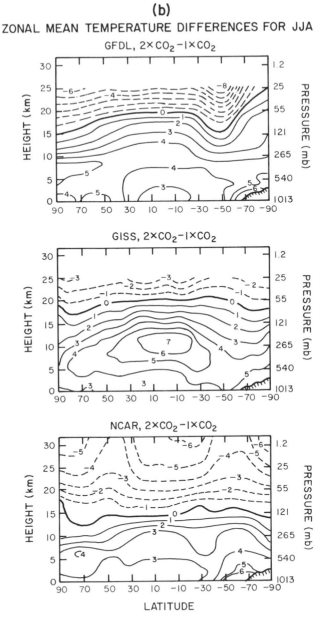

Fig. 6.10 (b) June/July/August zonal mean air temperature differences (in K) between 2 × CO₂ experiments and control experiments (1 × CO₂) done with the GFDL, GISS, and NCAR models. Results are plotted as functions of latitude and height. [From Schlesinger and Mitchell (1985).]

large. The climatological statistical fluctuations or "noise" make finding the CO_2 signal in the atmosphere even more difficult since the system is always responding as well to factors other than CO_2. This indicates one of the advantages of numerical modeling: in the model studies all prescribed parameters other than CO_2 can be kept unchanged, thereby allowing a better estimate of signal-to-noise patterns. Chervin (1981) and others have established a statistical significance methodology for GCM experiments that quantitatively determines whether differences between the results of a control run and modifications can be attributed to the prescribed change. Because of the differences in inherent variability, the statistical significance of geographical patterns is harder to establish than that of zonal patterns. In order to obtain statistical significance in the geographical patterns, one either has to run the models longer or to be content with a lower degree of confidence in the results. It probably will be some time before the simulated geographical patterns can be relied upon to give highly significant results since they differ considerably among models. The hope in the meantime is that the present simulations will provide valuable insights into important issues, not that they will provide final answers.

Manabe and Bryan (1985) have recently used a coupled atmosphere/ocean/ice model to simulate results for conditions with atmospheric carbon dioxide increased to two, four, and eight times its present value and decreased to one-half its present value. The results are particularly interesting in that they show features of the general circulation that remain relatively stable even over this large range in carbon dioxide amounts.

The simulated response of the Arctic ice pack to an atmospheric warming, whether CO_2 induced or otherwise, has been examined by Budyko (1966, 1974) with a heat balance model and by Parkinson and Kellogg (1979) with the thermodynamic/dynamic sea ice model of Parkinson and Washington (1979) discussed in Chapter 5. Budyko simulated a disappearance of the ice pack with a 4 K rise in summer temperatures and conjectured that under such conditions the ice would not reform in winter. In the Parkinson and Kellogg study, a prescribed uniform 5 K temperature increase resulted in an ice-free Arctic Ocean in August and September, but the ice reappeared in the central Arctic in mid-fall. Parkinson and Kellogg performed three experiments: (1) increasing temperatures uniformly by 5 K (Fig. 6.11); (2) increasing temperatures by 6–9 K, the largest increases occurring in winter, along with increasing cloud cover by 5% from May through September and by 10% from October through April; (3) retaining the changes of Experiment 2

(a) RESULTS USING MEAN CLIMATOLOGICAL DATA

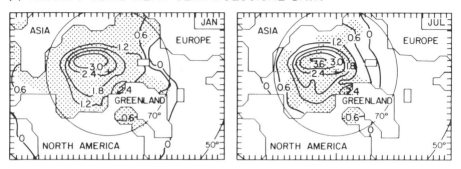

(b) RESULTS WITH ATMOSPHERIC TEMPERATURES INCREASED BY 5 K

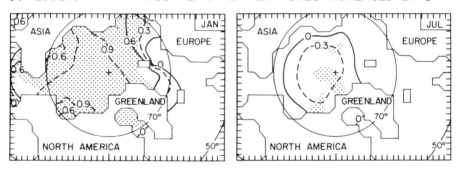

Fig. 6.11 January and July ice thicknesses simulated by an Arctic sea ice model with prescribed atmospheric forcing from (a) mean monthly climatological atmospheric data and (b) the same data set as in (a) except with all temperatures uniformly increased by 5 K. [Redrawn from Parkinson and Kellogg (1979).]

while also increasing the ocean heat flux to the bottom of the sea ice from 2 to 25 W m^{-2}. Even in the final case, with atmospheric temperature increases of 6–9 K and greater than an order of magnitude increase in the upward ocean heat flux, the ice still reappeared in the simulated late fall and winter (Parkinson and Kellogg, 1979). The prescribed atmospheric temperature increases used in this study were based conservatively on values below the 10 K polar surface temperature increase calculated by Manabe and Wetherald (1975) and within the vicinity of the 7–8 K increase calculated by Manabe and Wetherald (1980) for a doubling of atmospheric CO_2. Parkinson and Kellogg examined the first-order effects of such an atmospheric temperature increase on the sea ice and not the resultant effects of the change of ice on the atmosphere and ocean. The latter, interactive effects should be examined with a coupled ocean/ice/atmosphere model.

In a study similar to that of Parkinson and Kellogg for the Arctic, Parkinson and Bindschadler (1984) examined the impact of uniform atmospheric temperature changes on a simulated Antarctic sea ice cover. They too used the Parkinson and Washington (1979) thermodynamic/dynamic sea ice model, although with the following revisions: (1) more realistic initial conditions for ice concentrations and ocean temperatures; (2) elimination of snowfall to avoid the complications of a snow-induced multiyear cycle in regions with perennial sea ice; (3) an increase from 1 to 10 cm of the prescribed thickness of ice upon initial freezing in previously ice-free grid squares. The third revision has little impact beyond the first few time steps after ice formation.

Parkinson and Bindschadler compare results of the model's standard case, which uses mean-monthly climatological atmospheric inputs, and results of four perturbed cases with atmospheric temperatures and dew points altered uniformly in space and time by adding $-1, +1, +3$, and $+5$ K, respectively. Although somewhat unrealistic, the uniformity in the temperature changes has the advantage of removing ambiguities in the interpretation of results and allowing calculations of the rate of latitudinal retreat of the ice edge per unit temperature increase. With a 5 K atmospheric warming, the modeled Antarctic winter ice cover is reduced by about 50%, and the modeled summer ice cover vanishes everywhere except in one grid square in the Amundsen Sea (Figs. 6.12 and 6.13). Even with a 3 K temperature increase, summer ice is restricted to the Amundsen Sea and the southwestern portion of the Weddell Sea. Parkinson and Bindschadler calculate a hemispherically averaged latitudinal retreat rate of the ice edge of 1.4°S per 1 K increase in temperature. However, as was the case with the Parkinson and Kellogg study for the Arctic, this Antarctic experiment was constrained in not allowing ice/ocean/atmosphere feedbacks.

Manabe and Stouffer (1980) use a coupled ocean/ice/atmosphere model to compare results for a control case with results for a case with the concentration of atmospheric CO_2 increased to four times its present value. The Antarctic sea ice in the control case is too sparse as a consequence of the excessively warm simulated surface air temperatures in the region of the Southern Ocean. Thus, although Antarctic sea ice vanishes for much of the year in the $4 \times CO_2$ case, the simulated contrast in the two cases may still be too small. By contrast, in the Arctic the simulated sea ice in the control case is too extensive, which may contribute there to a misleadingly large temperature increase due to the CO_2 quadrupling. Nonetheless, in spite of the problems, by including sea ice Manabe and Stouffer have allowed

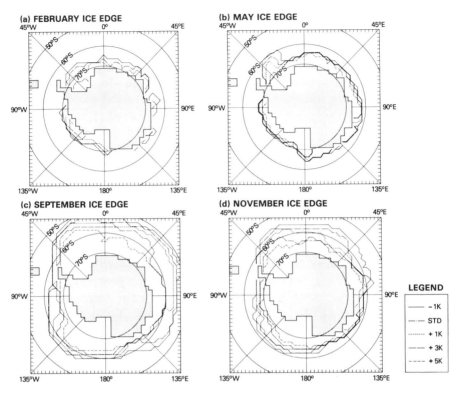

Fig. 6.12 Simulated seasonal cycle of the Southern Ocean sea ice edge with five different prescribed atmospheric forcings. In the standard case (labeled STD) the prescribed temperatures and dew points are mean monthly climatological values, while in the four perturbed cases all atmospheric temperatures and dew points have been uniformly adjusted by adding -1, $+1$, $+3$, and $+5$ K, respectively. Results are presented for the following months: (a) February, (b) May, (c) September, and (d) November. [From Parkinson and Bindschadler (1984).]

the lessened sea ice amounts resulting from increased CO_2 to feed back to the atmospheric calculations, thereby incorporating in their simulations more of the important interactions of the climate system. Hansen et al. (1984) and Washington and Meehl (1984) obtain somewhat different results with similar models, although naturally they too show a retreat of sea ice associated with the CO_2-induced surface warming at high latitudes.

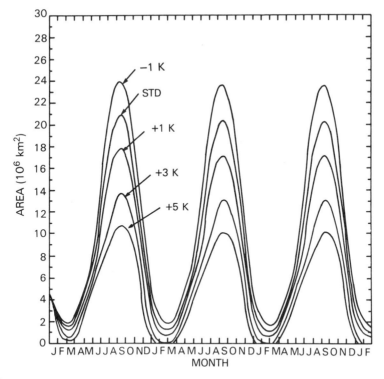

Fig. 6.13 Simulated time sequences of the areal extent of sea ice in the Southern Ocean using the same five sets of prescribed atmospheric forcing fields as in Fig. 6.12; namely, with mean monthly climatological data for the standard case (STD) and with the climatological temperatures and dew points uniformly altered by −1, +1, +3, and +5 K for the four perturbed cases. [From Parkinson and Bindschadler (1984).]

Possible Climatic Effects Due to Nuclear War

One of the more controversial uses of climate models is for the investigation of possible climatic effects of global nuclear war. Aside from the political issues involved, there are many sources of uncertainty in the modeling itself, as pointed out in a recent series of studies by the National Academy of Sciences (NAS, 1985). Nonetheless, several three-dimensional model simulations have been performed, and these suggest that under different sets of assumptions the climatic effects of nuclear war can be substantial (Covey et al., 1984; Aleksandrov and Stenchikov, 1983; Thompson et al., 1984; Covey et al., 1985). We present here an example to illustrate the application of a climate model to this issue and to indicate where some of the modeling uncertainties lie. In particular, we examine the climatic effect of the injection

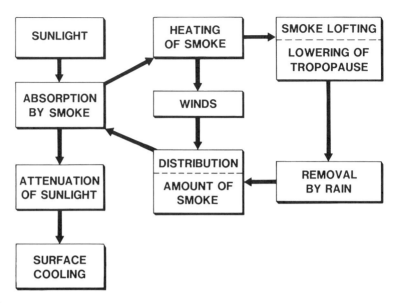

Fig. 6.14 Schematic of the interactions of solar radiation, smoke, and atmospheric processes. [From Malone et al. (1986).]

into the atmosphere of massive amounts of smoke from the fires speculated by Crutzen and Birks (1982) and Turco et al. (1983) to be likely during and after a nuclear war.

Figure 6.14 shows a depiction by Malone et al. (1986) of the principal atmospheric processes controlling the expected climatic effects from smoke caused by large-scale nuclear war. The radiative effects are shown on the left-hand side of the figure, with incoming sunlight absorbed by smoke particles, and this in turn causing a diminution or attenuation of sunlight lower in the atmosphere. This attenuation of sunlight results in less solar absorption at the ground and thus a cooling of the surface. There is, of course, much uncertainty about how much smoke would occur or what would be the precise radiative effects on the sunlight, and several alternative numerical treatments of these effects have been suggested. Most radiation experts agree, however, that if the optical depth (see Chapter 3) is sufficiently large, then the details of the radiative treatment are not critically important. The boxes in the center of Fig. 6.14 indicate the interactions of smoke with the wind, or atmospheric dynamics, which in an interactive model would greatly alter the smoke distribution. Maximum solar heating is believed likely to occur in the lower stratosphere (at an altitude of about 15 km in the tropics, 6–10 km in higher latitudes), and to cause a lowering of the tropopause and a vertical "lofting" of a hot smoke-filled "cloud" by buoyancy forces. One of the most uncertain aspects of the problem concerns the box in the lower right of Fig. 6.14, which deals with the removal of the smoke particles by precipitation processes.

Malone et al. (1986) have used a general circulation model of the Los Alamos National Laboratory (LANL) to simulate the effect of a nuclear war on atmospheric circulation. The LANL model uses a variant of the NCAR Community Climate Model. The model was modified for the nuclear-war study to include sources of smoke, transport of smoke, solar heating of smoke particles, removal of smoke particles by precipitation processes, and gravitational sedimentation.

Malone et al. (1986) simulate for January and July and for various amounts and vertical distributions of smoke. In the July experiment discussed here, 170 Tg (1 Tg $= 10^9$ kg) of smoke are injected into the troposphere over North America, Europe, and the western Soviet Union over a 7-day period, with a constant density of smoke from the surface to 9 km. In the model simulation, the mid-latitude westerlies transport the smoke rapidly eastward, especially at high levels, while the low-level smoke cloud is strongly scavenged by precipitation and gravitational sedimentation. Within a few weeks the smoke is fairly uniformly mixed in the Northern Hemisphere and has spread somewhat into the Southern Hemisphere. Figure 6.15 illustrates the effect of the smoke on the July zonal temperature structure by contrasting the temperature structures for (a) the long-term average for an unperturbed atmosphere, and (b) the average for days 15–20 of the simulation in which the atmosphere is perturbed by injecting smoke during days 1–7. The smoke injection causes a considerable lowering of the tropopause, which forms in the perturbed simulation under the region of maximum heating. The heating is caused by the absorption of solar radiation by the smoke and the consequent smoke lofting. It reaches a maximum at altitudes of about 10–20 km, where the atmospheric temperatures warm by 60–100°C.

The geographic changes in surface temperature resulting from the smoke injection are shown in Fig. 6.16. The deviation between the average for days 5–10 of the perturbed simulation and the long-term average of the unperturbed simulation is shown in panel (a), while the deviation between the average for days 35–40 of the perturbed simulation and the unperturbed long-term average is shown in panel (b). Only temperature changes exceeding 5 K in absolute value are indicated. For days 5–10, most of the temperature changes are concentrated over the continents where the smoke is being injected; by days 35–40,

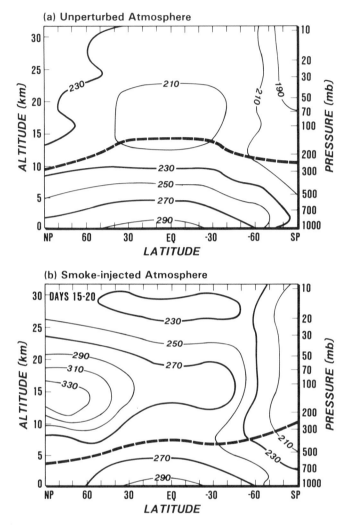

Fig. 6.15 The simulated height/latitude distributions of the zonally averaged temperature (K) for July for the following cases: (a) the long-term average for an unperturbed atmosphere and (b) the average for days 15–20 in a simulation in which 170×10^9 kg of smoke are injected into the atmosphere during days 1–7. The heavy dashed line is the approximate tropopause position. [From Malone et al. (1986).]

however, the Northern Hemisphere changes are somewhat more uniform latitudinally. The uniformity is even stronger in the upper atmosphere (not shown), where strong zonal winds spread the smoke zonally around the globe. The temperature changes near the edge of Antarctica visible in Fig. 6.16 are not related to the injection of smoke in the Northern Hemisphere but instead

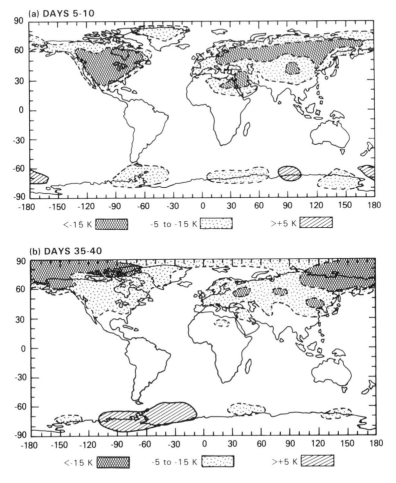

Fig. 6.16 Global distributions of the difference between the simulated July surface air temperature for a smoke-injected atmosphere and the simulated July surface air temperature for an unperturbed atmosphere. The long-term average of the unperturbed case is subtracted from the averages for the perturbed case over the following days: (a) days 5–10, (b) days 35–40. Changes near Antarctica are caused by individual storms, and are unconnected to the injection of smoke in the Northern Hemisphere. [Relabeled from Malone et al. (1986).]

reflect expected deviations of averages over only a few days from the long-term average.

In the corresponding January simulations done by Malone et al. (1986), the smoke spreads across the Northern Hemisphere more quickly than in the July simulations, because of the more intense winter atmospheric circulations. Also, the smoke stays closer to its initial injection height in January than in July, because of lesser solar heating and hence lower buoyancy. As a result of the lower buoyancy, less of the smoke moves into the stratosphere, where it tends to have considerably longer residence times, and hence the depletion of the smoke from the atmosphere is much more rapid in January than in July (Malone et al., 1986).

Overview of Climate Sensitivity Studies

Numerical modeling can serve a useful purpose in evaluating hypothetical scenarios of the future involving one or more of the various physical parameters and interactions in the climate system. Although certain experiments, such as increasing the atmospheric carbon dioxide content, might inadvertently be performed in the real world, allowing direct observation of the consequences, it is less risky and much quicker to perform such experiments numerically. Also, in the real world other complicating factors, such as volcanic eruptions, might confuse the issues. However, one must keep in mind the flaws of the numerical approach and particularly that the response of a model is not necessarily the response that the real world would have, especially in view of the large number of interactive processes essentially ignored in the simulations (Kellogg, 1975). Bach (1984) shows a schematic, reproduced in Fig. 6.17, of various climate components and possible interactions among them. No model yet incorporates all of these interactions. Nevertheless, model results can provide an estimate of possible effects of various altered conditions. As long as they are used with caution and researchers understand and communicate their limitations, models can be useful tools to suggest how the climate system might respond to changes and what feedback mechanisms might be important under various circumstances. They cannot be relied upon to give *definitive* answers, but they can contribute to increasing our understanding of past, present, and future climates.

As an example, modeling results support other indications that continued carbon dioxide insertion into the atmosphere could cause dramatic climatic consequences worldwide. Neither our numerical models nor our general understanding are sufficiently sophisticated to provide unrestrained confidence that the real world would respond in the manner that the models predict. However, in view of the possibility of dramatic and perhaps irreversible consequences, the warnings inherent in the model results should not be ignored. No model yet incorporates the full range of climate processes and feedbacks in the real world (or even the

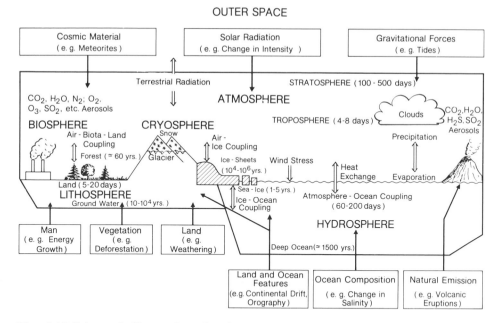

Fig. 6.17 Schematic illustration of various components and interactions in the climate system. [From Bach (1984).]

full range of those depicted in Fig. 6.17), though eventually climate models should improve and our confidence in them increase.

In this chapter only a few examples of climate sensitivity experiments have been addressed. Many other experiments have been completed or are presently underway with three-dimensional models, and more can be expected in the future. Some of the research topics presently being examined by various modeling groups around the world are:

1. Climates of past geologic periods ($10^2 - 10^6$ years).

2. Causes of ice-age climates.

3. Effects of alternative parameterizations of various physical processes.

4. Interactions of atmospheric circulations with photochemistry.

5. Interactions of the atmosphere and oceans with sea ice and continental ice sheets.

6. El Niño, the Southern Oscillation, and the Walker Circulation.

7. Interannual and intra-annual variability.

8. Vegetation changes (for example, those produced by deforestation).

9. Causes of changes in monsoon circulations.

10. Causes of droughts and the formation of deserts.

11. Effects of volcanic eruptions.

12. Effects of CO_2 and other trace gases.

13. Climate predictions on monthly, seasonal, and longer time scales.

14. Nuclear winter scenarios.

Results from present numerical models can provide insights into each of the above topics, and as the models continue to be improved, the insights they provide and our confidence in them should improve as well.

CHAPTER 7

Outlook for
Future Developments

Numerical climate modeling is a relatively new and rapidly developing discipline. Electronic computers have been in operation only since the late 1940s, and although atmospheric simulations were among the first applications of computer calculations, four decades are insufficient for establishing a mature science. Much progress has been made, however, as the discussions and examples provided in this book illustrate.

Historically, modeling of the atmospheric component of the climate system has received the most attention, largely because on a day-to-day basis the overwhelming majority of mankind is influenced much more directly by events in the atmosphere than by events in the oceans or sea ice. A serious, although areally-limited, numerical atmospheric model appeared even before the development of sophisticated computers, through the work of L. F. Richardson during World War I. However, because of the enormous number of calculations involved and the failure to forecast the atmospheric variables correctly, Richardson's goal of forecasting weather through numerical computations was viewed as impractical until the advent of electronic computers. Usable numerical weather prediction models were developed in the 1950s, and the first general atmospheric models were evolved from them; global atmospheric models were developed in the 1960s. These models have increased in complexity and overall capabilities as the availability of computers has become greater and computer storage capacity and speed of computation have increased.

A wide variety of three-dimensional atmospheric models has been created over the past three decades. Some are based on finite differences, others on spectral techniques; some use a vertical coordinate system based on pressure, others use one based on height; some incorporate a full set of "primitive equations," others use a reduced system of equations; some include very sophisticated parameterizations of physical phenomena, others employ much more simplistic parameterizations; and some use a very coarse horizontal resolution, while others use a finer horizontal resolution. Of the models tested over the years, many remain in use today. The diversity of model formulations reflects both the continuing search for improved models and the fact that there can be no single set of "ideal" parameterizations for all purposes, since indeed the "best" combination of parameterizations will depend on the intended uses of the model. For instance, models for very long-term paleoclimatic simulations will have different requirements than models meant to test the impact of a particular volcanic eruption on the climate of the next several months to years.

Efforts at numerical global modeling of the oceans and of sea ice began later than the efforts at global modeling of the atmosphere and have tended to involve far fewer researchers. Nonetheless, ocean and ice modeling have expanded greatly over the past two decades. Because of the geographic boundary conditions, both fields lend themselves more readily to the modeling of a single ocean or region than to global modeling, which is more natural to atmospheric fields.

As with atmospheric modeling, a variety of models has been formulated for the oceans and sea ice, with differing horizontal and vertical resolutions and differing levels of sophistication in the various parameterizations. Some ocean models use a full set of primitive equations, others use quasi-geostrophic or quasi-analytic equations; some deal with the entire depth of the ocean, others deal with only the upper levels; some are largely wind driven, while others are largely thermally driven. Similarly, some sea ice models emphasize thermodynamic calculations, others emphasize dynamic calculations; and some include the internal ice stress explicitly as a term in the momentum equation, while others incorporate it indirectly.

Although many of the basic equations are similar for all three climatic components (especially for the two fluids, ocean and atmosphere), there are distinct differences that have to be addressed in the modeling programs because of the basic physical differences in the respective fields: The atmosphere is composed largely of a compressible gas which, for the bulk of the atmosphere, has no horizontal boundaries; the ocean is composed largely of a nearly incompressible liquid with boundary constraints, often awkward to handle numerically, in every ocean, sea, and bay; and sea ice is a solid of variable extent and thickness that can disappear altogether, as it does over vast areas each summer. Furthermore, the density structure of the oceans, especially in the polar regions, is heavily dependent on salinity as well as temperature, so that the equation of state used in ocean modeling tends to be more involved than that used in atmospheric modeling, while the assumption of incompressibility in the oceans allows other equations to be less complicated.

In addition, the modeling of the three media is affected by the differing time and space scales involved in important energy transfers and other processes. Typically the time scales of response of the atmosphere to changes in surrounding conditions are much smaller than the time scales of response of the ocean, especially the deep ocean, and of the land surface. The sea ice responds much faster than the deep oceans but generally slower than the atmosphere. Therefore, time steps for atmospheric simulations tend to be smaller than those needed for oceanic or sea ice simulations. On the other hand, the space scales of the important eddies in the oceans tend to be much smaller than those in the atmosphere, so that finer horizontal resolution is required in the ocean models if these eddies are to be resolved.

Numerical models have successfully simulated large-scale features in all three media; among these are the easterly trade winds, mid-latitude westerlies, and jet streams in the atmosphere, the Gulf Stream and other major currents in the oceans, and the approximate annual cycle of sea ice areal extent and concentration in both polar regions. However, many features are well simulated (especially in terms of their intensity and geographic locations) only after testing various ranges of poorly known parameters such as heat and momentum transfer coefficients and selecting the value that produces the best results. For models to be applied with confidence to situations far different from those in which we can adjust selected parameters for specific results, we must continue to seek better understandings of why successful adjustments have worked and to develop the models further so that adjustments become less essential. A thorough documentation of model sensitivities is also needed.

As computing power has increased, certain improvements in numerical models have developed naturally, such as improved horizontal and vertical resolution, greater detail in oceanic bottom topography, and an increase in the number of variables that are computed rather than prescribed. Other improvements have come from deeper understandings of the climate system and of the numerics of the intertwined set of equations used to model it. It is now well understood that, given the limiting restrictions of individual situations to be modeled, the set of equations that is most complete and incorporates the most detail on the physical processes is not necessarily the set that will yield the most realistic results. For example, significant improvements are realized for day-to-day weather forecasts when undesired gravity waves are filtered out of the equations. Additional model enhancements have derived from improved data sets, both for initializing the models and for providing boundary conditions as well as for model verification. Some of the improved data sets come from intensive ground-based international efforts such as the Global Atmospheric Research Program (GARP), the Tropical Ocean and Global Atmosphere Program (TOGA), the World Ocean Circulation Experiment (WOCE) (see Webster, 1984), the First Implementation Plan for the World Climate Research Programme (see Mason, 1985), and the Arctic Ice Dynamics Joint Experiment (AIDJEX), but of even more potential importance for global data collection is the continuing development of satellite technology. Among the crucial variables that can be derived from satellite data on a routine basis are the radiation budget at the top of the atmosphere, cloud cover, atmospheric temperatures, sea surface temperatures, inferred surface winds over the oceans, sea level height, sea ice extents, and sea ice concentrations.

Each of the above three categories of model improvements—those deriving from increased computing power, from deeper understandings, and from better data sets—is expected to continue to yield improvements in the years and decades ahead.

The climate system is intricately coupled, with the four main climatic components of atmosphere, oceans, ice, and land surfaces each influencing the others. The modeling of these four components has reached a stage advanced enough that increasing efforts are being expended to include calculations for all of the components in combined models. To simulate the details of future events near the ocean/atmosphere interface, the changes in both media must be determined; if the calculations are being done in the polar regions, then changes in the sea ice cover must be determined as well; and for atmospheric calculations over land, major changes in the vegetation cover, snow cover, and glacial ice can be significant.

In the long term, proper coupling should yield improved model results for all media, whether atmosphere, ocean, ice, or land. However, initial efforts at coupling do not necessarily produce distributions of the modeled variables that are improved over the distributions produced by stand-alone models. Errors in the calculated fields of the other climatic components in the coupled models are likely to affect adversely the calculations for any given component. For example, an ocean model that had used carefully prescribed, realistic atmospheric fields is almost certain to have its oceanic results worsened rather than improved when first coupled with an atmospheric model, as the realistic prescribed fields are replaced by calculated atmospheric fields with their various imperfections. Prescribing all the atmospheric variables or ocean variables or sea ice variables, however, is not acceptable for long-term simulations under conditions more and more divergent from the present. The coupling efforts are therefore essential even though initial results may prove discouraging.

The coupling of ocean, ice, and atmosphere models should soon yield improved simulations of many phenomena known to derive at least in part from an interaction between two or all three of these climate components. Among these phenomena are the following: ocean downwelling and bottom water formation, influenced in several key regions at least in part by the rejection of salt during sea ice formation; wind-induced deepening of the oceanic mixed layer; all surface heat and moisture fluxes; the El Niño phenomenon; and low-level cyclogenesis.

As coupling efforts proceed, among the issues receiving increased attention are the differing typical time and space scales important for simulating the atmosphere, oceans, ice, and land surfaces. Because on an overall basis the atmosphere tends to respond much faster than other media, schemes continue to be devised, although with limited success, for asynchronous couplings, which allow much longer time steps for the calculations done for the less rapidly responding media than for the more rapidly responding media, thereby conserving computer time. It may turn out, however, that synchronous coupling will be the only viable method of effective coupling, since the upper ocean responds quickly to changes in wind forcing (especially in the tropics) and both the upper ocean and the deep ocean are influenced by bottom water formation, the onset of which is often a short time scale response to sea ice freezing in the high latitudes.

More attention needs to be directed to land surface parameters, such as vegetation, soil moisture, snow cover, mountain glaciers, and continental ice sheets. Vegetation is important for surface albedo, and soil moisture for surface moisture fluxes. For regions with snow, snow cover is particularly important for all time scales since it is highly variable and significantly increases the surface albedo. Proper simulation of snow cover, however, requires proper simulation of snowfall and hence of clouds and cloud processes, which are generally subgrid-scale phenomena. Proper treatment of snow cover also requires parameterization of the dependence of snow albedo on such factors as terrain, snow age, and snow depth. Grid-cell averages of snow-cover properties are therefore difficult to obtain in a reliable way. These and other subgrid-scale processes and parameterizations also are expected to receive more attention in the years ahead.

For very long-term simulations, ice sheet and ice shelf models should be coupled with models of the other climatic components. Ice sheets destroy the vegetation cover, greatly affect surface albedo, and force adjustments in atmospheric flow. Their modeling is easier in some ways than atmospheric and oceanic modeling because of simplifications permitted by the very high viscosity of the ice, but their flow equations are similar and are affected by the three-dimensional temperature and density structure within the ice. Large-scale ice sheet models have been constructed (for instance, see Budd and Smith, 1982), but they have not yet been coupled with atmospheric and oceanic general circulation models and have not been described in this book.

Detailed knowledge of many of the above-mentioned processes is severely limited by lack of observations. Numerous climate-oriented studies by various national and international bodies are addressing this situation, and eventually the knowledge gained from such studies should help to verify or lead to improved model parameterizations. The overwhelming majority of the existing data sets for rainfall, temperature, cloudiness, sea level pressure, and other variables is based upon observations from no more than the past several decades. Data sets on soil moisture, snow cover, surface albedo, and many other variables are virtually nonexistent for large regions of the globe; but this situation is dramatically improving, as satellite coverage of the earth is allowing the collection of consistent global data sets never before possible.

Numerical simulations should continue to increase our understanding of the earth's climate system and to provide insight into many factors that may influence climate change. As the models improve, our confidence in their results should increase. The potential benefits for mankind are multifold, as more accurate prediction of future states over both short and long time periods should assist decisions on future planning in many areas of human activity. One of the real merits of such models is that they will allow mankind to evaluate possible changes in a quantitative manner.

We close this chapter with a still-applicable quotation meant both to encourage and to caution those entering the research field of general circulation modeling. It was made by Joseph Smagorinsky, one of the leaders in the early development of general circulation models, and at the time head of the Geophysical Fluid Dynamics Laboratory at Princeton University. At a Joint Conference of the Royal Meteorological Society and the American Meteorological Society in 1969, Smagorinsky (1969) stated:

> Those research groups that have ventured into the field of general circulation modelling quickly come to realize the enormity of the undertaking—not merely the computation bulk but also the disciplinary scope.... As models become more complex, the starting obligation of time and talent becomes greater—and I am afraid this demand may deter the impatient and discourage the timid. Fortunately the latent rewards of discovery will continue to entice.
>
> ...a scientist entering this field must have a capacity for deep involvement and a temperament of unrelenting commitment; also he must have or develop a multi-disciplinary consciousness and an ability to function as a member of a team. In return, his institution must fulfil his human need for individual recognition. Although such demands are relatively new in meteorology, other fields, such as space exploration, have evidently learned to cope on a far larger scale.
>
> When one is in the midst of such long-term projects, months and even years can slip by without encountering a natural publication milestone. Yet in the prevailing environment of publish or perish, there is an impatience on the part of the scientist as well as the institution that supports him. ...Having been a young scientist myself, I can testify that it was worth the gamble. The satisfactions of treading new ground, of encountering new vistas from the shoulders of our predecessors are opportunities that must be preserved and strengthened. New advances will not come easily—some may come to think that much of the cream has already been skimmed from the top. Although the apparently easy things have already been done, at the time of their accomplishment the difficulties seemed insurmountable.

Vector Calculus

A vector is a quantity that has both magnitude and direction. It commonly is represented pictorially by a directed line segment with length scaled to the magnitude of the vector and notationally by boldface letters. Examples of vector quantities relevant to this book include velocity, force, and wind. In particular, a vector wind, V, can be represented in three-dimensional space either as

$$V = (V_x, V_y, V_z) \qquad (A.1)$$

or as

$$V = iV_x + jV_y + kV_z \qquad (A.2)$$

where V_x, V_y, and V_z are the speeds of the wind along three perpendicular axes with unit vectors i, j, and k, respectively. The speed of the wind itself (or the magnitude of the vector) is, by the Pythagorean theorem in elementary geometry,

$$|V| = \sqrt{V_x^2 + V_y^2 + V_z^2} \qquad (A.3)$$

In studies in fluid dynamics, it is conventional to symbolize the components of velocity in the x, y, and z directions as u, v, and w, respectively; thus (A.1) and (A.2) become

$$V = (u, v, w) \qquad (A.4)$$

and

$$V = iu + jv + kw \tag{A.5}$$

Since climate models are based on the terrestrial globe, for the purposes of this book u is defined along a latitude circle, positive to the east, v is defined along a meridian of longitude, positive to the north, and w is in the vertical direction, defined along a radius from the center of the earth, positive outward. Thus, although at each individual point the unit vectors in the three primary directions are perpendicular, these unit vectors are not the same (or parallel) from one point to another, as they would be in a Cartesian coordinate system. For clarity, however, in this appendix we first describe vector operations in a Cartesian coordinate system, before turning to the case of generalized coordinates and then to the spherical coordinates of most relevance to global climate modeling.

Vector Operations in a Cartesian Coordinate System

Vector addition and subtraction

By definition, two vectors $V_1 = (u_1, v_1, w_1)$ and $V_2 = (u_2, v_2, w_2)$ are added and subtracted vis-à-vis their components, so that

$$V_1 + V_2 = (u_1 + u_2, v_1 + v_2, w_1 + w_2) \tag{A.6}$$

and

$$V_1 - V_2 = (u_1 - u_2, v_1 - v_2, w_1 - w_2) \tag{A.7}$$

Vector multiplication

Mathematicians have defined two separate vector multiplications. The first, called a dot or scalar product, yields a scalar, and the second, called a cross product, yields a vector. By definition, the dot product of vectors V and B is the product of the magnitudes of the two vectors and the cosine of the angle between them:

$$V \cdot B \equiv |V||B| \cos(V, B) \tag{A.8}$$

In the case of two vectors at right angles to each other, the cosine of the angle between them (90°) is zero and hence the dot or scalar product is zero. The angle between two vectors that are parallel or overlap is 0, with a cosine of 1, and hence the dot product of two such vectors is the product of their magnitudes. More generally, the dot product of any two three-dimensional vectors $V = (u, v, w)$ and $B = (B_x, B_y, B_z)$ defined with the same unit vectors in Cartesian coordinates can be shown to be

$$V \cdot B = uB_x + vB_y + wB_z \tag{A.9}$$

The cross product, $V \times B$, by definition is O if either V or B has magnitude 0 or V and B are collinear. Otherwise the cross product is the vector with magnitude

$$|V \times B| \equiv |V| |B| \sin(V, B) \tag{A.10}$$

and direction perpendicular to the plane determined by V and B and such that V, B, and $V \times B$ form a right-handed triple. The term $\sin(V, B)$ is the sine of the angle between V and B. The cross product in Cartesian coordinates can be shown to be

$$V \times B = (vB_z - wB_y)i + (wB_x - uB_z)j + (uB_y - vB_x)k \tag{A.11}$$

(for instance, see Morrey, 1962, p. 531). For those familiar with determinants, this is equivalent to

$$V \times B = \begin{vmatrix} i & j & k \\ u & v & w \\ B_x & B_y & B_z \end{vmatrix} \tag{A.12}$$

In several places in this book products involving three vector quantities are used. Among the useful relationships derivable from the elementary definitions of dot and cross products for three arbitrary vectors A, B, and C and scalars m and n are the following:

$$A \times (mB + nC) = mA \times B + nA \times C \tag{A.13}$$

$$(A \times B) \cdot C = A \cdot (B \times C) = \begin{vmatrix} A_x & A_y & A_z \\ B_x & B_y & B_z \\ C_x & C_y & C_z \end{vmatrix} \tag{A.14}$$

$$(A \times B) \times C = (C \cdot A)B - (C \cdot B)A \tag{A.15}$$

$$A \times (B \times C) = (A \cdot C)B - (A \cdot B)C \tag{A.16}$$

In this context it should also be noted that

$$A \cdot B = B \cdot A \tag{A.17}$$

but

$$A \times B = -B \times A \qquad (A.18)$$

so that the dot product is commutative but the vector product is not. The reader should be able to verify (A.13)–(A.18) by expanding each of the terms (using (A.9) and (A.11)).

Vector differentiation

If a vector in Cartesian space is a function, for instance $V(t) = u(t)i + v(t)j + w(t)k$, then the derivative of the vector can be obtained by differentiating component by component, that is,

$$\frac{dV(t)}{dt} = \frac{du(t)}{dt}i + \frac{dv(t)}{dt}j + \frac{dw(t)}{dt}k \qquad (A.19)$$

Gradient (del) operator

The three-dimensional gradient operator, ∇_3, is the vector differential operator defined as

$$\nabla_3 = i\frac{\partial}{\partial x} + j\frac{\partial}{\partial y} + k\frac{\partial}{\partial z} \qquad (A.20)$$

Thus the gradient of a function $f(x, y, z)$ is the vector whose components in the direction of each of the unit vectors i, j, and k are the respective partial derivatives of the function:

$$\nabla_3 (f(x, y, z)) = i\frac{\partial f}{\partial x} + j\frac{\partial f}{\partial y} + k\frac{\partial f}{\partial z} \qquad (A.21)$$

The two-dimensional gradient operator, ∇, is simply the two-dimensional portion of ∇_3:

$$\nabla = i\frac{\partial}{\partial x} + j\frac{\partial}{\partial y} \qquad (A.22)$$

It is useful to note that for any scalar a and three-dimensional vector $B = (B_x, B_y, B_z)$,

$$\nabla_3 \cdot (aB) = a\nabla_3 \cdot B + B \cdot \nabla_3 a, \qquad (A.23)$$

or, for a two-dimensional vector $B = (B_x, B_y)$,

$$\nabla \cdot (aB) = a\nabla \cdot B + B \cdot \nabla a \qquad (A.24)$$

a relationship that is used, for instance, in deriving the energy conservation equation in Appendix C. The reader should be able to verify (A.23) and (A.24) easily by expanding each of the terms.

Vector Operations in Generalized and Spherical Coordinates

Following Haltiner and Williams (1980), we define various operations in general three-dimensional orthogonal coordinates with unit vectors $\boldsymbol{a}_1$, $\boldsymbol{a}_2$, and $\boldsymbol{a}_3$ in the x_1, x_2, x_3 directions, respectively, and metric coefficients h_1, h_2, h_3 such that if s_1, s_2, s_3 are the scalar curvilinear distances in the three respective directions then $ds_1 = h_1 dx_1$, $ds_2 = h_2 dx_2$, and $ds_3 = h_3 dx_3$. Specifically, the three-dimensional gradient operator $\boldsymbol{\nabla}_3(\ )$ acts on a scalar A as follows:

$$\boldsymbol{\nabla}_3 A = \frac{\boldsymbol{a}_1}{h_1}\frac{\partial A}{\partial x_1} + \frac{\boldsymbol{a}_2}{h_2}\frac{\partial A}{\partial x_2} + \frac{\boldsymbol{a}_3}{h_3}\frac{\partial A}{\partial x_3} \qquad (A.25)$$

The dot product of $\boldsymbol{\nabla}_3$ with a vector $\boldsymbol{T}_3 = (T_1, T_2, T_3)$, termed the divergence of $\boldsymbol{T}_3$, is:

$$\boldsymbol{\nabla}_3 \cdot \boldsymbol{T}_3 = \frac{1}{h_1 h_2 h_3}\left[\frac{\partial}{\partial x_1}(h_2 h_3 T_1)\right.$$
$$\left. + \frac{\partial}{\partial x_2}(h_1 h_3 T_2) + \frac{\partial}{\partial x_3}(h_1 h_2 T_3)\right] \qquad (A.26)$$

The Laplacian $\boldsymbol{\nabla}_3^2 A$ is

$$\boldsymbol{\nabla}_3^2 A = \frac{1}{h_1 h_2 h_3}\left[\frac{\partial}{\partial x_1}\left(\frac{h_2 h_3}{h_1}\frac{\partial A}{\partial x_1}\right)\right.$$
$$\left. + \frac{\partial}{\partial x_2}\left(\frac{h_1 h_3}{h_2}\frac{\partial A}{\partial x_2}\right) + \frac{\partial}{\partial x_3}\left(\frac{h_1 h_2}{h_3}\frac{\partial A}{\partial x_3}\right)\right] \qquad (A.27)$$

The curl of $\boldsymbol{T}_3$ is

$$\boldsymbol{\nabla}_3 \times \boldsymbol{T}_3 = \frac{\boldsymbol{a}_1}{h_2 h_3}\left[\frac{\partial}{\partial x_2}(h_3 T_3) - \frac{\partial}{\partial x_3}(h_2 T_2)\right]$$
$$+ \frac{\boldsymbol{a}_2}{h_1 h_3}\left[\frac{\partial}{\partial x_3}(h_1 T_1) - \frac{\partial}{\partial x_1}(h_3 T_3)\right]$$
$$+ \frac{\boldsymbol{a}_3}{h_1 h_2}\left[\frac{\partial}{\partial x_1}(h_2 T_2) - \frac{\partial}{\partial x_2}(h_1 T_1)\right] \qquad (A.28)$$

In Cartesian coordinates, the three directions x_1, x_2, x_3 are x, y, z with corresponding unit vectors i, j, k and metric coefficients $h_1 = h_2 = h_3 = 1$. Hence, for Cartesian coordinates (A.25)–(A.28) reduce to:

$$\nabla_3 A = i\frac{\partial A}{\partial x} + j\frac{\partial A}{\partial y} + k\frac{\partial A}{\partial z} \tag{A.29}$$

$$\nabla_3 \cdot T_3 = \frac{\partial}{\partial x}(T_1) + \frac{\partial}{\partial y}(T_2) + \frac{\partial}{\partial z}(T_3) \tag{A.30}$$

$$\nabla_3^2 A = \frac{\partial}{\partial x}\left(\frac{\partial A}{\partial x}\right) + \frac{\partial}{\partial y}\left(\frac{\partial A}{\partial y}\right) + \frac{\partial}{\partial z}\left(\frac{\partial A}{\partial z}\right)$$

$$= \frac{\partial^2 A}{\partial x^2} + \frac{\partial^2 A}{\partial y^2} + \frac{\partial^2 A}{\partial z^2} = \nabla_3 \cdot \nabla_3 A \tag{A.31}$$

$$\nabla_3 \times T_3 = i\left[\frac{\partial}{\partial y}(T_3) - \frac{\partial}{\partial z}(T_2)\right] + j\left[\frac{\partial}{\partial z}(T_1) - \frac{\partial}{\partial x}(T_3)\right]$$

$$+ k\left[\frac{\partial}{\partial x}(T_2) - \frac{\partial}{\partial y}(T_1)\right]$$

$$= \begin{vmatrix} i & j & k \\ \frac{\partial}{\partial x} & \frac{\partial}{\partial y} & \frac{\partial}{\partial z} \\ T_1 & T_2 & T_3 \end{vmatrix} \tag{A.32}$$

These can be verified against (A.21), (A.9), and (A.12), with the appropriate notational substitutions.

In spherical coordinates, the three primary directions x_1, x_2, x_3 are longitude λ, latitude ϕ, and distance r from the sphere's center, with the corresponding metric coefficients being $h_1 = r\cos\phi$, $h_2 = r, h_3 = 1$. As in the case of Cartesian coordinates, we let i, j, and k symbolize the unit vectors in the three primary directions. However, for spherical coordinates, these unit vectors do not remain identical in direction from one location to another, and the operations defined above are not reducible to as simple a form as for Cartesian coordinates. Direct substitution shows that (A.25)–(A.28) instead become

$$\nabla_3 A = \frac{i}{r\cos\phi}\frac{\partial A}{\partial \lambda} + \frac{j}{r}\frac{\partial A}{\partial \phi} + k\frac{\partial A}{\partial r} \tag{A.33}$$

$$\nabla_3 \cdot \boldsymbol{T_3} = \frac{1}{r^2 \cos \phi} \left[\frac{\partial}{\partial \lambda}(rT_1) + \frac{\partial}{\partial \phi}(r \cos \phi \, T_2) \right.$$

$$\left. + \frac{\partial}{\partial r}(r^2 \cos \phi \, T_3) \right]$$

$$= \frac{1}{r \cos \phi} \frac{\partial T_1}{\partial \lambda} + \frac{1}{r \cos \phi} \frac{\partial(\cos \phi \, T_2)}{\partial \phi}$$

$$+ \frac{1}{r^2} \frac{\partial(r^2 T_3)}{\partial r} \tag{A.34}$$

$$\nabla_3^2 A = \frac{1}{r^2 \cos \phi} \left[\frac{\partial}{\partial \lambda} \left(\frac{1}{\cos \phi} \frac{\partial A}{\partial \lambda} \right) + \frac{\partial}{\partial \phi} \left(\cos \phi \frac{\partial A}{\partial \phi} \right) \right.$$

$$\left. + \frac{\partial}{\partial r} \left(r^2 \cos \phi \frac{\partial A}{\partial r} \right) \right]$$

$$= \frac{1}{r^2 \cos^2 \phi} \frac{\partial^2 A}{\partial \lambda^2} + \frac{1}{r^2 \cos \phi} \frac{\partial}{\partial \phi} \left(\cos \phi \frac{\partial A}{\partial \phi} \right)$$

$$+ \frac{1}{r^2} \frac{\partial}{\partial r} \left(r^2 \frac{\partial A}{\partial r} \right) \tag{A.35}$$

$$\nabla_3 \times \boldsymbol{T_3} = \frac{\boldsymbol{i}}{r} \left[\frac{\partial}{\partial \phi}(T_3) - \frac{\partial}{\partial r}(rT_2) \right]$$

$$+ \frac{\boldsymbol{j}}{r \cos \phi} \left[\frac{\partial}{\partial r}(r \cos \phi \, T_1) - \frac{\partial}{\partial \lambda}(T_3) \right]$$

$$+ \frac{\boldsymbol{k}}{r^2 \cos \phi} \left[\frac{\partial}{\partial \lambda}(rT_2) - \frac{\partial}{\partial \phi}(r \cos \phi \, T_1) \right] \tag{A.36}$$

Narrowing specifically to the spherical system of our interest, the earth, we make several additional adjustments to convert (A.33)–(A.36) to the form used in climate modeling. Since the distance r from the center of the earth is the sum of the radius of the earth, a, and the height above the earth's surface, z (with the earth being approximated as a perfect sphere, with a constant),

$$r = z + a \tag{A.37}$$

and

$$\frac{\partial}{\partial r} = \frac{\partial}{\partial z} \tag{A.38}$$

Also, in view of the shallowness of the atmosphere with respect to the earth's radius, $z \ll a$, so that the approximation $r = a$ is often made. Using this approximation, (A.33)–(A.36) reduce to:

$$\nabla_3 A = \frac{i}{a \cos \phi} \frac{\partial A}{\partial \lambda} + \frac{j}{a} \frac{\partial A}{\partial \phi} + k \frac{\partial A}{\partial z} \qquad \text{(A.39)}$$

$$\nabla_3 \cdot T_3 = \frac{1}{a \cos \phi} \frac{\partial T_1}{\partial \lambda} + \frac{1}{a \cos \phi} \frac{\partial (\cos \phi \, T_2)}{\partial \phi} + \frac{\partial T_3}{\partial z} \qquad \text{(A.40)}$$

$$\nabla_3^2 A = \frac{1}{a^2 \cos^2 \phi} \frac{\partial^2 A}{\partial \lambda^2} + \frac{1}{a^2 \cos \phi} \frac{\partial}{\partial \phi} \left(\cos \phi \, \frac{\partial A}{\partial \phi} \right)$$
$$+ \frac{\partial^2 A}{\partial z^2} \qquad \text{(A.41)}$$

$$\nabla_3 \times T_3 = i \left[\frac{1}{a} \frac{\partial T_3}{\partial \phi} - \frac{\partial T_2}{\partial z} \right] + j \left[\frac{\partial T_1}{\partial z} - \frac{1}{a \cos \phi} \frac{\partial T_3}{\partial \lambda} \right]$$
$$+ k \left[\frac{1}{a \cos \phi} \frac{\partial T_2}{\partial \lambda} - \frac{1}{a \cos \phi} \frac{\partial (\cos \phi \, T_1)}{\partial \phi} \right] \qquad \text{(A.42)}$$

These are the equations used in Chapter 3 for the spherical coordinate system appropriate to atmosphere and ocean modeling.

The vertical component of vorticity $\varsigma = k \cdot (\nabla_3 \times V_3)$ with $V_3 = (u, v, w)$, then becomes, in spherical coordinates,

$$\varsigma = k \cdot (\nabla_3 \times V_3) = \frac{1}{a \cos \phi} \frac{\partial v}{\partial \lambda} - \frac{1}{a \cos \phi} \frac{\partial (u \cos \phi)}{\partial \phi}$$
$$= \frac{1}{a \cos \phi} \frac{\partial v}{\partial \lambda} - \frac{1}{a} \frac{\partial u}{\partial \phi} + \frac{u}{a} \tan \phi \qquad \text{(A.43)}$$

and the three-dimensional divergence $D_3 = \nabla_3 \cdot V_3$ becomes

$$D_3 = \nabla_3 \cdot V_3 = \frac{1}{a \cos \phi} \frac{\partial u}{\partial \lambda} + \frac{1}{a} \frac{\partial v}{\partial \phi} - \frac{v}{a} \tan \phi + \frac{\partial w}{\partial z} \qquad \text{(A.44)}$$

The two-dimensional divergence D is

$$D = \nabla \cdot V = \frac{1}{a \cos \phi} \frac{\partial u}{\partial \lambda} + \frac{1}{a} \frac{\partial v}{\partial \phi} - \frac{v}{a} \tan \phi \qquad \text{(A.45)}$$

Vectors on a Rotating Sphere

An additional complication arises in modeling climate because the motions of both the atmosphere and ocean take place not just on a sphere, making a spherical coordinate system more appropriate than the simpler Cartesian coordinate system, but also on a rotating sphere. A full mathematical treatment of these motions must take the rotation into account.

A vector A associated with the earth system can be defined either with respect to a fixed frame of reference, such as the relatively "fixed" stars, or with respect to a rotating frame of reference, such as the earth itself. Let i, j, and k be unit orthogonal vectors in the absolute frame of reference and i', j', and k' unit orthogonal vectors in the rotating frame of reference. Then the arbitrary vector A can be expressed as

$$A = A_x i + A_y j + A_z k \qquad (A.46)$$

where (A_x, A_y, A_z) are its coordinates in the absolute frame of reference, or as

$$A = A'_x i' + A'_y j' + A'_z k' \qquad (A.47)$$

where (A'_x, A'_y, A'_z) are its coordinates in the rotating frame of reference.

The time derivative of A in the absolute system, symbolized $d_a A / dt$, is

$$\frac{d_a A}{dt} = \frac{dA_x}{dt} i + \frac{dA_y}{dt} j + \frac{dA_z}{dt} k$$

$$= \frac{dA'_x}{dt} i' + \frac{dA'_y}{dt} j' + \frac{dA'_z}{dt} k' + A'_x \frac{di'}{dt}$$

$$+ A'_y \frac{dj'}{dt} + A'_z \frac{dk'}{dt} \qquad (A.48)$$

The first three terms following the last equal sign of (A.48) constitute the time derivative of A in the rotating system, symbolized $d_r A / dt$. Let Ω be the angular velocity vector with which the rotating system, in our case the earth, is rotating with respect to the fixed system. Then

$$\frac{di'}{dt} = \Omega \times i', \quad \frac{dj'}{dt} = \Omega \times j', \quad \text{and} \quad \frac{dk'}{dt} = \Omega \times k' \qquad (A.49)$$

from which it follows from (A.13) that

$$\Omega \times A = \Omega \times \left(A'_x i' + A'_y j' + A'_z k' \right)$$

$$= A'_x \frac{di'}{dt} + A'_y \frac{dj'}{dt} + A'_z \frac{dk'}{dt} \qquad (A.50)$$

Substituting (A.50) into (A.48), we obtain

$$\frac{d_a A}{dt} = \frac{d_r A}{dt} + \boldsymbol{\Omega} \times \boldsymbol{A} \qquad (A.51)$$

Alternatively, (A.51) can be written as

$$\frac{dA}{dt}\bigg|_{\text{fixed}} = \frac{dA}{dt}\bigg|_{\text{rotating}} + \boldsymbol{\Omega} \times \boldsymbol{A} \qquad (A.52)$$

In order to apply (A.52) to the earth/atmosphere system, let the arbitrary vector $\boldsymbol{A}$ in (A.52) be a position vector $\boldsymbol{r}$ of a particle measured from the earth's center, and symbolize the velocity of that particle with respect to the absolute frame of reference by $\boldsymbol{V}_a$ and the velocity with respect to the rotating earth by $\boldsymbol{V}$. Then from (A.52),

$$\boldsymbol{V}_a = \boldsymbol{V} + \boldsymbol{\Omega} \times \boldsymbol{r} \qquad (A.53)$$

This expresses the absolute velocity as the sum of the velocity relative to the earth and the velocity due to the rotation of the earth. Abbreviating the notation d_r/dt by d/dt and substituting $\boldsymbol{V}_a$ for the arbitrary vector $\boldsymbol{A}$ in (A.51) we obtain

$$\frac{d_a \boldsymbol{V}_a}{dt} = \frac{d \boldsymbol{V}_a}{dt} + \boldsymbol{\Omega} \times \boldsymbol{V}_a \qquad (A.54)$$

Substitution of (A.53) into the terms on the right-hand side of (A.54) yields

$$\frac{d_a \boldsymbol{V}_a}{dt} = \frac{d(\boldsymbol{V} + \boldsymbol{\Omega} \times \boldsymbol{r})}{dt} + \boldsymbol{\Omega} \times (\boldsymbol{V} + \boldsymbol{\Omega} \times \boldsymbol{r}) \qquad (A.55)$$

In view of the constancy of $\boldsymbol{\Omega}$, (A.55) can be rewritten as

$$\frac{d_a \boldsymbol{V}_a}{dt} = \frac{d\boldsymbol{V}}{dt} + 2\boldsymbol{\Omega} \times \boldsymbol{V} + \boldsymbol{\Omega} \times (\boldsymbol{\Omega} \times \boldsymbol{r}) \qquad (A.56)$$

Since Newton's second law states that the acceleration of a body equals the sum of the forces acting on the body per unit mass (see Chapter 3), and since the relevant forces in the atmosphere and ocean systems are the pressure gradient, gravitation, $\boldsymbol{g}_a$, and friction, $\boldsymbol{F}$, (A.56) can be used with Newton's second law to obtain

$$\frac{d\boldsymbol{V}}{dt} = -\frac{1}{\rho}\nabla p - 2\boldsymbol{\Omega} \times \boldsymbol{V} + \boldsymbol{g} + \boldsymbol{F} \qquad (A.57)$$

where ρ is density, p is pressure, and the gravitational force is combined with the centrifugal force per unit mass $(\boldsymbol{\Omega} \times (\boldsymbol{\Omega} \times \boldsymbol{r}))$, i.e.,

$$\boldsymbol{g} = \boldsymbol{g}_a - \boldsymbol{\Omega} \times (\boldsymbol{\Omega} \times \boldsymbol{r}) \qquad (A.58)$$

The second term on the right-hand side of (A.57) can be written

$$2\mathbf{\Omega} \times \mathbf{V} = 2 \begin{vmatrix} \mathbf{i} & \mathbf{j} & \mathbf{k} \\ 0 & \Omega \cos \phi & \Omega \sin \phi \\ u & v & w \end{vmatrix}$$

$$= 2\mathbf{i}(w\Omega \cos \phi - v\Omega \sin \phi) + 2\mathbf{j}(u\Omega \sin \phi)$$

$$- 2\mathbf{k}(u\Omega \cos \phi)$$

$$= \mathbf{j}(fu) - \mathbf{i}(fv) + \mathbf{i}(2w\Omega \cos \phi)$$

$$- \mathbf{k}(2u\Omega \cos \phi) \tag{A.59}$$

where $f = 2\Omega \sin \phi$ is the Coriolis parameter. The terms involving the Coriolis parameter are of paramount importance in atmosphere, ocean, and ice motions and are explained in Chapter 3. The final terms in (A.59) generally are relatively small for large-scale atmospheric and oceanic motions in extratropical latitudes and usually can be ignored.

Legendre Polynomials
and Gaussian Quadrature

Because of their convenient properties for treating the latitude variation on a sphere, Legendre polynomials are often used in global atmospheric models. These polynomials are defined as

$$P_\ell(\mu) = \frac{1}{2^\ell \ell!} \frac{d^\ell[(\mu^2 - 1)^\ell]}{d\mu^\ell} \qquad \ell = 0, 1, 2, \ldots \qquad \text{(B.1)}$$

and have the following properties:

$$P_\ell(1) = 1 \qquad \ell = 0, 1, 2, \ldots \qquad \text{(B.2)}$$

$$P_\ell(-1) = (-1)^\ell \qquad \ell = 0, 1, 2, \ldots \qquad \text{(B.3)}$$

$$\int_{-1}^{1} P_\ell(\mu) Q_k(\mu) \, d\mu = 0 \qquad k < \ell \qquad \text{(B.4)}$$

where Q_k is any polynomial of degree k less than ℓ. [See Demidovich and Maron (1976) for further details.] They also have ℓ distinct real roots in the interval $(-1, 1)$. The first five Legendre polynomials obtained from (B.1) are

$$P_0(\mu) = 1 \qquad \text{(B.5)}$$

$$P_1(\mu) = \mu \qquad \text{(B.6)}$$

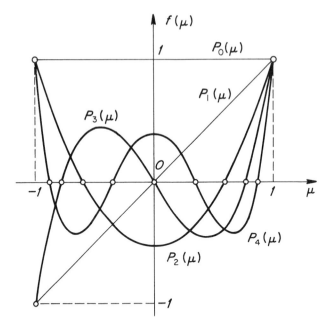

Fig. B.1 The first five Legendre polynomials $P_\ell(\mu)$, $\ell = 0, 1, 2, 3, 4$, plotted as a function of μ over the interval $(-1, 1)$. [Redrawn and relabeled from Demidovich and Maron (1976).]

$$P_2(\mu) = \tfrac{1}{2}(3\mu^2 - 1) \tag{B.7}$$

$$P_3(\mu) = \tfrac{1}{2}(5\mu^3 - 3\mu) \tag{B.8}$$

$$P_4(\mu) = \tfrac{1}{8}(35\mu^4 - 30\mu^2 + 3) \tag{B.9}$$

These polynomials are plotted in Fig. B.1 on the interval $(-1, 1)$. One of the reasons Legendre polynomials are so convenient for calculations on a sphere is that $|P_\ell(\mu)| = 1$ at the poles if μ is defined as $\sin \phi$, where ϕ is latitude. This can readily be confirmed for the first five Legendre polynomials by substituting $\mu = \pm 1$ into (B.5)–(B.9).

In order to solve for the coefficients of the spherical harmonics needed in Chapter 4, integrals over latitude must be computed. The usual technique used to solve these integrals is called the Gaussian quadrature formula.

Given a polynomial function $f(\mu)$ of degree equal to or less than $2\ell - 1$, the integral of $f(\mu)$ can be computed over the range -1 to $+1$ as the following sum, with ℓ properly selected coefficients A_i, at ℓ discrete points μ_i:

$$\int_{-1}^{1} f(\mu) \, d\mu = \sum_{i=1}^{\ell} A_i f(\mu_i) \tag{B.10}$$

In order to determine the points μ_i and coefficients A_i needed to make (B.10) generally valid, it is necessary and sufficient to ensure its validity for the single-term polynomials $f(\mu) = 1, \mu, \mu^2, \ldots, \mu^{2\ell-1}$. Hence the problem is reduced to solving the following system of 2ℓ equations:

$$\int_{-1}^{1} \mu^k \, d\mu = \sum_{i=1}^{\ell} A_i \mu_i^k \quad \text{for} \quad k = 0, 1, 2, \ldots, 2\ell - 1 \tag{B.11}$$

for the unknowns A_i and μ_i. The left-hand side of (B.11) can be integrated as

$$\int_{-1}^{1} \mu^k \, d\mu = \frac{1 - (-1)^{k+1}}{k+1} = \begin{cases} 2/(k+1) & k \text{ even} \\ 0 & k \text{ odd} \end{cases} \tag{B.12}$$

and hence equations (B.11) can be reduced to

$$\sum_{i=1}^{\ell} A_i = 2 \tag{B.13}$$

$$\sum_{i=1}^{\ell} A_i \mu_i = 0 \tag{B.14}$$

$$\sum_{i=1}^{\ell} A_i \mu_i^2 = \frac{2}{3} \tag{B.15}$$

$$\cdots\cdots\cdots\cdots\cdots\cdots$$

$$\sum_{i=1}^{\ell} A_i \mu_i^{2\ell-2} = \frac{2}{2\ell - 1} \tag{B.16}$$

$$\sum_{i=1}^{\ell} A_i \mu_i^{2\ell-1} = 0 \tag{B.17}$$

This set of 2ℓ equations is nonlinear, because the μ_i are unknown as well as the A_i. However, it can be simplified to a set of linear equations in the A_i once the points μ_i are determined. This is done by using Legendre polynomials.

Let $P_\ell(\mu)$ be the Legendre polynomial of degree ℓ, and form the ℓ polynomials

$$f_k(\mu) = \mu^k P_\ell(\mu) \tag{B.18}$$

where $k = 0, 1, ..., \ell - 1$. By property (B.4) for Legendre polynomials,

$$\int_{-1}^{1} \mu^k P_\ell(\mu) \, d\mu = 0 \qquad (k = 0, 1, 2, ... \ell - 1) \tag{B.19}$$

However, if (B.10) is to be satisfied in general, then

$$\int_{-1}^{1} \mu^k P_\ell(\mu) \, d\mu = \sum_{i=1}^{\ell} A_i \mu_i^k P_\ell(\mu_i) \tag{B.20}$$

for the correct choice of coefficients A_i and points μ_i. From (B.19) and (B.20), we then have that

$$\sum_{i=1}^{\ell} A_i \mu_i^k P_\ell(\mu_i) = 0 \tag{B.21}$$

To ensure that equations (B.21) are satisfied, the μ_i need simply be selected as the roots of $P_\ell(\mu)$. Note the use here of the important property of Legendre polynomials that the ℓ roots are distinct. Inserting these roots into (B.13)–(B.17) reduces that set of equations to a system of linear equations in the coefficients A_i. This system can then be solved, completing the determination of a set of coefficients A_i and points μ_i. These values of A_i and μ_i can then be shown to satisfy (B.10).

As an example of how this works, consider the case of $\ell = 3$ (see Fig. B.1). If the Legendre polynomial of degree three, expressed in (B.8), is set equal to zero, it can readily be seen to have the three roots

$$\mu_1 = -\sqrt{\tfrac{3}{5}} \simeq -0.774597 \tag{B.22}$$

$$\mu_2 = 0 \tag{B.23}$$

$$\mu_3 = \sqrt{\tfrac{3}{5}} \simeq 0.774597 \tag{B.24}$$

which correspond to the three crossings of the μ axis for $P_3(\mu)$ on Fig. B.1. From (B.13)–(B.15), the A_i must satisfy

$$A_1 + A_2 + A_3 = 2 \tag{B.25}$$

$$-\sqrt{\tfrac{3}{5}}\, A_1 + \sqrt{\tfrac{3}{5}}\, A_3 = 0 \tag{B.26}$$

$$\tfrac{3}{5} A_1 + \tfrac{3}{5} A_3 = \tfrac{2}{3} \tag{B.27}$$

This linear system of three equations in A_1, A_2, and A_3 can be solved for

$$A_1 = \tfrac{5}{9}, \quad A_2 = \tfrac{8}{9}, \quad \text{and} \quad A_3 = \tfrac{5}{9} \tag{B.28}$$

Substituting the just-determined values of A_i and μ_i into (B.10), we obtain

$$\int_{-1}^{1} f(\mu)\, d\mu = \tfrac{1}{9}\left[5f\left(-\sqrt{\tfrac{3}{5}}\right) + 8f(0) + 5f\left(\sqrt{\tfrac{3}{5}}\right)\right] \tag{B.29}$$

so that if $f(\mu)$ is known at the three points $-\sqrt{.6}$, 0, and $+\sqrt{.6}$, then the integral of $f(\mu)$ from -1 to $+1$ can be determined. For climate modeling applications, the three points are latitude points, commonly referred to as Gaussian latitudes, i.e., $\mu = \sin\phi$, and the range of the integral extends from the south pole at $\phi = -\pi/2$ to the north pole at $\phi = \pi/2$.

The above discussion of Gaussian quadrature follows closely the discussion in Demidovich and Maron (1976). The reader is referred to Machenhauer and Daley (1972), Bourke (1972), Hoskins and Simmons (1975), and Machenhauer (1979) for further discussions of the properties of Legendre polynomials and their applications to atmospheric global modeling.

The two-dimensional Laplacian operator appears throughout the spectral system of equations (for example, in Chapter 4 in (4.107), (4.148), and (4.149)). If scalar functions such as the streamfunction and velocity potential are expressed as orthogonal surface spherical harmonics, then the eigenfunctions, Y_ℓ^m, of the two-dimensional Laplacian operator are solutions of the following equation:

$$\nabla^2 Y_\ell^m + \ell(\ell+1)Y_\ell^m = 0 \tag{B.30}$$

where

$$\nabla^2(\ ) = \frac{\partial}{\partial\mu}\left[(1-\mu^2)\frac{\partial}{\partial\mu}\right] + \frac{1}{1-\mu^2}\frac{\partial^2}{\partial\lambda^2} \tag{B.31}$$

Substitution of (B.31) and

$$Y_\ell^m = P_\ell^m(\mu)e^{im\lambda} \tag{B.32}$$

into (B.30) yield

$$\frac{\partial}{\partial \mu}\left[(1-\mu^2)\frac{\partial}{\partial \mu}P_\ell^m(\mu)\right]e^{im\lambda}$$

$$+\left[\ell(\ell+1)-\frac{m^2}{1-\mu^2}\right]P_\ell^m(\mu)e^{im\lambda}=0 \quad \text{(B.33)}$$

This can be further simplified to:

$$(1-\mu^2)\frac{\partial^2}{\partial \mu^2}P_\ell^m(\mu)-2\mu\frac{\partial P_\ell^m(\mu)}{\partial \mu}$$

$$+\left[\ell(\ell+1)-\frac{m^2}{1-\mu^2}\right]P_\ell^m(\mu)=0 \qquad \text{(B.34)}$$

This equation is the well-known associated Legendre equation, so named because the solutions to (B.34) are the associated Legendre polynomials, $P_\ell^m(\mu)$, defined as:

$$P_\ell^m(\mu)=(1-\mu^2)^{m/2}\frac{d^m}{d\mu^m}P_\ell(\mu) \qquad \text{(B.35)}$$

If $m=0$ then the associated polynomials become the Legendre polynomials given in (B.5)–(B.9). The first few associated Legendre polynomials with $m>0$ are

$$P_1^1(\mu)=(1-\mu^2)^{1/2} \qquad \text{(B.36)}$$

$$P_2^1(\mu)=3\mu(1-\mu^2)^{1/2} \qquad \text{(B.37)}$$

$$P_2^2(\mu)=3(1-\mu^2) \qquad \text{(B.38)}$$

We now illustrate with a few examples the fact that the associated polynomials satisfy (B.34). Substitution of the first four Legendre polynomials into (B.34) yields the following:

$$0-0+0=0 \qquad \text{(B.39)}$$

for $m=0$, $\ell=0$;

$$-2\mu+2\mu=0 \qquad \text{(B.40)}$$

for $m=0$, $\ell=1$;

$$(1-\mu^2)3-6\mu^2+3(3\mu^2-1)=0 \qquad \text{(B.41)}$$

for $m=0$, $\ell=2$; and

$$(1 - \mu^2)15\mu - 2\mu \left(\frac{15}{2}\mu^2 - \frac{3}{2}\right) + 12\left(\frac{1}{2}\right)(5\mu^3 - 3\mu) = 0 \quad \text{(B.42)}$$

for $m = 0$, $\ell = 3$. Since (B.39)–(B.42) are each immediately seen to be identities, this verifies that the first four Legendre polynomials satisfy (B.34). Similarly, substitution of $m = 2$, $\ell = 2$, and $P_2^2(\mu)$ from (B.38) into (B.34) yields

$$(1 - \mu^2)(-6) - 2\mu(-6\mu) + \left[6 - \frac{4}{1 - \mu^2}\right](3)(1 - \mu^2) = 0 \quad \text{(B.43)}$$

Since this also is readily seen to be an identity, this verifies that the associated Legendre polynomial $P_2^2(\mu)$ is a solution to (B.34). The reader can find detailed derivations and solutions of (B.34) in various mathematical physics texts, including those by Arfken (1966), Sommerfeld (1949), and Dettman (1969).

The associated Legendre polynomials used in Chapter 4, and in the spectral modeling literature in general, are normalized to unity through division by the square root of the integral

$$\int_{-1}^{1} [P_\ell^m(\mu)]^2 \, du = \frac{2}{2\ell + 1} \frac{(\ell + m)!}{(\ell - m)!} \quad \text{(B.44)}$$

which is obtained with the help of the orthogonality property satisfied by associated Legendre polynomials.

Among the useful recursion relationships involving P_ℓ^m and Y_ℓ^m are the following:

$$\epsilon_{\ell+1}^m P_{\ell+1}^m = \mu P_\ell^m - \epsilon_\ell^m P_{\ell-1}^m \quad \text{(B.45)}$$

$$(1 - \mu^2)\frac{d}{d\mu} Y_\ell^m = -\ell \epsilon_{\ell+1}^m Y_{\ell+1}^m + (\ell + 1)\epsilon_\ell^m Y_{\ell-1}^m \quad \text{(B.46)}$$

where

$$\epsilon_\ell^m = \left\{\frac{(\ell^2 - m^2)}{(4\ell^2 - 1)}\right\}^{1/2} \quad \text{(B.47)}$$

Derivation of
Energy Equations

The total energy of any closed physical system should be at least approximately conserved, and the changes in total energy of any physical system, whether closed or open, should be balanced by the net inputs and outputs to the system. The reason that total energy may not be conserved exactly in a climate model, even though the original differential equation for energy conservation may be exact, is that by necessity various numerical approximations in space and time are made to the original equation, which may prevent exact conservation. Experimental tests indicate that simulated conversions from one form of energy to another are performed reasonably well by most major climate models, so that energy is approximately conserved.

Following the treatment of Kasahara (1974), the equation for kinetic energy for a hydrostatic system can be derived by multiplying the equations of motion (3.51), (3.52), and (3.53), by u, v, and w, respectively, and then adding to give

$$\frac{d}{dt}\left[\frac{1}{2}(u^2 + v^2) + gz\right] = -\frac{1}{\rho}\left(\boldsymbol{V} \cdot \boldsymbol{\nabla} p + w\frac{\partial p}{\partial z}\right) + \boldsymbol{V} \cdot \boldsymbol{F} \qquad \text{(C.1)}$$

where $\boldsymbol{V}$ is the horizontal velocity vector, $\boldsymbol{F} = F_\lambda \boldsymbol{i} + F_\phi \boldsymbol{j}$, and $\boldsymbol{V} \cdot \boldsymbol{F}$ is the frictional dissipation:

$$V \cdot F = uF_\lambda + vF_\phi \qquad \text{(C.2)}$$

The two terms in brackets on the left-hand side of (C.1) are the kinetic energy per unit mass, $(u^2 + v^2)/2$, and the potential energy per unit mass, gz. The terms on the right-hand side of (C.1) represent the work done by pressure and frictional forces which allow for conversions between kinetic and potential energy.

From the first law of thermodynamics and the equation of continuity, the following equation can be derived:

$$C_v \frac{dT}{dt} + \frac{p}{\rho}\left(\nabla \cdot V + \frac{\partial w}{\partial z} \right) = Q \qquad \text{(C.3)}$$

If (C.3) is added to (C.1) then

$$\frac{d}{dt}\left[\frac{1}{2}(u^2 + v^2) + gz + C_v T \right] = -\frac{1}{\rho}\left[V \cdot \nabla p + p\nabla \cdot V \right.$$

$$\left. + p\frac{\partial w}{\partial z} + w\frac{\partial p}{\partial z} \right] + V \cdot F + Q \qquad \text{(C.4)}$$

where $C_v T$ is the internal energy per unit mass and C_v is assumed constant.

The total energy, e, is the sum of the kinetic, potential, and internal energies (here each is divided by density):

$$e = \tfrac{1}{2}(u^2 + v^2) + gz + C_v T \qquad \text{(C.5)}$$

so that, regrouping the terms in the brackets on the right-hand side of (C.4) and using (A.24), (C.4) becomes

$$\frac{de}{dt} = -\frac{1}{\rho}\left[\nabla \cdot (pV) + \frac{\partial}{\partial z}(pw) \right] + V \cdot F + Q \qquad \text{(C.6)}$$

The equation of continuity (3.40) can be written as follows:

$$\frac{d\rho}{dt} = -\rho\left[\nabla \cdot V + \frac{\partial w}{\partial z} \right] \qquad \text{(C.7)}$$

If (C.6) is multiplied by ρ and (C.7) is multiplied by e, the sum is

$$\frac{d}{dt}(\rho e) = -\left[\nabla \cdot (pV) + \frac{\partial}{\partial z}(pw) \right] - \rho e\left[\nabla \cdot V + \frac{\partial w}{\partial z} \right]$$

$$+ \rho(V \cdot F + Q) \qquad \text{(C.8)}$$

Expanding the left-hand side of (C.8) and regrouping terms reduce (C.8) to

$$\frac{\partial(\rho e)}{\partial t} + \nabla \cdot (\rho e \mathbf{V}) + \frac{\partial}{\partial z}(\rho e w) = - \left[\nabla \cdot p\mathbf{V} + \frac{\partial}{\partial z}(pw) \right]$$

$$+ \rho(\mathbf{V} \cdot \mathbf{F} + Q) \qquad \text{(C.9)}$$

Integration of (C.9) over the entire globe with boundary conditions of $w = 0$ at the top ($z = z_T$) and bottom ($z = 0$) of the atmosphere yields

$$\frac{\partial}{\partial t} \int_{z=0}^{z=z_T} \int_0^{2\pi} \int_{-\pi/2}^{\pi/2} \rho e a^2 \cos \phi \, d\phi \, d\lambda \, dz$$

$$= \int_{z=0}^{z=z_T} \int_0^{2\pi} \int_{-\pi/2}^{\pi/2} [Q + \mathbf{V} \cdot \mathbf{F}] a^2 \cos \phi \, d\phi \, d\lambda \, dz \quad \text{(C.10)}$$

If there are no sources of frictional dissipation or nonadiabatic (diabatic) heating, Q, then the right-hand side of (C.10) becomes 0, so that energy is conserved exactly within the system as a whole. If the earth's orography is included, (C.10) should be modified by changing the lower limit on the integral to be a function of latitude and longitude, i.e., $z = H(\lambda, \phi)$.

Finite Difference
Barotropic Forecast Model

The finite difference numerical solution of the barotropic forecast model is fairly straightforward. It will be described in this
appendix, following the discussion in Holton (1979), and a computer code for its solution will be provided at the end of the
appendix. As mentioned in Chapter 3, the barotropic forecast
model was the first model solved with the electronic computer
in the late 1940s and early 1950s. It is based on a fundamental
equation in atmosphere and ocean dynamics, and its numerical
solution is a useful demonstration of the finite difference method
for the vorticity-type equation (see Chapters 3 and 4).

The barotropic vorticity equation can be written as

$$\frac{\partial \varsigma}{\partial t} = -\boldsymbol{V}_\psi \cdot \boldsymbol{\nabla}(\varsigma + f) = -u_\psi \frac{\partial}{\partial x}(\varsigma + f) - v_\psi \frac{\partial}{\partial y}(\varsigma + f) \quad \text{(D.1)}$$

where

$$\varsigma = \boldsymbol{\nabla}^2 \psi, \quad u_\psi = -\frac{\partial \psi}{\partial y}, \quad v_\psi = \frac{\partial \psi}{\partial x}, \quad \text{and} \quad \beta = \frac{\partial f}{\partial y} \quad \text{(D.2)}$$

Setting

$$F(x, y, t) = \frac{\partial \psi}{\partial x} \frac{\partial \nabla^2 \psi}{\partial y} - \frac{\partial \psi}{\partial y} \frac{\partial \nabla^2 \psi}{\partial x} + \beta \frac{\partial \psi}{\partial x} \qquad \text{(D.3)}$$

(D.1) can be rewritten as

$$\nabla^2 \left(\frac{\partial \psi}{\partial t} \right) + F(x, y, t) = 0 \qquad \text{(D.4)}$$

since $\partial f / \partial x = 0$ and $\partial (\nabla^2 \psi)/\partial t = \nabla^2 \partial \psi / \partial t$. If $F(x, y, t)$ is known in two-dimensional space then $(\partial \psi / \partial t)$ can be found from (D.4) at each point of that two-dimensional space. By using $\partial \psi / \partial t$ and ψ at some instant in time, a new value of ψ can be obtained. The computation of (D.4) can be repeated indefinitely into the future, albeit with some practical limitations. We will now obtain a finite difference form of (D.4) in order to allow the computations to be carried out.

Following Holton's notation, one could imagine an x-y two-dimensional grid structure where x has discrete values ranging from 0 to Md in the east-west direction and y has discrete values ranging from 0 to Nd in the north-south direction. Specifically, $x = md$, where d is the spacing between gridpoints and $m = 0, 1, 2, \ldots, M$; and $y = nd$, where $n = 0, 1, 2, \ldots, N$. The number of gridpoints in two dimensions is $(M + 1) \times (N + 1)$. The velocities in (D.1) are expressed as centered differences of the streamfunction:

$$u_\psi = -\frac{\partial \psi}{\partial y} \simeq -\frac{\psi_{m,n+1} - \psi_{m,n-1}}{2d} \qquad \text{(D.5)}$$

$$v_\psi = \frac{\partial \psi}{\partial x} \simeq \frac{\psi_{m+1,n} - \psi_{m-1,n}}{2d} \qquad \text{(D.6)}$$

and the vorticity can be put into finite difference form by taking uncentered approximations to the first derivatives of the streamfunction, for example $(\psi_{m+1,n} - \psi_{m,n})/d$ and $(\psi_{m,n} - \psi_{m-1,n})/d$, and then approximating the second derivatives by taking the second differences:

$$\varsigma = \nabla^2 \psi$$

$$= \frac{\partial^2 \psi}{\partial x^2} + \frac{\partial^2 \psi}{\partial y^2} \simeq \frac{\psi_{m+1,n} - 2\psi_{m,n} + \psi_{m-1,n}}{d^2}$$

$$+ \frac{\psi_{m,n+1} - 2\psi_{m,n} + \psi_{m,n-1}}{d^2}$$

$$= \frac{\psi_{m+1,n} + \psi_{m,n+1} + \psi_{m-1,n} + \psi_{m,n-1} - 4\psi_{m,n}}{d^2}$$

$$= \overset{2}{\nabla} \psi_{m,n} \tag{D.7}$$

We have followed Holton in using the notation $\overset{2}{\nabla} \psi_{m,n}$ for the finite difference form of $\nabla^2 \psi$ at point (m, n). This scheme has second-order accuracy (see Chapter 4).

As discussed in Chapter 4, a given function may have several different finite difference approximations, some of which may have specific advantages over others. In the case of $F(x, y, t)$ it is desirable to obtain a finite difference form that conserves both energy and average vorticity. This can be done by rewriting (D.3) as

$$F(x, y, t) = \frac{\partial}{\partial y}\left(\frac{\partial \psi}{\partial x}\nabla^2 \psi\right) - \frac{\partial}{\partial x}\left(\frac{\partial \psi}{\partial y}\nabla^2 \psi\right) + \beta\frac{\partial \psi}{\partial x} \tag{D.8}$$

and then taking centered differences about the gridpoint (m, n) to obtain the finite difference form

$$F_{m,n} = \frac{1}{4d^2}\Big[(\psi_{m+1,n+1} - \psi_{m-1,n+1})(\overset{2}{\nabla}\psi)_{m,n+1}$$

$$- (\psi_{m+1,n-1} - \psi_{m-1,n-1})(\overset{2}{\nabla}\psi)_{m,n-1}$$

$$- (\psi_{m+1,n+1} - \psi_{m+1,n-1})(\overset{2}{\nabla}\psi)_{m+1,n}$$

$$+ (\psi_{m-1,n+1} - \psi_{m-1,n-1})(\overset{2}{\nabla}\psi)_{m-1,n}\Big]$$

$$+ \frac{\beta}{2d}(\psi_{m+1,n} - \psi_{m-1,n}) \tag{D.9}$$

The finite difference form of (D.4) thus becomes

$$\left(\nabla^2\left(\frac{\partial\psi}{\partial t}\right)\right)_{m,n} + F_{m,n} = 0 \tag{D.10}$$

The solution of (D.10) requires inverting the finite difference Laplacian operator. Laplacian equations are common in physics and engineering and are solved by a wide variety of numerical techniques. The method used here is the so-called relaxation method, which is an iterative procedure in that a first guess is made and then the solution is improved by recursive repetition of the iteration until the solution satisfies (D.10) to some predetermined accuracy. The method begins by making an initial guess $(\partial\psi/\partial t)^\circ$ and substituting it into the left-hand side of (D.10) to obtain a residual:

$$\frac{1}{d^2}R_{m,n}^\circ = \nabla^2\left(\frac{\partial\psi}{\partial t}\right)^\circ + F_{m,n} \tag{D.11}$$

If $R_{m,n}^\circ$ is zero everywhere in m,n space, then (D.10) is satisfied exactly and the solution is $(\partial\psi/\partial t)^\circ$. Usually $R_{m,n}^\circ$ is not zero, however, and consequently it is reduced or "relaxed" by taking

$$\left(\frac{\partial\psi}{\partial t}\right)^1 = \left(\frac{\partial\psi}{\partial t}\right)^\circ + \frac{1}{4}R_{m,n}^\circ \tag{D.12}$$

as a second guess to the solution (this guess being selected because if $(\partial\psi/\partial t)^\circ$ were replaced by $(\partial\psi/\partial t)^1$ at the single point (m,n) while all other partial derivatives retained their $(\partial\psi/\partial t)^\circ$ values then (D.10) would be satisfied exactly). This guess is then substituted into (D.10) and a new residual $R_{m,n}^1$ is obtained. For subsequent iterations

$$\left(\frac{\partial\psi}{\partial t}\right)^{\nu+1} = \left(\frac{\partial\psi}{\partial t}\right)^\nu + \frac{1}{4}R_{m,n}^\nu \tag{D.13}$$

where the iteration cycle or counter is ν and each new residual $R_{m,n}^\nu$ is obtained from $(\partial\psi/\partial t)^\nu$ by

$$R_{m,n}^\nu = d^2\nabla^2\left(\frac{\partial\psi}{\partial t}\right)^\nu + d^2 F_{m,n} \tag{D.14}$$

The time derivatives in (D.14) are calculated with centered approximations as discussed in Chapter 4, that is,

$$\frac{\partial \psi}{\partial t} \simeq \frac{\psi^{t+\Delta t} - \psi^{t-\Delta t}}{2\Delta t} \tag{D.15}$$

where Δt is the time increment. Note that two time levels must be retained in (D.15). The Courant-Friedrichs-Levy restriction on the length of the time step (see Chapter 4) makes it important that the time step Δt be such that $(M\Delta t)/d \leq 1$, where M is the maximum of either $|u_\psi|$ or $|v_\psi|$ within the region.

The rate of convergence of the solution can be increased by converting from the method of simultaneous relaxation described above to a method of successive relaxation, as discussed in Holton (1979). Another technique to accelerate the convergence is to increase the 1/4 relaxation factor in (D.13). This latter modification is called over-relaxation. However, if the over-relaxation factor is too large the method will not converge to the true solution.

The following coding of the numerical barotropic forecast model is a modified version of a program provided by John E. Kutzbach of the University of Wisconsin, Department of Meteorology, the modifications having been made by Thomas W. Bettge of NCAR. The program is written in the BASIC programming language and is appropriate for IBM/IBM compatible desktop and other size computers using BASIC. The model requires initial inputs of the speed of the westerly flow, the east-west wave number, and the amplitude of the wave, and it can generate a 24-hour forecast within a few minutes of computing time on a small personal computer.

An IBM personal computer floppy diskette or a Macintosh micro floppy disk containing this program and the computer programs described in Appendices E and F can be purchased as explained at the end of Appendix F. The version of this numerical barotropic forecast model found on the Macintosh/Macintosh compatible floppy disk is a slightly altered form of the IBM code that follows.

```
10 LPRINT "BAROTROPIC FCST MODEL,based upon Holton, developed for Meteorology 32
3 at UNIV. of WISCONSIN by J.E.Kutzbach":LPRINT "-----------------------------
---------------------------------------------"
20 LPRINT "Barotropic model, mid-lat,4000km by 10000km,dx=500km,8 by 20 grid.":
   LPRINT "User sets initial stream function; model calculates the streamfunctio
n tendency and the new streamfunction.":LPRINT "-----------------------------
-------------"
30 LPRINT "1)specify e-w wavenumber (for example,2)"
40 LPRINT "2)specify n-s streamfunction gradient(for ex.,50 gives 20mps at 45N)"
50 LPRINT "3)specify ratio of n-s to e-w wind at wave inflection point(for ex.,
   0.1 gives a ratio of approx. 1 to 10 for wavenumber 2)"
60 LPRINT "4)specify time step(for ex.,6 gives 6 hour time step)"
65 LPRINT "5)specify the number of time steps"
70 LPRINT "---------------------------------":LPRINT "date of run="DATE$
80 LPRINT "--------Order of output------------":LPRINT "1)characteristics of ini
tial field":LPRINT "2)initial streamfunction field"
90 LPRINT "3)streamfunction tendency field":LPRINT "4)forecast streamfunction fi
eld":LPRINT "5)forecasts continue until specified number of time steps are compl
eted or until control-break is executed to terminate
100 LPRINT "----------------characteristics of initial field----------------"
110 DIM P(20,8),LP(20,8),F(20,8),GPT(20,8),R(20,8):REM P is psi (streamfunction)
   field;LP is Laplacian of psi;F is the vorticity advection term;GPT is   guessed
   psi tendency-which,upon iteration,becomes the psi tendency;R is residual
115 DIM SAV(20,8):REM SAVE is the psi field at T-1 used in centered time diff.
120 INPUT "what is the east-west wavenumber";K
130 INPUT "what is the n-s gradient";S:INPUT "what is the n-s/e-w ratio";R
140 INPUT "what is the time step";DT:TSTEP=DT*3600
145 INPUT "how many time steps";TSTEPS
150 BETA=.16*10^-10:DX=5*10^5:T=0
160 REM calculate psi field and other characteristic values of field
170 FOR M=0 TO 20
180 FOR N=0 TO 8
190 P(M,N)=S+S*COS(2*3.1416*N/16)+S*R*COS(K*2*3.1416*M/20)*SIN(2*3.1416*N/16)
200 NEXT N
210 NEXT M
220 U4= INT((P(0,3)-P(0,5))/2):V4= INT(S*R/(10/K)):KK=(2*3.1416*K/10^7)^2+(2*3.1
416/(8*10^6))^2:C=INT(U4-  BETA/KK):DKM=INT(C*24*3600/1000)
230 LPRINT "u-component at lat=45N is equal to  "U4"  meters/sec"
240 LPRINT "E-W wavenumber="K,10000/K "km wavelength"
250 LPRINT "Rossby wave speed=  "C" m/s":LPRINT "Rossby wave would move  "DKM"
   km in 24 hours."
260 LPRINT "avg. v-component btwn trough and ridge=  "V4"  meters/sec":LPRINT "-
----------------------------------------------------------------":LPRIN
T "stream function field, T=0, initial conditions":LPRINT "made at time="TIME$
270 LPRINT "************STREAMFUNCTION FIELD****************************"
280 FOR N=8 TO 0 STEP -1
290 FOR M=0 TO 20
300 LPRINT USING "###";P(M,N)/(S/5),
310 NEXT M
320 LPRINT " lat= " N*5+25" N"
330 NEXT N
340 LPRINT "************LONGITUDE****************************************"
350 REM compute Laplacian of psi
```

```
360 FOR M=1 TO 19
370 FOR N=1 TO 7
380 LP(M,N)=P(M+1,N)+P(M,N+1)+P(M-1,N)+P(M,N-1)-4*P(M,N):REM eq.8.30,Holton
390 NEXT N
400 NEXT M
410 REM compute vorticity advection
420 FOR M=1 TO 19
430 FOR N=1 TO 7
440 F(M,N)=(P(M+1,N+1)-P(M-1,N+1))*LP(M,N+1)-(P(M+1,N-1)-P(M-1,N-1))*LP(M,N-1)-
        (P(M+1,N+1)-P(M+1,N-1))*LP(M+1,N)+(P(M-1,N+1)-P(M-1,N-1))*LP(M-1,N)+
        2*DX*BETA*(P(M+1,N)-P(M-1,N))
450 F(M,N)=.25*F(M,N):F(M,N)=F(M,N)/DX:REM scaling to same as initial field-
    lines 450-460 are eq. 8.32 in Holton
460 NEXT N
470 NEXT M
480 IF T>0 THEN GOTO 550 ELSE GOTO 490:REM if T>0 use previous guess
490 REM set initial psi tendency guess to zero everywhere
500 FOR M=0 TO 20
510 FOR N=0 TO 8
520 GPT(M,N)=0
530 NEXT N
540 NEXT M
550 REM iterate to get relaxation solution-sequential relaxation,see Holton
560 REM g= guess number
570 G=0
580 G=G+1
590 REM compute residual
600 FOR N=1 TO 7
610 FOR M=1 TO 19
620 R(M,N)=F(M,N)+GPT(M+1,N)+GPT(M,N+1)+GPT(M-1,N)+GPT(M,N-1)-4*GPT(M,N):REM
    eq. 8.38,Holton
630 GPT(M,N)=GPT(M,N)+.25*R(M,N):REM eq. 8.37,Holton
640 NEXT M
650 NEXT N
660 REM print residuals
670 PRINT USING "###";G,
680 REM check size of residuals
690 FOR M=1 TO 19
700 FOR N=1 TO 7
710 IF ABS(.25*R(M,N))>9.999999E-06 THEN GOTO 750 ELSE GOTO 720
720 NEXT N
730 NEXT M
740 PRINT G: GOTO 770
750 REM not good enough
760 GOTO 580
770 REM convergence
780 REM boundary values-establishes cyclic e-w boundary condition
790 FOR N=1 TO 7
800 GPT(20,N)=GPT(19,N)
810 GPT(0,N)=GPT(19,N)
820 NEXT N
830 REM print tendencies
840 LPRINT "Relaxation solution required "G"  iterations for convergence"
850 LPRINT "streamfunction tendencies for next "DT" hours(10X actual values)" :
```

```
LPRINT "----------------- tendency field ----------------------------"
860 FOR N=8 TO 0 STEP -1
870 FOR M=0 TO 20
880 GPT(M,N)=TSTEP*GPT(M,N):LPRINT USING "###";GPT(M,N)*10/(S/5),
890 NEXT M
900 LPRINT " lat="N*5+25" N"
910 NEXT N
920 LPRINT "-----------longitude-------------------------------------"
930 REM new streamfunction  field-page 194, Holton (use forward time
       differencing on first step, centered differencing thereafter)
931 IF T=0 THEN GOTO 939 ELSE GOTO 982
939 REM forward time differencing
940 FOR N=0 TO 8
950 FOR M=0 TO 20
951 SAV(M,N)=P(M,N)
960 P(M,N)=P(M,N)+GPT(M,N):GPT(M,N)=GPT(M,N)/TSTEP:REM resets gpt
970 NEXT M
980 NEXT N
981 GOTO 991
982 REM centered time differencing
983 FOR N=0 TO 8
984 FOR M=0 TO 20
985 XX=P(M,N)
986 P(M,N)=SAV(M,N)+2*GPT(M,N)
987 GPT(M,N)=GPT(M,N)/TSTEP
988 SAV(M,N)=XX
989 NEXT M
990 NEXT N
991 REM iterate
1000 T=T+1
1010 LPRINT "streamfunction field forecast for   "T*DT"  hours":LPRINT "made at
       time="TIME$
1020 IF T>TSTEPS THEN GOTO 1060 ELSE GOTO 270
1030 REM this is a test
1040 REM Note from study of line 190 and 220 that P(psi) is 1/dx of true value.
However, the calculation of F(vorticity advection) in line 440 is correct for
true value of psi because (1/dx)^2 is deleted from Laplacian calculation (line
380).
1050 REM Similarly, in line 620, F is not multiplied by dx^2 because it was not
divided by dx^2 in line 450.  Finally, in the unnumbered statement after line
450, division by dx is required to keep scale of psi the same as defined in line
190.
1060 PRINT " Normal Termination of Barotropic Forecast Model"
1065 LPRINT " Normal Termination of Barotropic Forecast Model"
1070 END
```

Spectral Transform
Technique

This appendix describes and provides a computer listing for the solution to a barotropic forecast model using a spectral transform technique instead of the finite difference technique of Appendix D. The program was written by Joseph Tribbia of the National Center for Atmospheric Research and makes use of the nondivergent barotropic vorticity equation described in Chapter 3. As discussed earlier, this equation has been useful for understanding the basic dynamics of idealized atmospheric and oceanic flows. Examination of the computer code should be useful in understanding the spectral method and its associated transforms. The program is self-contained and can be run on small or large computers to generate computer simulations that can be used for forecasts or otherwise.

The basic prediction equation is given by

$$\frac{\partial \varsigma}{\partial t} = -\mathbf{V} \cdot \mathbf{\nabla}(\varsigma + f) \tag{E.1}$$

where ς is vorticity, f is the Coriolis parameter, $\mathbf{V} = \mathbf{i}u + \mathbf{j}v$ is the horizontal wind vector, and u and v are eastward and northward components of the wind, respectively, as discussed in Chapters 3 and 4 and Appendix A. As pointed out in Chapter 4, Robert (1966) suggested redefining the wind vector components as

$$U = u \cos \phi \qquad \text{and} \qquad V = v \cos \phi \tag{E.2}$$

in order to make them more suitable scalar fields that vanish at the poles [see (4.119) and (4.120) in Chapter 4].

The velocity vector involves only the streamfunction since the flow is assumed to be nondivergent. Thus, by eliminating the divergence terms from (4.131), (4.137), and (4.138),

$$\boldsymbol{V} = \boldsymbol{k} \times \boldsymbol{\nabla}\psi \tag{E.3}$$

with components

$$U = -\frac{\cos\phi}{a}\frac{\partial\psi}{\partial\phi} \quad \text{and} \quad V = \frac{1}{a}\frac{\partial\psi}{\partial\lambda} \tag{E.4}$$

As in (4.132), the vorticity is

$$\varsigma = \boldsymbol{k}\cdot(\boldsymbol{\nabla}\times\boldsymbol{V}) = \boldsymbol{\nabla}^2\psi \tag{E.5}$$

Furthermore, by eliminating the divergence term in (4.134), the spherical form of the nondivergent barotropic vorticity equation becomes

$$\frac{\partial}{\partial t}\boldsymbol{\nabla}^2\psi = -\frac{1}{a\cos^2\phi}\left[\frac{\partial}{\partial\lambda}(U\boldsymbol{\nabla}^2\psi) + \cos\phi\frac{\partial}{\partial\phi}(V\boldsymbol{\nabla}^2\psi)\right] - 2\Omega\frac{V}{a} \tag{E.6}$$

where Ω is the rotation rate of the earth and a is the mean radius of the earth. Although there are other forms of (E.6), this form is convenient for the transform technique.

The spectral representations of U, V, and ψ are given as

$$U = a\sum_{m=-J}^{+J}\sum_{\ell=|m|}^{|m|+J+1}U_\ell^m\,Y_\ell^m \tag{E.7}$$

$$V = a\sum_{m=-J}^{+J}\sum_{\ell=|m|}^{|m|+J+1}V_\ell^m\,Y_\ell^m \tag{E.8}$$

$$\psi = a^2\sum_{m=-J}^{+J}\sum_{\ell=|m|}^{|m|+J}\psi_\ell^m\,Y_\ell^m \tag{E.9}$$

where

$$Y_\ell^m = P_\ell^m(\sin\phi)e^{im\lambda} \tag{E.10}$$

P_ℓ^m is an associated Legendre polynomial of the first kind normalized to 1 (see Chapter 4 and Appendix B). The coefficients ψ_ℓ^m, U_ℓ^m, and V_ℓ^m are time-dependent expansion coefficients which are complex. Note that since ψ is differentiated to obtain U and V by (E.4), the truncated series for U and V have one more term than the series for ψ. After using recursive relationships of spherical harmonics (as described in Abramowitz and Stegun, 1964), substitution of (E.7) and (E.8) into (E.5) yields the following:

$$U_\ell^m = (\ell - 1)\epsilon_\ell^m \psi_{\ell-1}^m - (\ell + 2)\epsilon_{\ell+1}^m \psi_{\ell+1}^m \qquad \text{(E.11)}$$

and

$$V_\ell^m = im\psi_\ell^m \qquad \text{(E.12)}$$

where

$$\epsilon_\ell^m = [(\ell^2 - m)/(4\ell^2 - 1)]^{1/2} \qquad \text{(E.13)}$$

The spectral fields of U, V, and $\nabla^2 \psi$ are transformed to a grid on the sphere, and products in parentheses in (E.6) can then be represented as truncated Fourier series at each latitude circle as

$$U\nabla^2\psi = a \sum_{m=-J}^{+J} A_m e^{im\lambda} \qquad \text{(E.14)}$$

$$V\nabla^2\psi = a \sum_{m=-J}^{+J} B_m e^{im\lambda} \qquad \text{(E.15)}$$

The inverse or backward transform of (E.14) and (E.15) gives the values of A_m and B_m.

Returning to the original forecast equation, substitution of (E.9) into (E.6) yields

$$-\ell(\ell + 1)\left(\frac{\partial \psi_\ell^m}{\partial t}\right) = C_\ell^m - 2i\Omega m\psi_\ell^m \qquad \text{(E.16)}$$

where C_ℓ^m is the nonlinear term denoted by square brackets in (E.6). If (E.14)–(E.15) and (E.9) are substituted into (E.16) and the orthogonality relationship is used, then

$$C_\ell^m = \int_{-\pi/2}^{+\pi/2} \frac{1}{\cos^2\phi}\left(imA_m + \cos\phi\frac{\partial B_m}{\partial\phi}\right) P_\ell^m(\sin\phi)\cos\phi\,d\phi \qquad \text{(E.17)}$$

The second term in parentheses can be integrated by parts with the boundary condition $B_m(\pm\pi/2) = 0$, which results from V being zero at both poles; thus

$$C_\ell^m = \int_{-\pi/2}^{+\pi/2} \frac{1}{\cos^2\phi}\left[imA_mP_\ell^m(\sin\phi)\right.$$

$$\left. - B_m\cos\phi\frac{\partial P_\ell^m}{\partial\phi}(\sin\phi)\right]\cos\phi\,d\phi \qquad (\text{E.18})$$

Equation (E.18) is the Legendre transform in the meridional direction. Its solution is made easy if A_m and $P_\ell^m(\sin\phi)$ are known at Gaussian latitudes (see Appendix B). There are some restrictions on the resolution of the spherical harmonics and the gridpoint resolution in order to obtain alias-free computations. Generally, there must be $3J - 1$ gridpoints in the longitudinal direction to obtain alias-free results, where J is the rhomboidal wave number truncation. In the latitudinal direction the number of gridpoints should be $> (5J - 1)/2$.

The above system (E.7) to (E.18), developed by Bourke (1972), is one of the most often used methods for solving the primitive equations by the spectral technique. However, Machenhauer (1979) has developed a variant of this method for the barotropic vorticity equation, and it is this variant that has been programmed by J. Tribbia and is presented below. For convenience in the subsequent FORTRAN program of the spectral model, (E.6) is changed to a nondimensionalized equation by using the radius of the earth as the length scale and $(2\Omega)^{-1}$ as the time scale so that the nondimensional form of (E.6) on the sphere becomes

$$\nabla^2\psi_t = -(\psi_\lambda\nabla^2\psi_\mu - \psi_\mu\nabla^2\psi_\lambda) - \psi_\lambda \qquad (\text{E.19})$$

and ς is represented as a spectral expansion for rhomboidal $J = 15$:

$$\varsigma(\lambda,\mu,t) = \sum_{m=-15}^{15}\sum_{\ell=|m|}^{15+|m|} V(\ell,m,t)e^{im\lambda}P_\ell^m(\mu) \qquad (\text{E.20})$$

where $\mu = \sin\phi$ and V is the vorticity coefficient. For ease of indexing and vectorization, the double index (ℓ,m) is converted to a single index, k, with $k = (m)\cdot 16 + (\ell - m) + 1$. For a given

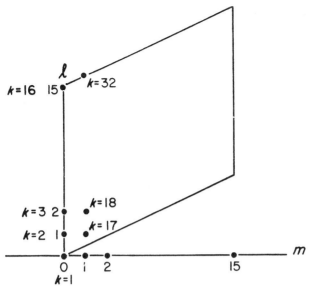

Fig. E.1 Diagram showing the relationship in the computer program between the longitudinal planetary wave number m, the meridional wave number $\ell - m$, and the index k, where $k = (m) \cdot 16 + (\ell - m) + 1$.

k, m_k and ℓ_k are calculated sequentially in integer arithmetic as $m_k = (k - 1)/16$ and $\ell_k = k - 16 \cdot m_k - 1$. Figure E.1 shows a schematic of the index, where m ranges from 0 to 15. Using this double index k, (E.20) becomes

$$\zeta(\lambda, \mu, t) = \sum_{k=1}^{kmax} [V(k,t)Y_k + (1 - \delta_{m_k,o})V^*(k,t)Y_k^*] \quad \text{(E.21)}$$

while (E.10) becomes

$$Y_k \equiv Y_k(\lambda, \mu) \equiv P_{\ell_k}^{m_k}(\mu) e^{im_k \lambda} \quad \text{(E.22)}$$

and $*$ denotes the complex conjugate. [Since $\zeta(\lambda, \mu, t)$ is real, only the $m_k \geq 0$ components need be predicted; $V(-|m_k|, \ell_k) = V^*(+|m_k|, \ell_k)$]. The streamfunction $\psi(\lambda, \mu, t)$ is related to $\zeta(\lambda, \mu, t)$ through (E.5). Spectrally this relationship is given as

$$\psi(\lambda, \mu, t) = \sum_{k=1}^{kmax} [PS(k)Y_k + (1 - \delta_{m_k,o})PS^*(k)Y_k^*]$$

$$= \sum_{k=0}^{kmax} CA(k)[V(k)Y_k + (1 - \delta_{m_k,o})V^*(k)Y_k^*] \quad \text{(E.23)}$$

where PS is the streamfunction and $CA(k) = -1/(\ell_k(\ell_k + 1))$.

Description of the FORTRAN Computer Program
—Barotropic Vorticity Equation (BVE)

Main program—BVE

Purpose: Set up a time integration of the spectral form of the governing equation (E.19)

$$\frac{dV(k)}{dt} = F(k) - BF(k) * PS(k) \qquad \text{(E.24)}$$

where

$$\frac{dV(k)}{dt} = (\nabla^2 \psi_t)_k \qquad \text{(E.25)}$$

The longitudinal and meridional spectral transform of the Jacobian is

$$F(k) = (J(\varsigma, \psi))_k = (\psi_\mu \varsigma_\lambda - \psi_\lambda \varsigma_\mu)_k \qquad \text{(E.26)}$$

and

$$BF(k) * PS(k) = im_k PS(k) \qquad \text{(E.27)}$$

is the spectral transform of ψ_λ, in which $BF(k) = im_k$.

Program scalars

For rhomboidal truncation 15 $(J = 15)$, with alias-free 40 × 48 latitude-longitude grid, the parameter identifications are:

IMAX	–	number of gridpoints in λ-direction (48 plus 2 additional wrap-around points at the ends of the grid)
JMAX	–	number of gridpoints in the meridional direction (40 Gaussian latitudes)
KMAX	–	total number of complex modes ($16^2 = 256$)
MMAX	–	array size of Fourier modes in longitude direction (25, although not all are used)

Program scalars—continued

MP	–	number of Fourier modes in longitude direction where MP < 2 MMAX/3 is required for alias-free calculation
NMAX	–	number of complex modes
IM2	–	number of nonredundant gridpoints in the λ direction (48, where there are no wrap-around points)
DT	–	time step (1 hr)
KPL	–	plotting or printout every KPL time step
KRS	–	time step parameter to determine frequency of "smooth start"
KTMAX	–	maximum number of time steps
NF	–	number of prime factors in FFT

Program vectors and arrays

AZ(I,J), AR(J,I), CZL(I,J), CZM(I,J), DZL(I,J), DZM(I,J), FZ(I,J)		
	–	Work space storage arrays for gridpoint fields (Note that AR(J,I) has reverse indexes)
GLATS(J)	–	GAUSSIAN latitudes
GWTS(J)	–	GAUSSIAN weights
P(J,K)	–	associated Legendre polynomial where $m = m_k$ and $\ell = \ell_k$ at μ_j
DP(J,F)	–	derivative of the associated Legendre polynomial with respect to μ at μ_j
CA(K)	–	$-1/(\ell_k(\ell_k + 1))$
V(K)	–	current time coefficients of vorticity
VM(K)	–	coefficients of vorticity at $t - \Delta t$
VS(K)	–	switching array between V(K) and VM(K) at the two time levels
PS(K)	–	current time coefficients of streamfunction
F(K)	–	coefficients of the Jacobian $J(\varsigma, \psi)$
BF(K)	–	im_k, Beta term (scaled)
IFAC(NF)	–	array of prime factors in FFT
NL(M)	–	number of complex coefficients

Subroutines and functions

VORPSI(V,PS...)	–	gives transform of $\psi(PS(K))$ from the transform of $\varsigma(V(K))$
JAC(V,PS,F...)	–	the transform of $J(\varsigma, \psi)(F(K))$ from the transform of $\varsigma(V(K))$ and the transform of $\psi(PS(K))$
COFG(V,AZ...)	–	calculates the model grid representation of $\varsigma(AZ(I,J))$ from the coefficients of $\varsigma(V(K))$
PSIVOR(PS,V...)	–	calculates coefficients of $\varsigma(V(K))$ from the coefficients of $\psi(PS(K))$
DDM(A,AZ...)	–	computes the gridpoint values of the meridional derivative, $(\partial/\partial\mu)$, of a field from its spectral coefficients $(A(K))$ and stores the field's meridional derivative in AZ(I,J)
DDL(A,AZ...)	–	computes the gridpoint values of the longitudinal derivatives $(\partial/\partial\lambda)$ of a field from its spectral coefficients $(A(K))$ and stores the longitudinal derivative in AZ(I,J)
GCOF(AZ,A...)	–	calculates the coefficients of a field $(A(K))$ from the gridpoint values of that field AZ(I,J)
PLOT	–	produces a latitude-longitude printout of any field Z(IMAX,JMAX)
LGTST	–	main initial subroutine that calls subsequent subroutines that compute Gaussian latitudes and weights, Legendre polynomials, and derivatives of Legendre polynomials
LEGTBL	–	computes Gaussian latitudes
LGNDR	–	computes Legendre polynomials
DYDMU	–	computes derivative Legendre polynomials
GAUSL1	–	computes Gaussian weights and abscissas
GAUSL2	–	used by GAUSL1 to compute Gaussian weights and abscissas
GAUSL3	–	used by GAUSL1 to compute Gaussian weights and abscissas
FFTIN	–	initial Fast Fourier Transform (FFT) subroutine call to obtain prime factors of number of points to be transformed

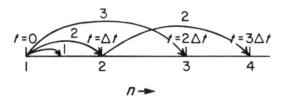

Fig. E.2 Schematic of the time differencing scheme for the "smooth start" method.

Subroutines and functions—continued

SFFT	–	computes forward FFT (i.e., grid data to spectral data) or backward FFT (i.e., spectral data to grid data)
PASSG	–	performs forward and backward transform
DCF	–	solves (E.13)

The time-stepping algorithm is centered with the so-called "smooth start method" as shown schematically in Fig. E.2. (The smooth restart is performed once each day, and there are 25 iterations per day.) The first time step is really a forward 1/2 time step, the results of which are used in the second time step to obtain values at $N = 2$. After that, centered time steps of 1 hour are used.

The FORTRAN program that follows has been run on several computers. The Fast Fourier Transform (FFT) used in (E.14)–(E.15) is a standard mathematical technique that is available in some form on many computers, ranging from supercomputers to home computers. Swarztrauber (1984) compares several FORTRAN programs that compute FFTs including the simple one used in the accompanying program. Users who have available FFT subroutines that are more efficient on their particular computers can substitute the alternative subroutine to reduce the overall computer time.

Once the user has programmed the model into his computer, there are many experiments that can be done, such as changing the wave number in the east-west direction, m, and the north-south direction, $(\ell - m)$, to study Rossby wave propagation, or adding forcing terms to simulate mountain and heating/cooling effects on upper tropospheric flows. On a minicomputer the running time for the example shown takes about 15 minutes per day of simulation. The parameters in the program can be set at almost any configuration. For illustrative purposes the initial condition in the current version assumes that there are five Rossby waves in the east-west direction and one wave in the north-south direction; thus $m = 5$, $\ell - m = 1$, and $k = 82$ with an amplitude of 0.1. It is further assumed that the Rossby wave is superimposed on a basic flow, with $m = 0$, $\ell - m = 1$, and $k = 2$ with a scaling of $4/84$. Figures E.3a, b, c show the streamfunction after day 1, day 2, and day 3. The wave is propagating to the west. Figures E.3d, e, f show another pattern in which a Rossby five-wave pattern with $k = 82$ and amplitude $= 0.9$ is superimposed on a Rossby one-wave pattern with $m = 1, \ell - m = 1$, $k = 18$, and amplitude $= 0.05$. The basic flow has a constant scaling of $8/84$.

An IBM personal computer diskette or a Macintosh micro floppy disk containing this and the computer programs of Appendices D and F can be purchased as explained at the end of Appendix F.

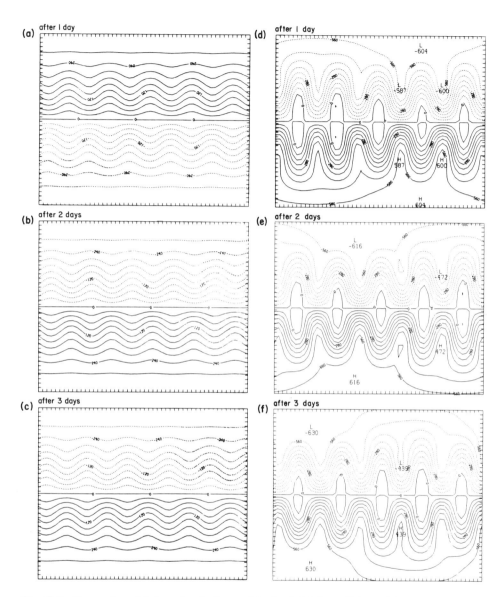

Fig. E.3 Examples of 3-day simulations with the initial condition of five Rossby waves in the east-west direction and one wave in the north-south direction $(m = 5, \ell - m = 1)$: (a-c) the streamfunction after day 1, day 2, and day 3, respectively, showing a weak westerly basic flow; (d-f) the streamfunction after day 1, day 2, and day 3 for a superposition of a Rossby five-wave pattern and a Rossby one-wave pattern with a strong basic flow.

```
      PROGRAM BVE
C *** BVE IS A BAROTROPIC SPECTRAL MODEL PROGRAMMED BY J. TRIBBIA, NCAR
      PARAMETER(IMAX=50,JMAX=40,KMAX=256,MMAX=25,MP=16,IM2=48,NF=3)
C *** NOTE: IF CHANGES ARE MADE IN THE RESOLUTION PARAMETER, ALSO MAKE
C *** CHANGES IN SUBROUTINE LGTST
      DIMENSION AZ(IMAX,JMAX),AR(JMAX,IMAX),P(JMAX,KMAX)
      DIMENSION DP(JMAX,KMAX),NL(MP),WK(IMAX,JMAX)
      DIMENSION GUSW(JMAX),GUSL(JMAX)
      DIMENSION CA(KMAX),PDP(JMAX,MP)
      DIMENSION CZL(IMAX,JMAX),CZM(IMAX,JMAX),DZL(IMAX,JMAX)
      DIMENSION DZM(IMAX,JMAX),FZ(IMAX,JMAX),IFAC(3)
      COMPLEX V(KMAX),PS(KMAX),F(KMAX),AKP(KMAX)
      COMPLEX AP(JMAX,MMAX),CT(IM2),W(IM2*3),DTF
      COMPLEX VM(KMAX),VS(KMAX),BF(KMAX)
      CHARACTER IDAY
      IU=10
      DO 9 K=1,KMAX
      V(K)=(0.,0.)
    9 CONTINUE
      DO 10 M=1,MP
      NL(M)=MP
   10 CONTINUE
C *** FIRST COMPUTE LEGENDRE POLYNOMIAL, GAUSSIAN WEIGHTS, AND LATITUDES
      CALL LGTST(GUSL,GUSW,CA,P,DP)
      CALL FFTIN(NF,IFAC)
      K=0
      DO 19 M=1,MP
      XM2=FLOAT(M-1)
      NLM=NL(M)
      DO 19 N=1,NLM
      K=K+1
      BF(K)=CMPLX(0.,XM2)
   19 CONTINUE
C *** SET INITIAL CONDITION FOR 5 EAST-WEST ROSSBY WAVES, V(82),
C *** ON A CONSTANT WESTERLY FLOW, V(2)
      S=4./84.
C *** NORMALIZE WITH TOTAL NUMBER OF POINTS FOR FFT
      V(2)=CMPLX(S,0.)*FLOAT(IM2)
      V(82)=CMPLX(.1,0.)*FLOAT(IM2)
      PI=4.*ATAN(1.)
      CALL VORPSI(V,PS,CA,KMAX)
      DT=PI/6.
C *** TO MAKE 20-DAY RUN, KTMAX=25*20
      KTMAX=25*1
C *** TO MAKE PLOT EVERY 5 DAYS, KINC=5
      KINC=1
      KPL=25*KINC
      KRS=25
      DTF=CMPLX(.5*DT,0.)
      DO 95 K=1,KMAX
      VM(K)=V(K)
   95 CONTINUE
      DO 96 KT=1,KTMAX
      KR=MOD(KT,KRS)
```

```
      CALL JAC(V,PS,F,CZL,DZL,CZM,DZM,FZ,AKP,AP,AR,P,DP,NL,WK,GUSW,
     +IMAX,JMAX,KMAX,MMAX,MP,IM2,CT,NF,IFAC,W)
      DO 99 K=1,KMAX
      VS(K)=V(K)
      V(K)=VM(K)+DTF*(F(K)-BF(K)*PS(K))
   99 CONTINUE
      IF(KR.LT.3) DTF=CMPLX(2.,0.)*DTF
      IF(KR.LT.3) GO TO 97
      DO 98 KK=1,KMAX
      VM(KK)=VS(KK)
   98 CONTINUE
   97 CONTINUE
      KPZ=MOD(KT,KPL)
      CALL VORPSI(V,PS,CA,KMAX)
      IF(KR.NE.0) GO TO 81
      DO 80 KK=1,KMAX
      VM(KK)=V(KK)
   80 CONTINUE
      DTF=CMPLX(.5*DT,0.)
   81 CONTINUE
C
      WRITE (6,1289) KPZ
 1289 FORMAT (1X,' KPZ=',I2)
C
      IF(KPZ.NE.0) GO TO 96
C
      CALL COFG(V,AZ,AP,AR,P,NL,WK,IMAX,JMAX,KMAX,MMAX,
     +MP,IM2,CT,NF,IFAC,W)
      DAY=(KT/KPL)*KINC
      DO 1010 JJ=1,JMAX
      AZ(IMAX-1,JJ)=AZ(1,JJ)
 1010 AZ(IMAX,JJ)=AZ(2,JJ)
      WRITE(6,2000)
 2000 FORMAT('0','VORTICITY VALUES:')
      CALL PLOT(AZ,IMAX,JMAX,KT,KTMAX,DAY)
C
      CALL COFG(PS,AZ,AP,AR,P,NL,WK,IMAX,JMAX,KMAX,MMAX,
     +MP,IM2,CT,NF,IFAC,W)
      DO 1020 JJ=1,JMAX
      AZ(IMAX-1,JJ)=AZ(1,JJ)
 1020 AZ(IMAX,JJ)=AZ(2,JJ)
      WRITE(6,2001)
 2001 FORMAT('0','STREAMFUNCTION VALUES:')
      CALL PLOT (AZ,IMAX,JMAX,KT,KTMAX,DAY)
C
   96 CONTINUE
   12 CONTINUE
   89 CONTINUE
      WRITE (6,2002)
 2002 FORMAT ('0',' ************* END OF PROGRAM BVE  *************')
      STOP
      END
      SUBROUTINE PLOT (Z,IMAX,JMAX,KT,KTMAX,DAY)
C *** SUBROUTINE TO MAKE TWO-DIMENSIONAL PLOTS OF FIELDS
```

```
      DIMENSION Z(IMAX,JMAX)
      DO 1 I1=1,IMAX,10
      I2=I1+9
      WRITE(6,100)KT,KTMAX,DAY,I1,I2
  100 FORMAT(1X,' KT=',I4,'   KTMAX=',I5,'   DAY=',F5.1,
     +'   I1=',I2,'    I2=',I2)
      DO 2 J=1,JMAX
      L=JMAX+1-J
    2 WRITE (6,400) (Z(I,L),I=I1,I2)
    1 CONTINUE
  400 FORMAT(1X,10F6.3)
      RETURN
      END
      SUBROUTINE PSIVOR(PS,V,CA,KMAX)
C *** SUBROUTINE TO COMPUTE LAPLACIAN OF STREAMFUNCTION
      COMPLEX V(KMAX),PS(KMAX)
      DIMENSION CA(KMAX)
      DO 10 K=2,KMAX
      V(K)=PS(K)/CMPLX(CA(K),0.)
   10 CONTINUE
      RETURN
      END
      SUBROUTINE VORPSI(V,PS,CA,KMAX)
C *** SUBROUTINE TO COMPUTE STREAMFUNCTION FROM THE LAPLACIAN
C *** OF STREAMFUNCTION
      COMPLEX V(KMAX),PS(KMAX)
      DIMENSION CA(KMAX)
      DO 10 K=1,KMAX
      PS(K)=CMPLX(CA(K),0.)*V(K)
   10 CONTINUE
      RETURN
      END
      SUBROUTINE DDM(A,AZ,AP,AR,DP,NL,WK,IMAX,JMAX,KMAX,MMAX,
     +MP,IM2,CT,NF,IFAC,W)
C *** SUBROUTINE TO COMPUTE LATITUDE DERIVATIVE
      COMPLEX A(KMAX),AP(JMAX,MMAX),CT(IM2),W(IM2*3)
      DIMENSION AZ(IMAX,JMAX),DP(JMAX,KMAX),NL(MP)
      DIMENSION WK(IMAX,JMAX),AR(JMAX,IMAX),IFAC(NF)
      DO 10 M=1,MMAX
      DO 10 J=1,JMAX
      AP(J,M)=CMPLX(0.,0.)
   10 CONTINUE
      K=0
      DO 20 M=1,MP
      NLM=NL(M)
      DO 20 N=1,NLM
      K=K+1
      DO 20 J=1,JMAX
      AP(J,M)=AP(J,M)+A(K)*CMPLX(DP(J,K),0.)
   20 CONTINUE
      DO 30 J=1,JMAX
      DO 24 I=1,IM2
      CT(I)=CMPLX(0.,0.)
   24 CONTINUE
```

```
      DO 25 I=1,MP
      CT(I)=AP(J,I)/FLOAT(IM2)
   25 CONTINUE
      CALL SFFT(1,IM2,NF,IFAC,CT,W)
      DO 26 I=1,IM2
      AZ(I,J)=REAL(CT(I))
   26 CONTINUE
   30 CONTINUE
      RETURN
      END
      SUBROUTINE GCOF(AZ,A,AP,AR,P,NL,WK,IMAX,JMAX,KMAX,MMAX,
     +MP,IM2,CT,NF,IFAC,W,GUSW)
C *** SUBROUTINE TO TRANSFORM GRIDPOINT VALUES TO SPECTRAL VALUES
      COMPLEX A(KMAX),AP(JMAX,MMAX),CT(IM2),W(IM2*3)
      DIMENSION AZ(IMAX,JMAX),AR(JMAX,IMAX),P(JMAX,KMAX)
      DIMENSION NL(MP),WK(IMAX,JMAX),GUSW(JMAX)
      DIMENSION IFAC(NF)
      DO 10 J=1,JMAX
      DO 15 I=1,IM2
      CT(I)=CMPLX(AZ(I,J),0.0)
   15 CONTINUE
      CALL SFFT(-1,IM2,NF,IFAC,CT,W)
      FAC=CMPLX(1.,0.)
      DO 16 M=1,MP
      IF(M.GT.1) FAC=CMPLX(2.,0.)
      AP(J,M)=CT(M)*FAC
   16 CONTINUE
   10 CONTINUE
      DO 26 K=1,KMAX
      A(K)=CMPLX(0.,0.)
   26 CONTINUE
      K=0
      DO 20 M=1,MP
      NLM=NL(M)
      DO 20 N=1,NLM
      K=K+1
      DO 20 J=1,JMAX
      A(K)=A(K)+AP(J,M)*CMPLX(P(J,K)*GUSW(J),0.)
   20 CONTINUE
      RETURN
      END
      SUBROUTINE DDL(A,AZ,AKP,AP,AR,P,NL,WK,IMAX,JMAX,KMAX,M
     +MAX,MP,IM2,CT,NF,IFAC,W)
C *** SUBROUTINE TO COMPUTE THE LONGITUDE DERIVATIVE
      COMPLEX AP(JMAX,MMAX),CT(IM2),W(IM2*3)
      COMPLEX A(KMAX),XIM,AKP(KMAX)
      DIMENSION AZ(IMAX,JMAX),NL(MP),P(JMAX,KMAX)
      DIMENSION WK(IMAX,JMAX),AR(JMAX,IMAX),IFAC(NF)
      K=0
      DO 10 M=1,MP
      XM=FLOAT(M-1)
      XIM=CMPLX(0.,XM)
      NLM=NL(M)
      DO 10 N=1,NLM
      K=K+1
      AKP(K)=XIM*A(K)
```

```
   10 CONTINUE
      CALL COFG(AKP,AZ,AP,AR,P,NL,WK,IMAX,JMAX,KMAX,MMAX,
     +MP,IM2,CT,NF,IFAC,W)
      RETURN
      END
      SUBROUTINE COFG(A,AZ,AP,AR,P,NL,WK,IMAX,JMAX,KMAX,MMAX,
     +MP,IM2,CT,NF,IFAC,W)
C *** SUBROUTINE TO TRANSFORM SPECTRAL VALUES TO GRIDPOINT VALUES
      COMPLEX A(KMAX),AP(JMAX,MMAX),CT(IM2),W(IM2*3)
      DIMENSION AZ(IMAX,JMAX),P(JMAX,KMAX),NL(MP)
      DIMENSION WK(IMAX,JMAX),AR(JMAX,IMAX),IFAC(NF)
      DO 10 M=1,MMAX
      DO 10 J=1,JMAX
      AP(J,M)=CMPLX(0.,0.)
   10 CONTINUE
      K=0
      DO 20 M=1,MP
      NLM=NL(M)
      DO 20 N=1,NLM
      K=K+1
      DO 20 J=1,JMAX
      AP(J,M)=AP(J,M)+A(K)*CMPLX(P(J,K),0.)
   20 CONTINUE
      DO 30 J=1,JMAX
      DO 24 I=1,IM2
   24 CT(I)=CMPLX(0.,0.)
      DO 25 I=1,MP
      CT(I)=AP(J,I)/FLOAT(IM2)
   25 CONTINUE
      CALL SFFT(1,IM2,NF,IFAC,CT,W)
      DO 26 I=1,IM2
      AZ(I,J)=REAL(CT(I))
   26 CONTINUE
   30 CONTINUE
      RETURN
      END
      SUBROUTINE JAC(C,D,F,CZL,DZL,CZM,DZM,FZ,AKP,AP,AR,P,DP,NL,WK,GUSW,
     +IMAX,JMAX,KMAX,MMAX,MP,IM2,CT,NF,IFAC,W)
C *** SUBROUTINE TO COMPUTE THE JACOBIAN TERM
      COMPLEX C(KMAX),D(KMAX),F(KMAX),CT(IM2),W(IM2*3)
      DIMENSION CZL(IMAX,JMAX),CZM(IMAX,JMAX),DZL(IMAX,JMAX),DZM(IMAX,JM
     +AX),FZ(IMAX,JMAX)
      COMPLEX AP(JMAX,MMAX),AKP(KMAX)
      DIMENSION AR(JMAX,IMAX),P(JMAX,KMAX),DP(JMAX,KMAX),NL(MP)
      DIMENSION WK(IMAX,JMAX),GUSW(JMAX),IFAC(NF)
      CALL DDM(C,CZM,AP,AR,DP,NL,WK,IMAX,JMAX,KMAX,MMAX,
     +MP,IM2,CT,NF,IFAC,W)
      CALL DDM(D,DZM,AP,AR,DP,NL,WK,IMAX,JMAX,KMAX,MMAX,
     +MP,IM2,CT,NF,IFAC,W)
      CALL DDL(C,CZL,AKP,AP,AR,P,NL,WK,IMAX,JMAX,KMAX,M
     +MAX,MP,IM2,CT,NF,IFAC,W)
      CALL DDL(D,DZL,AKP,AP,AR,P,NL,WK,IMAX,JMAX,KMAX,M
     +MAX,MP,IM2,CT,NF,IFAC,W)
      DO 10 J=1,JMAX
```

```
      DO 10 I=1,IM2
      FZ(I,J)=CZL(I,J)*DZM(I,J)-CZM(I,J)*DZL(I,J)
   10 CONTINUE
      CALL GCOF(FZ,F,AP,AR,P,NL,WK,IMAX,JMAX,KMAX,MMAX,
     +MP,IM2,CT,NF,IFAC,W,GUSW)
      RETURN
      END
      SUBROUTINE LGTST(GLATS,GWTS,CA,P,DP)
C *** SUBROUTINE TO COMPUTE GAUSSIAN LATITUDES, WEIGHTS AND
C *** LEGENDRE POLYNOMIALS, AND THE DERIVATIVE LEGENDRE POLYNOMIALS
      PARAMETER(MMAX=16,NMAX=16,JMAX=40)
      DIMENSION GWTS(JMAX),GLATS(JMAX),P(JMAX,NMAX,MMAX),P1(JMAX,NMAX)
      DIMENSION P2(JMAX,NMAX),P3(JMAX,NMAX),COA(JMAX),SIA(JMAX)
      DIMENSION DP(JMAX,NMAX,MMAX)
      DIMENSION NN(NMAX)
      DIMENSION CA(NMAX,MMAX)
      CALL LEGTBL(MMAX,NMAX,JMAX,GWTS,GLATS,P,P1,P2,P3,COA,SIA)
      CALL DYDMU(MMAX,JMAX,NMAX,P,DP,COA,SIA)
      DO 15 M=1,MMAX
      MT=M
      DO 15 N=1,NMAX
      XC=FLOAT(MT+N-2)*FLOAT(MT+N-1)
      IF(XC.LE.0.00001) XC=1.E10
      CA(N,M)=-1./XC
   15 CONTINUE
C *** NMAX,MMAX = RHOMBOIDAL TRUNCATION (E.G., R15+1)
C *** FOR ALIAS-FREE LATITUDE/LONGITUDE GRID, JMAX=40
C *** GWTS IS GAUSSIAN WEIGHTS AND GLATS IS GAUSSIAN LATITUDES
C *** P IS LEGENDRE POLYNOMIALS
C *** DP IS DIFFERENTIATED LEGENDRE POLYNOMIALS
      WRITE(6,401)
      WRITE(6,402)GLATS
  401 FORMAT(1X,'GAUSSIAN LATITUDES')
  402 FORMAT(1X,5E15.8)
      WRITE(6,403)
      WRITE(6,402)GWTS
  403 FORMAT(1X,'GAUSSIAN WEIGHTS')
      RETURN
      END
      SUBROUTINE LEGTBL(MMAX,NMAX,JMAX,GWTS,GLATS,P,P1,P2,P3,COA,SIA)
C *** SUBROUTINE TO COMPUTE GAUSSIAN LATITUDES
      DIMENSION GWTS(JMAX),GLATS(JMAX),P(JMAX,NMAX,MMAX),P1(JMAX,NMAX)
      DIMENSION P2(JMAX,NMAX),P3(JMAX,NMAX),COA(JMAX),SIA(JMAX)
      CALL GAUSL1(JMAX,-1.,1.,1,GWTS,GLATS,IER)
      DO 1 J=1,JMAX
      COA(J)=GLATS(J)
      SIA(J)=SQRT(1.-COA(J)*COA(J))
    1 CONTINUE
      MC=MMAX/3+1
      DO 10 M=1,MC
      MP=3*M-1
      MT=MP
      IF(MP-1.GT.MMAX) GO TO 10
      CALL LGNDR(JMAX,MT,NMAX,P1,P2,P3,COA,SIA)
```

```
      DO 5 N=1,NMAX
      DO 5 J=1,JMAX
      P(J,N,MP-1)=P1(J,N)
      IF(MP.GT.MMAX) GO TO 8
      P(J,N,MP)=P2(J,N)
      IF(MP+1.GT.MMAX) GO TO 8
      P(J,N,MP+1)=P3(J,N)
    8 CONTINUE
    5 CONTINUE
   10 CONTINUE
      RETURN
      END
      SUBROUTINE LGNDR(LATS,MMAX,NMAX,PNM1,PNM2,PNM3,COA,SIA)
C *** THIS SUBROUTINE IS BASED ON BELOUSOV'S
C *** "TABLES OF NORMALIZED ASSOCIATED LEGENDRE POLYNOMIALS,"
C *** MATHEMATICAL TABLES SER.,VOL.18,PERGAMON PRESS,N.Y.,1962,379PP.
C *** THE PROGRAM USES VARIOUS RECURRENCE FORMULAE TO CALCULATE
C *** NORMALIZED ASSOCIATED LEGENDRE POLYNOMIALS EVALUATED AT THE
C *** COSINES OF A SET OF SPECIFIED CO-LATITUDES FOR A GIVEN RANGE
C *** OF M AND (N-M)
C *** PARAMETER DEFINITIONS:
C *** LATS: NUMBER OF CO-LATITUDES
C *** MMAX: MAXIMUM DESIRED M-VALUE
C *** NMAX: NUMBER OF COMPONENTS IN N-DIRECTION,
C ***       FROM (N-M)=0 TO (N-M)=NMAX-1
C *** PNM1, PNM2, AND PNM3 (LATS,NMAX) ARE MATRICES CONTAINING VALUES AT
C ***       ALL CO-LATITUDES, WITH (N-M) RANGING FROM 0 TO (NMAX-1) FOR
C ***       MMAX-2, MMAX-1, AND M=MMAX, RESPECTIVELY
C
C *** NOTES:
C *** IN THE GENERAL CASE, THE ROUTINE FIRST CALCULATES ALL NMAX*LATS
C *** COMPONENTS FOR M=0 AND M=1.  THEN USING RECURRENCE FORMULAE WHICH
C *** REQUIRE THE PREVIOUS TWO SETS OF M AND (N-M) VALUES FOR A GIVEN M,
C *** THE ROUTINE STARTS AT M=2 AND KEEPS APPLYING THE RECURRENCE
C *** FORMULAE UNTIL IT GETS TO THE DESIRED VALUE OF M(MMAX).  THE
C *** ROUTINE "REMEMBERS" WHAT THE VALUES OF "LATS,MMAX,AND NMAX" WERE IN
C *** THE PREVIOUS CALL.  IF ON THE NEXT CALL "NMAX" IS .LE. ITS OLD
C *** VALUE AND "MMAX" IS .GE. ITS OLD VALUE, AND IF "LATS" IS UNCHANGED,
C *** THE RECURRENCE FORMULAE ARE RESTARTED AT THAT POINT RATHER THAN FROM
C *** THE BEGINNING.  THE VALUES IN PNM1,PNM2, AND PNM3 THEN ARE ASSUMED
C *** TO BE ~rTHE SAME AS THEY WERE AT THE LAST RETURN TO THE CALLING PROGRAM.
C
C *** MMAX MUST BE .GE. 2
C *** NMAX MUST BE .GE. 1
C *** LATS MUST BE .GE. 2
C *** IF "LATS" IS TO BE .GT. 31, SOME REDIMENSIONING MUST BE DONE
C
C *** FOR FASTEST RESULTS, CALLS TO LGNDRE SHOULD BE ARRANGED SO
C *** THAT NMAX(NEW) IS .LE. NMAX(OLD) AND MMAX IS AT LEAST 3 GREATER
C *** THAN IT WAS IN THE PREVIOUS CALL; OTHERWISE, THE RECURRENCE
C *** FORMULAE MUST BE REINITIALIZED.
C
      DIMENSION PNM1(LATS,NMAX),PNM2(LATS,NMAX),PNM3(LATS,NMAX)
      DIMENSION COA(LATS),SIA(LATS)
```

```
      DATA ILAT/-1/
C *** CHECK FOR ILLEGAL VALUES OF LATS,MMAX,!NMAX
      IF (LATS.LT.2.OR.MMAX.LT.2.OR.NMAX.LT.1) STOP
      IF (LATS.EQ.ILAT) GO TO 1
C *** INITIALIZE COA ! SIA
      ILAT=LATS
      PI=4.*ATAN(1.)
      ILAT1=ILAT-1
      DELTA=PI/(2.*ILAT1)
    2 CONTINUE
      SQ=2.
      SQR2=SQRT(SQ)
      GO TO 222
C *** CHECK TO SEE IF THE RECURRENCE FORMULAE MUST BE INITIALIZED
C *** UPON ENTRY:
C *** (NM+1)=LAST VALUE OF NMAX
C *** MM     =LAST VALUE OF MMAX
C *** MMIN   =MM+1
C *** ILAT   =LAST VALUE OF LATS (ILAT=-1 INITIALLY, SO THE BRANCH TO
C *** STATEMENT #1 WILL NEVER BE TAKEN THE FIRST TIME, AND HENCE, NM,MM,
C *** AND MMIN WILL BE INITIALIZED CORRECTLY THE FIRST TIME THROUGH.)
    1 IF (NMAX.GT.NM+1) GO TO 222
      IF (MMAX.GT.MM) GO TO 3
C *** THIS BRANCH (PREVIOUS STATEMENT) ASSUMES PNM1,PNM2,AND PNM3
C *** ARE THE SAME AS THEY WERE AT THE LAST RETURN FROM LGNDRE.
C *** CHECK TO SEE IF THIS CASE HAS JUST BEEN DONE;IF SO, RETURN.
      IF (MMAX.EQ.MM) RETURN
C *** START FROM THE BEGINNING
  222 MMIN=2
      NM=NMAX-1
C *** CALCULATE VALUES FOR M=O AND M=1
      DO 22 I=1,ILAT
      THETA=ACOS(COA(I))
      C1=SQR2
      PNM2(I,1)=1./C1
      DO 20 N=1,NMAX
      FN=N
      FN2=2.*FN
      FN2SQ=FN2*FN2
C *** BELOUSOV EQUATION 19
      C1=C1*SQRT(1.-1./FN2SQ)
C *** BELOUSOV EQUATION 21
      C3=C1/SQRT(FN*(FN+1.))
      ANG=FN*THETA
      S1=0.
      S2=0.
      C4=1.
      C5=FN
      A=-1.
      B=0.
      N1=N+1
      DO 27 KK=1,N1,2
      K=KK-1
      IF(K.EQ.N) C4=.5*C4
```

```
      S2=S2+C5*SIN(ANG)*C4
      S1=S1+C4*COS(ANG)
      A=A+2.
      B=B+1.
      FK=K
      ANG=THETA*(FN-FK-2.)
      C4=(A*(FN-B+1.)/(B*(FN2-A)))*C4
      C5=C5-2.
   27 CONTINUE
      IF(N-NMAX) 23,24,20
   23 PNM2(I,N1)=S1*C1
   24 IF(MMAX) 20,20,21
   21 PNM3(I,N)=S2*C3
   20 CONTINUE
   22 CONTINUE
C *** DO-LOOP 22 HAS SET UP LEGENDRE POLYNOMIALS FOR M=O AND M=1
    3 MM=MMAX
C *** NOW LOOP THROUGH THE RECURRENCE FORMULAE INCREASING "M" UNTIL
C *** IT REACHES THE DESIRED VALUE
      DO 4 M=MMIN,MM
C *** THE EXISTING VALUES MUST BE MOVED OVER FOR THE RECURRENCE
      DO 41 I=1,ILAT
      DO 41 N=1,NMAX
      PNM1(I,N)=PNM2(I,N)
   41 PNM2(I,N)=PNM3(I,N)
      FM=M
      FM1=FM-1.
      FM2=FM-2.
      FM3=FM-3.
      MM1=M-1
      C6=SQRT((2.*FM+1.)/(2.*FM))
  333 DO 39 I=1,ILAT
C *** BELOUSOV EQUATION 23
      PNM3(I,1)=C6*SIA(I)*PNM2(I,1)
      M1=M+1
      IF (NMAX.EQ.1) GO TO 39
      DO 40 N=2,NMAX
      FN=N+MM1
      NM1=N-1
      NM2=N-2
      NN1=N+1
      C7=(FN*2.+1.)/(FN*2.-1.)
      C8=(FM1+FN)/((FM+FN)*(FM2+FN))
      C=SQRT((FN*2.+1.)/(FN*2.-3.)*C8*(FM3+FN))
      D=SQRT(C7*C8*(FN-FM1))
      E=SQRT(C7*(FN-FM)/(FN+FM))
      IF(N-NMAX) 43,42,40
C *** BELOUSOV EQUATION 17
   43 PNM3(I,N)=C*PNM1(I,N)-D*PNM1(I,NN1)*COA(I)+E*PNM3(I,NM1)*COA(I)
      GO TO 40
C *** BELOUSOV EQUATION 11
   42 A=SQRT((FN*FN-.25)/(FN*FN-FM*FM))
      B=SQRT((2.*FN+1.)*(FN-FM-1.)*(FN+FM1)/((2.*FN-3.)*(FN-FM)*(FN+FM))
     +)
```

```
C *** BELOUSOV EQUATION 11 FOR ONE MISSING VALUE
      PNM3(I,N)=2.00*A*COA(I)*PNM3(I,NM1)-B*PNM3(I,NM2)
   40 CONTINUE
   39 CONTINUE
    4 CONTINUE
C *** C,D,E ARE BELOUSOV'S C,D,E
C *** PNM1(I,N)=P(N,M-2)(COS I)
C *** PNM2(I,N)=P(N,M-1)(COS I)
C *** PNM3(I,N)=P(N,M)(COS I)
C *** (COS I) IS THE COSINE OF THE ITH CO-LATITUDE
C *** N IS NOW THE MERIDIONAL WAVE NUMBER
C
C *** SET UP FOR POSSIBLE NEW ENTRY
      MMIN=MM+1
      RETURN
      END
      SUBROUTINE DYDMU(MMAX,JMAX,NMAX,P,DP,COA,SIA)
C *** SUBROUTINE TO COMPUTE DERIVATIVE OF LEGENDRE POLYNOMIAL
      DIMENSION P(JMAX,NMAX,MMAX),DP(JMAX,NMAX,MMAX),COA(JMAX),SIA(JMAX)
      LT=NMAX-1
      DO 20 M=1,MMAX
      MM=M-1
      DO 10 L=2,LT
      N=MM+L-1
      DM=DCF(MM,N)*FLOAT(N+1)
      DP1=DCF(MM,N+1)*FLOAT(N)
      DO 10 J=1,JMAX
      DP(J,L,M)=(DM*P(J,L-1,M)-DP1*P(J,L+1,M))/(SIA(J)*SIA(J))
   10 CONTINUE
      NL=MM+NMAX-1
      DO 30 J=1,JMAX
      DP(J,1,M)=-FLOAT(MM)*DCF(MM,MM+1)*P(J,2,M)/(SIA(J)*SIA(J))
      DP(J,NMAX,M)=(FLOAT(2*NL+1)*DCF(MM,NL)*P(J,NMAX-1,M)-FLOAT(NL)*COA
     +(J)*P(J,NMAX,M))/(SIA(J)*SIA(J))
   30 CONTINUE
   20 CONTINUE
      RETURN
      END
      FUNCTION DCF(M,N)
      X=FLOAT(M)
      Y=FLOAT(N)
      DCF=SQRT((Y*Y-X*X)/(4.*Y*Y-1.))
      RETURN
      END
      SUBROUTINE GAUSL1 (N,A,B,NSUB,CK,XK,IER)
C *** SUBROUTINE TO CALCULATE GAUSS-LEGENDRE WEIGHTS AND ABSCISSAE OF A
C *** GIVEN ORDER ON A GIVEN INTERVAL, OR TO COMPUTE THE INTEGRAL OF A
C *** FUNCTION ON A GIVEN INTERVAL USING GAUSS-LEGENDRE QUADRATURE
C *** FORMULAE.  (FOR OTHER GAUSSIAN QUADRATURE FORMULAE, USE THE NSSL
C *** PACKAGE, GAUSS.)
C *** GAUSS-LEGENDRE QUADRATURE CORRESPONDS TO A WEIGHT FUNCTION OF 1 ON
C *** A FINITE INTERVAL.  IT INVOLVES APPROXIMATING THE INTEGRAL OF A
C *** FUNCTION BY A SUM OF FUNCTIONAL VALUES MULTIPLIED BY APPROPRIATE
C *** WEIGHTS.  WITH N ABSCISSAE AND WEIGHTS, THE APPROXIMATION IS EXACT
```

```
C *** FOR ALL POLYNOMIALS OF DEGREE LESS THAN 2*N.  THIS IS THE HIGHEST
C *** PRECISION THAT CAN BE OBTAINED USING N POINTS.  GAUSS-LEGENDRE
C *** QUADRATURE FORMULAE HAVE THE ADDITIONAL FEATURE OF NOT USING THE
C *** FUNCTIONAL VALUES AT THE ENDPOINTS OF THE INTERVAL OF INTEGRATION.
C ***
C *** SUBROUTINE GAUSL1 - COMPUTES THE WEIGHTS AND ABSCISSAE
C *** FUNCTION GAUSL2   - CALCULATES THE GAUSS-LEGENDRE APPROXIMATION
C *** TO THE INTEGRAL OF A GIVEN FUNCTION USING WEIGHTS AND ABSCISSAE
C *** PRODUCED BY GAUSL1.  PROGRAMMED BY RUSSELL K. REW, NCAR, FROM A
C *** METHOD DESCRIBED IN "THE HANDBOOK OF MATHEMATICAL FUNCTIONS" BY
C *** M. ABRAMOWITZ AND I. STEGUN
C -------------------------------------------------------------------
C *** SUBROUTINE GAUSL1 (N,A,B,NSUB,CK,XK,IER)
C *** DIMENSION OF   CK(N*NSUB),XK(N*NSUB)
C *** USAGE          CALL GAUSL1 (N,A,B,NSUB,CK,XK,IER)
C *** ARGUMENTS:
C *** ON INPUT      N - ORDER OF QUADRATURE DESIRED ON EACH ORDER
C ***                   SUBINTERVAL.  N MUST BE A POSITIVE INTEGER.
C ***               A - LEFT ENDPOINT OF INTEGRATION INTERVAL
C ***               B - RIGHT ENDPOINT OF INTEGRATION INTERVAL
C ***            NSUB - NUMBER OF SUBINTERVALS INTO WHICH THE INTEGRA-
C ***                   TION INTERVAL IS TO BE SUBDIVIDED, WITH AN N
C ***                   POINT QUADRATURE RULE APPLIED TO EACH.  IF NSUB
C ***                   IS 1, THE N POINT QUADRATURE RULE IS CALCULATED
C ***                   FOR THE WHOLE INTERVAL (A,B)
C *** ON OUTPUT     CK - THE N*NSUB WEIGHTS RETURNED BY GAUSL1
C ***               XK - THE CORRESPONDING N*NSUB ABSCISSAE.  AN APPROX-
C ***                    IMATION TO THE INTEGRAL OF F(X) FROM A TO B IS
C ***                    GIVEN BY:
C ***                        CK(1)*F(XK(1)) + CK(2)*F(XK(2))
C ***                        + ... + CK(N*NSUB)*F(XK(N*NSUB))
C ***            IER = 0  IF NO ERRORS OCCURRED
C ***                = 1  IF GAUSL1 WAS CALLED WITH N OR NSUB .LE. 0
C ***                     (IF IER .NE. 0, A MESSAGE IS PRINTED)
C ***
      DIMENSION    CK(1)     ,XK(1)
C ***
C *** TEST FOR INVALID INPUT
      IF (N.LE.0 .OR. NSUB.LE.0) GO TO 103
      IER = 0
      DINT = (B-A)/FLOAT(NSUB)
      B1 = A+DINT
C *** GET WEIGHTS AND ABSCISSAE FOR FIRST SUBINTERVAL
      CALL GAUSL3 (N,A,B1,CK,XK)
      IF (NSUB .EQ. 1) GO TO 104
      NSUBM1 = NSUB-1
C *** GET WEIGHTS AND ABSCISSAE FOR REST OF SUBINTERVALS
      DO 102 ISUB=1,NSUBM1
         DISP = DINT*FLOAT(ISUB)
         NISUB = N*ISUB
         DO 101 J=1,N
            K = J+NISUB
            CK(K) = CK(J)
            XK(K) = XK(J)+DISP
```

```
      101    CONTINUE
      102 CONTINUE
          GO TO 104
      103 IER = 1
C *** ADD ERROR MESSAGE PRINT
          WRITE (6,174)
      174 FORMAT (1X,'ARGUMENTS IN GAULS1 ARE INCORRECT')
      104 RETURN
          END
          FUNCTION GAUSL2 (N,A,B,NSUB,F,WRK,IER)
C *** FUNCTION GAUSL2 (N,A,B,NSUB,F,WRK,IER)
C *** DIMENSION OF    WRK(2*N)
C *** USAGE           FINT = GAUSL2 (N,A,B,NSUB,F,WRK,IER)
C *** ARGUMENTS:
C *** ON INPUT        N - ORDER OF QUADRATURE DESIRED ON EACH SUBINTER-
C ***                     VAL.  N MUST BE A POSITIVE INTEGER.
C ***                 A - LEFT ENDPOINT OF INTEGRATION INTERVAL
C ***                 B - RIGHT ENDPOINT OF INTEGRATION INTERVAL
C ***              NSUB - NUMBER OF SUBINTERVALS INTO WHICH THE INTEGRA-
C ***                     TION INTERVAL IS TO BE SUBDIVIDED, WITH AN N
C ***                     POINT QUADRATURE RULE APPLIED TO EACH.  IF NSUB
C ***                     IS 1, THE N POINT QUADRATURE RULE IS CALCULATED
C ***                     FOR THE WHOLE INTERVAL (A,B)
C ***                 F - THE NAME OF THE USER-SUPPLIED INTEGRAND,
C ***                     WRITTEN AS A FUNCTION OF ONE VARIABLE AND
C ***                     DECLARED EXTERNAL IN THE CALLING ROUTINE
C ***               WRK - A WORK ARRAY WHICH MUST BE DIMENSIONED FOR AT
C ***                     LEAST 2*N
C *** ON OUTPUT     IER = 0 IF NO ERRORS OCCURRED
C ***                   = 1 IF GAUSL2 WAS CALLED WITH N OR NSUB .LE.0.
C ***                     (IF IER .NE. 0, A MESSAGE IS PRINTED BY THE
C ***                     INDICATING ERROR)
C ***            GAUSL2 - THE FUNCTION NAME CONTAINS THE VALUE OF THE
C ***                     INTEGRAL, CALCULATED AS THE SUM (FOR I=1 TO
C ***                     N*NSUB) OF CK(I)*F(XK(I)) WHERE THE FIRST N
C ***                     ELEMENTS OF THE CK AND XK ARRAYS ARE OBTAINED
C ***                     BY CALLING GAUSL1 AND ARE STORED IN THE WORK
C ***                     ARRAY, WRK.  FOR NSUB .GT.1, THE REST OF THE
C ***                     WEIGHTS AND ABSCISSAE ARE OBTAINED FROM THE
C ***                     FIRST N.
C ***
      DIMENSION    WRK(1)
      GAUSL2 = 0.
C
C *** TEST FOR INVALID INPUT
      IF (N.LE.0 .OR. NSUB.LE.0) GO TO 103
      IER = 0
      DINT = (B-A)/FLOAT(NSUB)
      B1 = A+DINT
C *** GET WEIGHTS AND ABSCISSAE FOR FIRST SUBINTERVAL
      CALL GAUSL3 (N,A,B1,WRK(1),WRK(N+1))
      DO 102 ISUB=1,NSUB
         DISP = DINT*FLOAT(ISUB-1)
C *** ADD CONTRIBUTION TO INTEGRAL OF THE ISUB SUBINTERVAL
```

```
      DO 101 J=1,N
         JPN = J+N
         GAUSL2 = GAUSL2+WRK(J)*F(WRK(JPN)+DISP)
  101    CONTINUE
  102 CONTINUE
      GO TO 104
  103 IER = 1
C *** WRITE ERROR MESSAGE
      WRITE (6,175)
  175 FORMAT (1X,'ARGUMENTS IN GAUSL2 ARE INCORRECT')
  104 RETURN
      END
      SUBROUTINE GAUSL3 (N,XA,XB,WT,AB)
C *** WEIGHTS AND ABSCISSAE FOR NTH ORDER GAUSSIAN QUADRATURE ON
C *** (XA,XB)
C *** ARGUMENTS:
C *** ON INPUT       N - THE ORDER DESIRED
C ***                XA - LEFT ENDPOINT OF THE INTERVAL OF INTEGRATION
C ***                XB - RIGHT ENDPOINT OF THE INTERVAL OF INTEGRATION
C *** ON OUTPUT      AB - N CALCULATED ABSCISSAE
C ***                WT - N CALCULATED WEIGHTS
C ***
      DOUBLE PRECISION            DZERO    ,DZERI      ,DP          ,
     1              DPM1    ,DPM2      ,DPPR      ,DP2PRI    ,
     2              DRAT    ,DPROD     ,DTMP      ,DDIF      ,
     3              DSUM    ,DDIFX
      DIMENSION     AB(N)     ,WT(N)
C ***
C *** MACHINE DEPENDENT CONSTANTS:
C *** TOL - CONVERGENCE CRITERION FOR DOUBLE PRECISION ITERATION
C *** PI  - GIVEN TO 15 SIGNIFICANT DIGITS
C *** C1  -  1/8               THESE ARE COEFFICIENTS IN MCMAHON'S
C *** C2  -  -31/(2*3*8**2)    EXPANSIONS OF THE KTH ZERO OF THE
C *** C3  -  3779/(2*3*5*8**3) BESSEL FUNCTION JO(X) (CF.
C *** C4  -  -6277237/(3*5*7*8**5)  ABRAMOWITZ, HANDBOOK OF MATHEMATI-
C *** U   -  (1-(2/PI)**2)/4   CAL FUNCTIONS)
C ***
      DATA TOL/1.E-22/,PI/3.14159265358979/,U/.148678816357662/
      DATA C1,C2,C3,C4/.125,-.080729166666667,.246028645833333,
     1                 -1.82443876720609 /
C *** MAXIMUM NUMBER OF ITERATIONS BEFORE GIVING UP ON CONVERGENCE
      DATA MAXIT /5/
C *** ARITHMETIC STATEMENT FUNCTION FOR CONVERTING INTEGER TO DOUBLE
C *** PRECISION
      DBLI(INTEGR) = DBLE(FLOAT(INTEGR))
      DDIF = .5D0*(DBLE(XB)-DBLE(XA))
      DSUM = .5D0*(DBLE(XB)+DBLE(XA))
      IF (N .GT. 1) GO TO 101
      AB(1) = DSUM
      WT(1) = 2.*DDIF
      GO TO 107
  101 CONTINUE
      NNP1 = N*(N+1)
      COND = 1./SQRT((.5+FLOAT(N))**2+U)
```

```
      LIM = N/2
      DO 105 K=1,LIM
         B = (FLOAT(K)-.25)*PI
         BISQ = 1./(B*B)
C *** ROOTBF APPROXIMATES THE KTH ZERO OF THE BESSEL FUNCTION JO(X)
         ROOTBF = B*(1.+BISQ*(C1+BISQ*(C2+BISQ*(C3+BISQ*C4))))
C *** INITIAL GUESS FOR KTH ROOT OF LEGENDRE POLYNOMIAL P-SUB-N(X)
         DZERO = COS(ROOTBF*COND)
         DO 103 I=1,MAXIT
            DPM2 = 1.DO
            DPM1 = DZERO
C *** RECURSION RELATION FOR LEGENDRE POLYNOMIALS
            DO 102 NN=2,N
               DP = (DBLI(2*NN-1)*DZERO*DPM1-DBLI(NN-1)*DPM2)/DBLI(NN)
               DPM2 = DPM1
               DPM1 = DP
  102       CONTINUE
            DTMP = 1.DO/(1.DO-DZERO*DZERO)
            DPPR = DBLI(N)*(DPM2-DZERO*DP)*DTMP
            DP2PRI = (2.DO*DZERO*DPPR-DBLI(NNP1)*DP)*DTMP
            DRAT = DP/DPPR
C *** CUBICALLY-CONVERGENT ITERATIVE IMPROVEMENT OF ROOT
            DZERI = DZERO-DRAT*(1.DO+DRAT*DP2PRI/(2.DO*DPPR))
            IF (ABS(SNGL(DZERI-DZERO)) .LE. TOL) GO TO 104
            DZERO = DZERI
  103       CONTINUE
  104       CONTINUE
            DDIFX = DDIF*DZERO
            AB(K) = DSUM-DDIFX
            WT(K) = 2.DO*(1.DO-DZERO*DZERO)/(DBLI(N)*DPM2)**2*DDIF
            I = N-K+1
            AB(I) = DSUM+DDIFX
            WT(I) = WT(K)
  105 CONTINUE
      IF (MOD(N,2) .EQ. 0) GO TO 107
      AB(LIM+1) = DSUM
      NM1 = N-1
      DPROD = N
      DO 106 K=1,NM1,2
         DPROD = DBLI(NM1-K)*DPROD/DBLI(N-K)
  106 CONTINUE
      WT(LIM+1) = 2.DO/DPROD**2*DDIF
  107 RETURN
      END
      SUBROUTINE FFTIN(NF,IFAC)
C *** SUBROUTINE SFFT
C *** THE ARRAY IFAC MUST BE INPUT WITH THE PRIME FACTORIZATION OF N,
C *** WHICH IS THE LENGTH OF THE SEQUENCE TO BE TRANSFORMED. E.G.,
C *** FOR N=48, THE ARRAY IFAC SHOULD CONTAIN IFAC(1)=3, IFAC(2)=4,
C *** IFAC(3)=4.  THE FORTRAN PROGRAM WAS WRITTEN BY P. SWARZTRAUBER,
C *** NCAR. (FOR ALGORITHM SEE "PARALLEL COMPUTING",1,(1984),45-63)
      DIMENSION IFAC(NF)
      IFAC(1)=3
      IFAC(2)=4
```

```
      IFAC(3)=4
      RETURN
      END
      SUBROUTINE SFFT(IS,N,NF,IFAC,C,CH)
C *** SUBROUTINE TO COMPUTE STOCKHAM FFT BOTH FORWARD AND REVERSE
C *** TRANSFORMS
      DIMENSION IFAC(NF),C(1),CH(1)
      COMMON /PASK/ OMEGA,TAU
      COMPLEX C,CH,OMEGA,TAU
C *** TRANSPOSED STOCKHAM FFT
C *** IS=-1 FOR FORWARD TRANSFORM AND IS=1 FOR BACKWARD TRANSFORM
      TPI = 8.*ATAN(1.)
      L1 = 1
      DO 1 K1=1,NF
      IP = IFAC(K1)
      ARG = FLOAT(IS)*TPI/FLOAT(IP)
      TAU = CMPLX(COS(ARG),SIN(ARG))
      L2 = IP*L1
      IDO = N/L2
      ARG = FLOAT(IS)*TPI/FLOAT(IP*IDO)
      OMEGA = CMPLX(COS(ARG),SIN(ARG))
      CALL PASSG(IDO,IP,L1,C,C,CH)
      L1=L2
    1 CONTINUE
      RETURN
      END
      SUBROUTINE PASSG(IDO,IP,L1,CC,C1,CH)
C *** SUBROUTINE NEEDED FOR SFFT IN SFFT
      DIMENSION CH(IDO,L1,IP),CC(IDO,IP,L1),C1(IDO,L1,IP)
      COMMON /PASK/ OMEGA,TAU
      COMPLEX OMEGA,TAU,CC,C1,CH
C *** GENERAL RADIX STOCKHAM FFT
      DO 2 I=1,IDO
      DO 2 K=1,L1
      DO 2 L=1,IP
      CH(I,K,L)=0.
      DO 2 J=1,IP
      CH(I,K,L)=CH(I,K,L)+TAU**((J-1)*(L-1))*CC(I,J,K)
    2 CONTINUE
      DO 1 J=1,IP
      DO 1 K=1,L1
      DO 1 I=1,IDO
      C1(I,K,J)=OMEGA**((I-1)*(J-1))*CH(I,K,J)
    1 CONTINUE
      RETURN
      END
```

Finite Difference Shallow Water Wave Equation Model

As discussed in Chapter 4, the finite difference primitive equation technique, with prediction equations for the velocity components u and v rather than for the vorticity ς as in the models of Appendices D and E, is in widespread use as part of many climate models, for example those of Arakawa and Lamb (1977) and Schlesinger et al. (1985) for the atmosphere, and those of Bryan (1969a) and Semtner (1974) for the ocean. At the end of this appendix a simplification of the primitive equation model in the form of a shallow water wave equation model is presented. The formulation of this model follows Sadourny (1975) as coded by Paul N. Swarztrauber of the National Center for Atmospheric Research (NCAR) for timing tests of supercomputers (see Hoffman et al., 1984). The code can be adapted to IBM-PC compatible software. By adding suitable forcing terms or initial conditions the model can be a very useful illustrative tool for students of both oceanographic and atmospheric sciences. In fact, if a limited closed domain is specified rather than a cyclic domain with identical east and west boundaries, and if stress terms are added to the momentum equations, the computer program could be adapted to a closed region such as the North Atlantic basin.

Following Sadourny (1975), we start with basic model equations that differ somewhat from the shallow water equations presented as (4.28)–(4.30) in Chapter 4. Here we ignore the Coriolis terms and employ potential vorticity in the advection terms:

$$\frac{\partial u}{\partial t} - \eta V + \frac{\partial H}{\partial x} = 0 \tag{F.1}$$

$$\frac{\partial v}{\partial t} + \eta U + \frac{\partial H}{\partial y} = 0 \tag{F.2}$$

$$\frac{\partial P}{\partial t} + \frac{\partial U}{\partial x} + \frac{\partial V}{\partial y} = 0 \tag{F.3}$$

where P is $gh (= \text{pressure/density})$ and u and v are east-west and north-south velocities, respectively,

$$U = Pu, \ V = Pv, \ H = P + \frac{1}{2}(u^2 + v^2) \tag{F.4}$$

and η is potential vorticity:

$$\eta = \frac{\partial v/\partial x - \partial u/\partial y}{P} \tag{F.5}$$

The Coriolis terms $-fv, +fu$ could be added to (F.1) and (F.2), respectively, to obtain a set of equations easily shown to be equivalent to (4.28)–(4.30).

The particular scheme that Sadourny uses to solve (F.1)–(F.5) is a potential-enstrophy conserving method similar to the energy conserving scheme in (4.57)–(4.59). His resulting set of equations, using the averaging operators $^{-x}$ and $^{-y}$ and the derivative operators δ_x and δ_y [see (4.53)–(4.56)], are

$$\frac{\partial u}{\partial t} - \overline{\eta}^y \overline{V}^{x}{}^{-y} + \delta_x H = 0 \tag{F.6}$$

$$\frac{\partial v}{\partial t} + \overline{\eta}^x \overline{U}^{y}{}^{-x} + \delta_y H = 0 \tag{F.7}$$

$$\frac{\partial P}{\partial t} + \delta_x U + \delta_y V = 0 \tag{F.8}$$

The grid structure is such that η is defined at the center of a grid volume, P is defined at the four corners, U is defined at the north-south walls, and V is defined at the east-west walls. If several grid cells are drawn adjacent to each other and the variables are placed at their respective locations, it should be clear where the averaging must take place to define the quantities needed for the finite difference equations. The initial conditions in the code assume a geostrophic balance between the wind and pressure fields. The time differencing method is centered with a small amount of smoothing between successive time steps to damp out high frequency oscillations.

Some of the key variables in the FORTRAN computer program are

DX	–	grid spacing in x
DY	–	grid spacing in y
DT	–	time step
M and N	–	number of gridpoints in x and y directions
PSI	–	initial streamfunction used for initial values of wind, i.e., $u_0 = -\partial\psi/\partial y$, $v_0 = \partial\psi/\partial x$
Alpha	–	time filtering parameter
CU	–	Pu
CV	–	Pv
Z	–	potential vorticity, η
H	–	$P + 1/2(u^2 + v^2)$
UNEW, VNEW, PNEW	–	variables at time step $N + 1$
UOLD, VOLD, POLD	–	variables at time step $N - 1$

The program below does not include a contouring routine, but there are many contour plotting subroutines available that can graphically display two-dimensional fields on small dot matrix printers, and these could profitably be used with this program to allow easy visualization of the simulated flow fields.

A single-sided double-density 5 1/4″ floppy diskette (IBM/IBM compatible) containing the computer models discussed in Appendix D (Finite Difference Barotropic Forecast Model), Appendix E (Barotropic Vorticity Spectral Model), and Appendix F (Finite Difference Shallow Water Wave Equation Model), can be purchased by sending one's name, address, telephone number, and a check or money order for $25.00, payable to LOGON/3, to the following address:

LOGON/3
P.O. Box 3031
Boulder, CO 80307–3031

The three models also are available at the same address on a single-sided 3 1/2″ micro floppy disk (Macintosh/Macintosh compatible) for $27.00 per disk. In either case, Colorado residents should add 5.75% sales tax. For overseas mailing, add $1.00 for postage to the base price of each diskette ordered.

Diskettes containing these three computer models, in both IBM and Macintosh formats, will be available from LOGON/3 as long as this book remains in print.

```
      PROGRAM SHALLOW
C ***
C *** BENCHMARK WEATHER PREDICTION PROGRAM FOR COMPARING THE PERFORMANCE
C *** OF CURRENT SUPERCOMPUTERS. THE MODEL IS BASED ON THE PAPER, "THE
C *** DYNAMICS OF FINITE-DIFFERENCE MODELS OF THE SHALLOW-WATER
C *** EQUATIONS," BY R. SADOURNY, J. ATM. SCI., VOL.32, NO.4, APRIL 1975
C *** CODE BY PAUL N. SWARZTRAUBER, NATIONAL CENTER FOR ATMOSPHERIC
C *** RESEARCH, BOULDER, COLORADO   OCTOBER 1984.
C ***
      DIMENSION U(32,16),V(32,16),P(32,16),UNEW(32,16),VNEW(32,16),
     1          PNEW(32,16),UOLD(32,16),VOLD(32,16),POLD(32,16),
     2          CU(32,16),CV(32,16),Z(32,16),H(32,16),PSI(32,16)
      CHARACTER*1 LABEL(3)
      LABEL(1)='P'
      LABEL(2)='U'
      LABEL(3)='V'
      IMAX = 32
      JMAX = 16
      A = 1.E6
      DT = 90.
      TDT = 90.
      TIME = 0.
      DX = 1.E5
      DY = 1.E5
      ALPHA = .001
      ITMAX = 120
      MPRINT = 120
      M = 31
      N = 15
      MP1 = M+1
      NP1 = N+1
      PI = 4.*ATAN(1.)
      TPI = PI+PI
      DI = TPI/FLOAT(M)
      DJ = TPI/FLOAT(N)
C *** INITIAL VALUES OF THE STREAMFUNCTION
      DO 50 J=1,NP1
      DO 50 I=1,MP1
      PSI(I,J) = A*SIN((FLOAT(I)-.5)*DI)*SIN((FLOAT(J)-.5)*DJ)
   50 CONTINUE
C *** INITIALIZE VELOCITIES
      DO 60 J=1,N
      DO 60 I=1,M
      U(I+1,J) = -(PSI(I+1,J+1)-PSI(I+1,J))/DY
      V(I,J+1) = (PSI(I+1,J+1)-PSI(I,J+1))/DX
   60 CONTINUE
C *** PERIODIC CONTINUATION
      DO 70 J=1,N
      U(1,J) = U(M+1,J)
      V(M+1,J+1) = V(1,J+1)
   70 CONTINUE
      DO 75 I=1,M
      U(I+1,N+1) = U(I+1,1)
      V(I,1) = V(I,N+1)
```

```
   75 CONTINUE
      U(1,N+1) = U(M+1,1)
      V(M+1,1) = V(1,N+1)
      DO 86 J=1,NP1
      DO 86 I=1,MP1
      UOLD(I,J) = U(I,J)
      VOLD(I,J) = V(I,J)
      POLD(I,J) = 50000.
      P(I,J) = 50000.
   86 CONTINUE
C *** PRINT INITIAL VALUES
      WRITE(*,390) N,M,DX,DY,DT,ALPHA
  390 FORMAT('1NUMBER OF POINTS IN THE X DIRECTION',I8/
     1        ' NUMBER OF POINTS IN THE Y DIRECTION',I8/
     2        ' GRID SPACING IN THE X DIRECTION   ',F8.0/
     3        ' GRID SPACING IN THE Y DIRECTION   ',F8.0/
     4        ' TIME STEP                         ',F8.0/
     5        ' TIME FILTER PARAMETER             ',F8.3)
      MNMIN = MINO(M,N)
      WRITE(*,391) (POLD(I,I),I=1,MNMIN)
  391 FORMAT(/' INITIAL DIAGONAL ELEMENTS OF P ' //,(8E16.6))
      WRITE(*,392) (UOLD(I,I),I=1,MNMIN)
  392 FORMAT(/' INITIAL DIAGONAL ELEMENTS OF U ' //,(8E16.6))
      WRITE(*,393) (VOLD(I,I),I=1,MNMIN)
  393 FORMAT(/' INITIAL DIAGONAL ELEMENTS OF V ' //,(8E16.6))
      NCYCLE = 0
   90 NCYCLE = NCYCLE + 1
C *** COMPUTE CAPITAL U, CAPITAL V, Z, AND H
      FSDX = 4./DX
      FSDY = 4./DY
      DO 100 J=1,N
      DO 100 I=1,M
      CU(I+1,J) = .5*(P(I+1,J)+P(I,J))*U(I+1,J)
      CV(I,J+1) = .5*(P(I,J+1)+P(I,J))*V(I,J+1)
      Z(I+1,J+1) = (FSDX*(V(I+1,J+1)-V(I,J+1))-FSDY*(U(I+1,J+1)
     1        -U(I+1,J)))/(P(I,J)+P(I+1,J)+P(I+1,J+1)+P(I,J+1))
      H(I,J) = P(I,J)+.25*(U(I+1,J)*U(I+1,J)+U(I,J)*U(I,J)
     1          +V(I,J+1)*V(I,J+1)+V(I,J)*V(I,J))
  100 CONTINUE
C *** PERIODIC CONTINUATION
      DO 110 J=1,N
      CU(1,J) = CU(M+1,J)
      CV(M+1,J+1) = CV(1,J+1)
      Z(1,J+1) = Z(M+1,J+1)
      H(M+1,J) = H(1,J)
  110 CONTINUE
      DO 115 I=1,M
      CU(I+1,N+1) = CU(I+1,1)
      CV(I,1) = CV(I,N+1)
      Z(I+1,1) = Z(I+1,N+1)
      H(I,N+1) = H(I,1)
  115 CONTINUE
      CU(1,N+1) = CU(M+1,1)
      CV(M+1,1) = CV(1,N+1)
```

```
      Z(1,1) = Z(M+1,N+1)
      H(M+1,N+1) = H(1,1)
C *** COMPUTE NEW VALUES U, V, AND P
      TDTS8 = TDT/8.
      TDTSDX = TDT/DX
      TDTSDY = TDT/DY
      DO 200 J=1,N
      DO 200 I=1,M
      UNEW(I+1,J) = UOLD(I+1,J)+
     1    TDTS8*(Z(I+1,J+1)+Z(I+1,J))*(CV(I+1,J+1)+CV(I,J+1)+CV(I,J)
     2    +CV(I+1,J))-TDTSDX*(H(I+1,J)-H(I,J))
      VNEW(I,J+1) = VOLD(I,J+1)-TDTS8*(Z(I+1,J+1)+Z(I,J+1))
     1    *(CU(I+1,J+1)+CU(I,J+1)+CU(I,J)+CU(I+1,J))
     2    -TDTSDY*(H(I,J+1)-H(I,J))
      PNEW(I,J) = POLD(I,J)-TDTSDX*(CU(I+1,J)-CU(I,J))
     1    -TDTSDY*(CV(I,J+1)-CV(I,J))
  200 CONTINUE
C *** PERIODIC CONTINUATION
      DO 210 J=1,N
      UNEW(1,J) = UNEW(M+1,J)
      VNEW(M+1,J+1) = VNEW(1,J+1)
      PNEW(M+1,J) = PNEW(1,J)
  210 CONTINUE
      DO 215 I=1,M
      UNEW(I+1,N+1) = UNEW(I+1,1)
      VNEW(I,1) = VNEW(I,N+1)
      PNEW(I,N+1) = PNEW(I,1)
  215 CONTINUE
      UNEW(1,N+1) = UNEW(M+1,1)
      VNEW(M+1,1) = VNEW(1,N+1)
      PNEW(M+1,N+1) = PNEW(1,1)
C
      IF(NCYCLE .GT. ITMAX) WRITE (*,220)
  220 FORMAT('0   ***** END OF PROGRAM SHALLOW *****')
C
      IF(NCYCLE .GT. ITMAX) STOP
      TIME = TIME + DT
      IF(MOD(NCYCLE,MPRINT) .NE. 0) GO TO 370
      PTIME = TIME/3600.
      WRITE(*,350) NCYCLE,PTIME
  350 FORMAT(//,' CYCLE NUMBER',I5,' MODEL TIME IN  HOURS',F6.2)
      WRITE(*,355) (PNEW(I,I),I=1,MNMIN)
  355 FORMAT(/,' DIAGONAL ELEMENTS OF P ' //,(8E16.6))
      WRITE(*,360) (UNEW(I,I),I=1,MNMIN)
  360 FORMAT(/,' DIAGONAL ELEMENTS OF U ' //,(8E16.6))
      WRITE(*,365) (VNEW(I,I),I=1,MNMIN)
  365 FORMAT(/,' DIAGONAL ELEMENTS OF V ' //,(8E16.6))
      CALL PLOT (P,IMAX,JMAX,NCYCLE,LABEL(1))
      CALL PLOT (U,IMAX,JMAX,NCYCLE,LABEL(2))
      CALL PLOT (V,IMAX,JMAX,NCYCLE,LABEL(3))
  370 IF(NCYCLE .LE. 1) GO TO 310
      DO 300 J=1,N
      DO 300 I=1,M
      UOLD(I,J) = U(I,J)+ALPHA*(UNEW(I,J)-2.*U(I,J)+UOLD(I,J))
```

```
        VOLD(I,J) = V(I,J)+ALPHA*(VNEW(I,J)-2.*V(I,J)+VOLD(I,J))
        POLD(I,J) = P(I,J)+ALPHA*(PNEW(I,J)-2.*P(I,J)+POLD(I,J))
        U(I,J) = UNEW(I,J)
        V(I,J) = VNEW(I,J)
        P(I,J) = PNEW(I,J)
  300 CONTINUE
C *** PERIODIC CONTINUATION
        DO 320 J=1,N
        UOLD(M+1,J) = UOLD(1,J)
        VOLD(M+1,J) = VOLD(1,J)
        POLD(M+1,J) = POLD(1,J)
        U(M+1,J) = U(1,J)
        V(M+1,J) = V(1,J)
        P(M+1,J) = P(1,J)
  320 CONTINUE
        DO 325 I=1,M
        UOLD(I,N+1) = UOLD(I,1)
        VOLD(I,N+1) = VOLD(I,1)
        POLD(I,N+1) = POLD(I,1)
        U(I,N+1) = U(I,1)
        V(I,N+1) = V(I,1)
        P(I,N+1) = P(I,1)
  325 CONTINUE
        UOLD(M+1,N+1) = UOLD(1,1)
        VOLD(M+1,N+1) = VOLD(1,1)
        POLD(M+1,N+1) = POLD(1,1)
        U(M+1,N+1) = U(1,1)
        V(M+1,N+1) = V(1,1)
        P(M+1,N+1) = P(1,1)
        GO TO 90
  310 TDT = TDT+TDT
        DO 400 J=1,NP1
        DO 400 I=1,MP1
        UOLD(I,J) = U(I,J)
        VOLD(I,J) = V(I,J)
        POLD(I,J) = P(I,J)
        U(I,J) = UNEW(I,J)
        V(I,J) = VNEW(I,J)
        P(I,J) = PNEW(I,J)
  400 CONTINUE
        GO TO 90
        END
C
        SUBROUTINE PLOT (FIELD,IMAX,JMAX,NCYCLE,FTITLE)
C *** SUBROUTINE TO MAKE TWO-DIMENSIONAL PLOTS OF FIELDS
        DIMENSION FIELD (32,16)
        CHARACTER*1 FTITLE
        DO 1  II=1,IMAX,8
        I1=II
        I2=I1+7
        WRITE (*,22) FTITLE,I1,I2,NCYCLE
   22 FORMAT ('OFIELD=',A1,'  I1=',I2,'  I2=',I2,'  NCYCLE=',I3)
        DO 2  J=1,JMAX
        L=JMAX+1-J
```

```
 2 WRITE (*,21) (FIELD(I,L),I=I1,I2)
 1 CONTINUE
21 FORMAT(1X,8E9.2)
   RETURN
   END
```

Atmospheric General Circulation Model Equations

The particular form of model equations for the atmosphere used by different modeling groups varies widely, but essentially all the major groups use the σ-system equations shown in Chapter 3. This appendix presents the complete set of equations used in the spectral atmospheric model developed by Bourke (1974) and Bourke et al. (1977). This model is widely used by many researchers and was the basis for an early version of the NCAR Community Climate Model (CCM) (Pitcher et al. (1983) and Ramanathan et al. (1983)). The description uses Bourke's notation to facilitate reference to other literature, although as a result some of the variable symbols differ from those found in Chapters 3 and 4.

Basic model equations

The horizontal equation of motion in vector form (using (3.61) and (3.62), and (3.53), (3.56), and (3.59)), is

$$\frac{d\boldsymbol{V}}{dt} = -f\boldsymbol{k} \times \boldsymbol{V} - \boldsymbol{\nabla}\Phi - RT\boldsymbol{\nabla}\ln p_s + \boldsymbol{F} \qquad (\mathrm{G.1})$$

The equation of continuity (modified from (3.69)) is

$$\frac{d \ln p_s}{dt} = -\nabla \cdot V - \frac{\partial \dot\sigma}{\partial \sigma} \tag{G.2}$$

The first law of thermodynamics must be written in a somewhat different form from (3.73) for the spectral method using semi-implicit numerical procedures. Starting with

$$C_p \frac{dT}{dt} - \frac{1}{\rho}\frac{dp}{dt} = H + C_p \left({}_h F_T + {}_v F_T \right) \tag{G.3}$$

where H is the diabatic term, insertion of the gas law (3.56) yields

$$\frac{dT}{dt} = \frac{RT}{C_p p}\frac{dp}{dt} + \frac{H}{C_p} + {}_h F_T + {}_v F_T \tag{G.4}$$

Using the definition of σ, i.e., $\sigma = p/p_s$,

$$\frac{dp}{dt} = \frac{d}{dt}(\sigma p_s) = \dot\sigma p_s + \sigma \frac{dp_s}{dt} \tag{G.5}$$

Combination of (G.4), (G.5), and (3.69) yields the form of the first law of thermodynamics most convenient for allowing a longer time step through a semi-implicit treatment of $\nabla \cdot V$:

$$\frac{dT}{dt} = \frac{RT}{C_p}\left(\frac{\dot\sigma}{\sigma} - \frac{\partial \dot\sigma}{\partial \sigma} - \nabla \cdot V \right) + \frac{H}{C_p} + {}_h F_T + {}_v F_T \tag{G.6}$$

From (3.63), (3.56), and the definition of σ, the hydrostatic equation becomes

$$\frac{\partial \Phi}{\partial \sigma} = -\frac{RT}{\sigma} \tag{G.7}$$

Finally, the moisture prediction equation in vector form (see (3.145)) is

$$\frac{\partial M}{\partial t} = -V \cdot \nabla M - \dot\sigma \frac{\partial M}{\partial \sigma} + C + {}_h F_M + {}_v F_M \tag{G.8}$$

The meanings of the symbols used in (G.1)–(G.8) are as follows: V is the horizontal wind vector with eastward and northward components u and v, respectively; T is the temperature; M

is the moisture mixing ratio; Φ is the geopotential height gz; f is the Coriolis parameter; R is the gas constant for dry air; $\boldsymbol{F}$ is the horizontal frictional force; $\boldsymbol{k}$ is the vertical unit vector; $\boldsymbol{\nabla}$ is the horizontal spherical gradient operator defined in Appendix A; $\dot{\sigma}$ is the total time derivative of σ; C_p is the specific heat at constant pressure for dry air; d/dt is the total time derivative in the σ-system; and H is the diabatic heating by condensation, convection, solar heating, and infrared heating and cooling. Terms $_hF_T$ and $_vF_T$ are the horizontal and vertical diffusions of sensible heat, respectively. The terms $_uF_M$ and $_vF_M$ are horizontal and vertical diffusions of moisture. C combines the sources and sinks of moisture connected with the precipitation process.

Prediction equations for the vertical component of the relative vorticity, $\varsigma = \boldsymbol{k} \cdot \boldsymbol{\nabla} \times \boldsymbol{V}$, and the horizontal divergence, $D = \boldsymbol{\nabla} \cdot \boldsymbol{V}$, can be obtained by performing the $\boldsymbol{\nabla} \times$ and $\boldsymbol{\nabla} \cdot$ operations on (G.1):

$$\frac{\partial \varsigma}{\partial t} = -\boldsymbol{\nabla} \cdot (\varsigma + f)\boldsymbol{V} - \boldsymbol{k} \cdot \boldsymbol{\nabla} \times \left(RT'\boldsymbol{\nabla}q + \dot{\sigma}\frac{\partial \boldsymbol{V}}{\partial \sigma} + \boldsymbol{F} \right) \text{ (G.9)}$$

$$\frac{\partial D}{\partial t} = \boldsymbol{k} \cdot \boldsymbol{\nabla} \times (\varsigma + f)\boldsymbol{V} - \boldsymbol{\nabla} \cdot \left(RT'\boldsymbol{\nabla}q + \dot{\sigma}\frac{\partial \boldsymbol{V}}{\partial \sigma} + \boldsymbol{F} \right)$$

$$- \boldsymbol{\nabla}^2 \left(\Phi' + RT_o q + \tfrac{1}{2}\boldsymbol{V} \cdot \boldsymbol{V} \right) \text{ (G.10)}$$

The first law of thermodynamics and the moisture prediction and continuity equations can be written as

$$\frac{\partial T}{\partial t} = -\boldsymbol{\nabla} \cdot \boldsymbol{V}T' + T'D + \dot{\sigma}\gamma - \frac{RT}{C_p}\left(D + \frac{\partial \dot{\sigma}}{\partial \sigma} \right)$$

$$+ \frac{H_c}{C_p} + {}_hF_T + {}_vF_T \text{ (G.11)}$$

$$\frac{\partial M}{\partial t} = -\boldsymbol{\nabla} \cdot \boldsymbol{V}M' + M'D - \dot{\sigma}\left(\frac{\partial M}{\partial \sigma} \right) + C$$

$$+ {}_hF_M + {}_vF_M \text{ (G.12)}$$

$$\frac{\partial q}{\partial t} = -D - \frac{\partial \dot{\sigma}}{\partial \sigma} - \boldsymbol{V} \cdot \boldsymbol{\nabla}q \text{ (G.13)}$$

where q denotes $\ln p_s$, and γ is the static stability defined as $RT/\sigma C_p - \partial T/\partial \sigma$. The reader should not confuse the use of q here with the mixing ratio in Chapter 3. The subscript zero denotes a horizontal mean value and a prime denotes the deviation from the mean.

The horizontal wind vector, V, can be defined in terms of a streamfunction, ψ, and a velocity potential, χ, as in (3.80):

$$V = k \times \nabla\psi + \nabla\chi \tag{G.14}$$

By performing the $k \cdot \nabla \times$ and $\nabla \cdot$ operations on (G.14), alternate expressions for vorticity and divergence are obtained (see (3.83) and (3.84)), i.e.,

$$\varsigma = k \cdot \nabla \times V = \nabla^2\psi \tag{G.15}$$

$$D = \nabla \cdot V \qquad = \nabla^2\chi \tag{G.16}$$

Integrating the continuity equation (G.13) in the vertical with boundary conditions $\dot\sigma = 0$ at $\sigma = 1$ and $\sigma = 0$ gives

$$\frac{\partial q}{\partial t} = \overline{D} + \overline{V} \cdot \nabla q \tag{G.17}$$

where

$$(\overline{}) = \int_{\sigma=1}^{0} ()\partial\sigma \quad \text{and} \quad (\overline{})^\sigma = \int_{\sigma=1}^{\sigma} ()\partial\sigma \tag{G.18}$$

The diagnostic equation for vertical velocity, $\dot\sigma$, is

$$\dot\sigma = \{(1-\sigma)\overline{D} - \overline{D}^\sigma\} + \{(1-\sigma)\overline{V} - \overline{V}^\sigma\} \cdot \nabla q \tag{G.19}$$

Redefining the horizontal wind components as $U = u\cos\phi$, $V = v\cos\phi$ as discussed in Chapter 4 ((4.119) and (4.120)), (G.9) to (G.13) may be expanded into spherical polar coordinates, where λ denotes longitude, ϕ denotes latitude, and a denotes the mean radius of the earth:

$$\frac{\partial}{\partial t}\varsigma = -\frac{1}{a\cos^2\phi}\left(\frac{\partial A}{\partial\lambda} + \cos\phi\frac{\partial B}{\partial\phi}\right) - 2\Omega\left(\sin\phi\, D + \frac{V}{a}\right)$$

$$+ k \cdot \nabla \times (_hF +_v F) \tag{G.20}$$

$$\frac{\partial}{\partial t} D = \frac{1}{a \cos^2 \phi} \left(\frac{\partial B}{\partial \lambda} - \cos \phi \frac{\partial A}{\partial \phi} \right) + 2\Omega \left(\sin \phi \zeta - \frac{U}{a} \right)$$

$$- \nabla^2 (E + \Phi' + RT_0 q) + \nabla \cdot ({}_h F + {}_v F) \tag{G.21}$$

$$\frac{\partial T}{\partial t} = -\frac{1}{a \cos^2 \phi} \left(\frac{\partial}{\partial \lambda} UT' + \cos \phi \frac{\partial}{\partial \phi} VT' \right) + T'D$$

$$+ \dot{\sigma}\gamma + \frac{RT}{C_p} \{ \overline{D} + (V + \overline{V}) \cdot \nabla q \} + \frac{H_c}{C_p}$$

$$+ {}_h F_T + {}_v F_T \tag{G.22}$$

$$\frac{\partial M}{\partial t} = -\frac{1}{a \cos^2 \phi} \left[\frac{\partial}{\partial \lambda} (UM') + \cos \phi \frac{\partial}{\partial \phi} (VM') \right]$$

$$+ M'D - \dot{\sigma} \left(\frac{\partial M}{\partial \sigma} \right) + C + {}_h F_M + {}_v F_M \tag{G.23}$$

$$\frac{\partial q}{\partial t} = \overline{V} \cdot \nabla q + \overline{D} \tag{G.24}$$

Here

$$A = \zeta U + \dot{\sigma} \frac{\partial V}{\partial \sigma} + \frac{RT'}{a} \cos \phi \frac{\partial q}{\partial \phi} \tag{G.25}$$

$$B = \zeta V - \dot{\sigma} \frac{\partial U}{\partial \sigma} - \frac{RT'}{a} \frac{\partial q}{\partial \lambda} \tag{G.26}$$

$$E = \frac{U^2 + V^2}{2 \cos^2 \phi} \tag{G.27}$$

The frictional force F in (G.9) and (G.10) has been divided into vertical and horizontal forces, ${}_v F$ and ${}_h F$, respectively.

The U and V appearing in (G.20) to (G.24) are obtained from the components of (G.14):

$$U = u \cos \phi = -\frac{\cos \phi}{a} \frac{\partial \psi}{\partial \phi} + \frac{1}{a} \frac{\partial \chi}{\partial \lambda} \tag{G.28}$$

$$V = v \cos \phi = \frac{1}{a} \frac{\partial \psi}{\partial \lambda} + \frac{\cos \phi}{a} \frac{\partial \chi}{\partial \phi} \tag{G.29}$$

The prognostic equations (G.20)–(G.24) and the diagnostic equations (G.7), (G.15), (G.16), (G.18), (G.28), and (G.29) constitute the predictive system for ψ, χ, T, M, q. However, to complete the set of equations, the diffusion, diabatic heating, and water vapor condensation must be calculated from predicted and diagnostic variables.

Horizontal and vertical diffusion

The horizontal diffusion terms $_hF$, $_hF_T$, and $_hF_M$ in (G.20), (G.21), (G.22), and (G.23) for the Bourke et al. (1977) model are given by the following equations:

$$\boldsymbol{k} \cdot \boldsymbol{\nabla} \times {}_hF = K_h[\boldsymbol{\nabla}^2\varsigma + 2(\varsigma/a^2)] \tag{G.30}$$

$$\boldsymbol{\nabla} \cdot {}_hF = K_h[\boldsymbol{\nabla}^2 D + 2(D/a^2)] \tag{G.31}$$

$${}_hF_T = K_h\boldsymbol{\nabla}^2 T \tag{G.32}$$

$${}_hF_M = K_h\boldsymbol{\nabla}^2 M \tag{G.33}$$

The vertical diffusion terms $_vF$, $_vF_T$, $_vF_M$ are given by:

$$\{{}_vF, {}_v F_T, {}_v F_M\} = (g/p_s)(\partial/\partial\sigma)\{\tau, \eta, \beta\} \tag{G.34}$$

where

$$\tau = \rho^2(g/p_s)K_v(\partial V/\partial\sigma)$$

$$\eta = \delta\rho^2(g/p_s)K_v(\partial\theta/\partial\sigma) \tag{G.35}$$

$$\beta = \rho^2(g/p_s)K_v(\partial M/\partial\sigma)$$

and θ is potential temperature. K_v is the vertical diffusion coefficient, which is expressed in terms of a mixing length, μ, and the magnitude of the wind shear:

$$K_v = \rho(g/p_s)\mu^2|\partial V/\partial\sigma| \tag{G.36}$$

where μ is assumed to be 30 m for $\sigma \leq 0.5$ and zero for $\sigma > 0.5$, so that vertical diffusion is used only in the lower part of the atmosphere. The V used in the K_v term involves only the global root mean square velocity; thus K_v is essentially a global constant throughout the lower atmosphere.

Heating—precipitation, radiation, and cumulus convection

Physical processes, such as precipitation, solar and infrared radiation, and cumulus convection, will not be described here in detail since the basic methods used in general circulation models have already been described in various subsections of Chapter 3. These processes enter the equations as time rate of change of temperature, i.e., dT/dt, or time rate of change of moisture, i.e., dM/dt.

Surface boundary conditions

The boundary stress, sensible heat flux, and moisture at the earth's surface are given by bulk formulas parameterizing surface flux mechanisms for momentum, τ_*, heat, η_*, and moisture, β_*:

$$\tau_* = \rho_1 C_D |V_1| V_1 \tag{G.37}$$

$$\eta_* = \delta \rho_1 C_D |V_1| (\theta_* - \theta_1) \tag{G.38}$$

$$\beta_* = \rho_1 C_D |V_1| (M_s(T_*) - M_1) \tag{G.39}$$

where the subscripts $_*$ and $_1$ indicate the values computed at the earth's surface and at the first model level above the surface. The potential temperature is related to temperature by (3.50); and

$$\delta = \left(\frac{P}{P_0}\right)^{R/c_p} \tag{G.40}$$

C_D is the drag coefficient used in (3.175)–(3.176) and β_* is multiplied by an evaporative wetness factor, C_W, that ranges between 0 and 1. C_W either can be specified, as in Bourke et al. (1977), or can be related to the soil moisture amount (for example, setting $C_W = W/W_c$ in (3.196)). The surface temperature T_* used in (G.39) can be specified or computed as shown, for example, in (3.193). Finally, M_s is the saturation mixing ratio, obtainable from the saturation vapor pressure, which in turn is a function of temperature alone (e.g., see Murray (1967), List (1951), and the precipitation and cloud processes subsection of Chapter 3).

The model equations can be expressed as truncated expansions of the spherical harmonics, Y_ℓ^m, at each vertical level (see Chapter 4 and Appendix B):

$$\{\psi, \chi, T, M, q, \Phi\} = \sum_{m=-J}^{+J} \sum_{\ell=|m|}^{|m|+J} \{a^2\psi_\ell^m, a^2\chi_\ell^m, T_\ell^m,$$

$$M_\ell^m, q_\ell^m, a^2\Phi_\ell^m\}Y_\ell^m \qquad (G.41)$$

and

$$\{U, V\} = a \sum_{m=-J}^{+J} \sum_{\ell=|m|}^{|m|+J+1} \{U_\ell^m, V_\ell^m\}Y_\ell^m \qquad (G.42)$$

Using (G.28) and (G.29) and the recursion relationship discussed in Appendix B

$$U_\ell^m = (\ell - 1)\epsilon_\ell^m \psi_{\ell-1}^m - (\ell + 2)\epsilon_{\ell+1}^m \psi_{\ell+1}^m + im\chi_\ell^m \qquad (G.43)$$

$$V_\ell^m = -(\ell - 1)\epsilon_\ell^m \chi_{\ell-1}^m + (\ell + 2)\epsilon_{\ell+1}^m \chi_{\ell+1}^m + im\psi_\ell^m \qquad (G.44)$$

where U and V are truncated at $\ell = |m| + J + 1$, which is one component more than that in (G.41). This yields an equivalence of representation between

$$\{\psi, \chi\} \quad \text{and} \quad \{U, V\} \qquad (G.45)$$

Using spherical harmonics, the hydrostatic equation (G.7) becomes

$$\partial\Phi_\ell^m/\partial\sigma = -\left(R/a^2\right)(T_\ell^m/\sigma) \qquad (G.46)$$

Many of the nonlinear terms in (G.20)–(G.24) involve products of spherical harmonics. Some of the complexities of these nonlinear interactions can be avoided by using the transform method described in Chapter 4 and Appendix E. The nonlinear products are defined at each discrete model level, and at each latitude circle, in terms of truncated Fourier series as follows:

$$\{A, B, E, UT', VT', UM', VM', H, I, \overline{V} \cdot \nabla_q\}$$

$$= \sum_{m=-J}^{+J} \{aA_m, aB_m, a^2 E_m, aF_m, aG_m, aP_m, aQ_m,$$

$$H_m, I_m, z_m\} e^{im\lambda} \tag{G.47}$$

The diagnostic vertical velocity $\dot{\sigma}$, via (G.19), is computed only in gridpoint space and thus is not represented spectrally.

The Legendre transform defined by

$$(\ )_\ell^m = \int_{-\pi/2}^{+\pi/2} (\ )_m P_\ell^m(\sin\phi) \cos\phi \, d\phi \tag{G.48}$$

is performed on the quantities E_m, H_m, I_m, and z_m, thereby yielding E_ℓ^m, H_ℓ^m, I_ℓ^m, and z_ℓ^m. Another operator, which allows the transformation of derivatives of nonlinear products, is defined as follows:

$$L_\ell^m(x, y) = \int_{-\pi/2}^{\pi/2} \frac{1}{\cos^2\phi} \{imx_m P_\ell^m(\sin\phi)$$

$$- y_m \cos\phi \frac{\partial P_\ell^m}{\partial \phi}(\sin\phi)\} \cos\phi \, d\phi \tag{G.49}$$

Using (G.41)–(G.49) in (G.20)–(G.24), the spectral prediction equations at each discrete model level are:

$$-\ell(\ell+1)\left(d\psi_\ell^m/dt\right) = -L_\ell^m(A, B) + 2\Omega\{\ell(\ell-1)\epsilon_\ell^m \chi_{\ell-1}^m$$

$$+ (\ell+1)(\ell+2)\epsilon_{\ell+1}^m \chi_{\ell+1}^m - V_\ell^m\}$$

$$+ \psi S_\ell^m \tag{G.50}$$

$$-\ell(\ell+1)\left(d\chi_\ell^m/dt\right) = L_\ell^m(B, -A) - 2\Omega\{\ell(\ell-1)\epsilon_\ell^m \psi_{\ell-1}^m$$

$$+ (\ell+1)(\ell+2)\epsilon_{\ell+1}^m \psi_{\ell+1}^m + U_\ell^m\}$$

$$+ \ell(\ell+1)\left[E_\ell^m + \Phi_\ell^m + \left(RT_0/a^2\right) q_\ell^m\right]$$

$$+ \chi S_\ell^m \tag{G.51}$$

$$dT_\ell^m/dt = -L_\ell^m(F,G) + H_\ell^m + {}_T S_\ell^m \qquad (G.52)$$

$$dM_\ell^m/dt = -L_\ell^m(P,Q) + I_\ell^m + {}_M S_\ell^m \qquad (G.53)$$

$$dq_\ell^m/dt = z_\ell^m - \ell(\ell+1)\overline{\chi_\ell^m} \qquad (G.54)$$

where the Laplacian operator $\nabla^2(\ \ )$ has been replaced throughout by $-\ell(\ell+1)$ (see Appendix B, Eq. (B.30)). The terms $_\psi S_\ell^m, {}_\chi S_\ell^m, {}_T S_\ell^m$, and $_M S_\ell^m$ are the spectral representations of the subtruncation scale diffusive terms; the vertical diffusion introduces coupling between the spectral tendencies at each level.

As described in Chapter 4, the nonlinear products in (G.47) are first evaluated at gridpoints, after being represented at Gaussian latitudes and longitudes using gridpoint values of $\nabla^2\psi, \nabla^2\chi, T, M, U, V, q, \partial q/\partial\lambda$, and $\cos\phi(\partial q/\partial\phi)$. While in gridpoint space, vertical integrals and differentials also are evaluated. Vertical finite differences at intermediate levels are obtained as the weighted averages based on the thickness of the σ layer. Vertical integrals are evaluated with one-sided, uncentered approximations both below the first full level and above the topmost full level; intermediate contributions to integrals assume the half-level value in a centered approximation to the integrand between full levels. Where appropriate, the products, such as $\dot\sigma(\partial U/\partial\sigma)$, are evaluated at half levels prior to weighted averaging to full levels to facilitate incorporation of the upper and lower boundary conditions for $\dot\sigma$.

The spectral amplitudes of the geopotential are computed from vertical integration of (G.46), and the topography is expressed spectrally $(\Phi_{*\ell}^m)$ for (G.7), the lower limit of the hydrostatic integration.

All the diabatic physical processes are computed in gridpoint space except for horizontal diffusion, which can be computed in spectral space. Thus, radiation, convection, and condensation are computed in gridpoint space using the adjustment scheme of Manabe et al. (1965) and lower boundary fluxes and are then transformed to spectral space as rates of change.

Many important details have been omitted from this brief description, such as the details of the semi-implicit method (see Chapter 4), which allows for a much longer time step, or the incorporation of the vertical diffusion terms, which requires the solving of a tridiagonal matrix. It is recommended that the reader consult Baede et al. (1979), Bourke (1974), Bourke et al. (1977), Gordon and Stern (1982), Hoskins and Simmons (1975), Sela (1980), Williamson (1983), and Williamson and

Swarztrauber (1984) for further details on the numerical aspects of the models.

Figure G.1 shows a schematic of the Australian spectral model for one time step. Note that two transforms are made from spectral space to gridpoint space, one to compute the physical processes in gridpoint space and the other to compute the nonlinear terms in (G.50)–(G.54). In some spectral models a single transform loop is used to compute all these quantities together. The advantages and disadvantages of either procedure are dependent largely upon the machine configuration. The possibility of ocean and sea ice calculations has been added to (G.1) for use in coupled atmosphere/ocean/sea ice models (e.g., Washington and VerPlank (1986)). In such a case, the ocean and sea ice calculations determine interactive ocean surface temperatures and sea ice distributions for the atmospheric calculations, and require from the atmospheric calculations such variables as the surface heat flux, momentum fluxes, and precipitation minus evaporation, which is needed for the salinity fluxes.

378

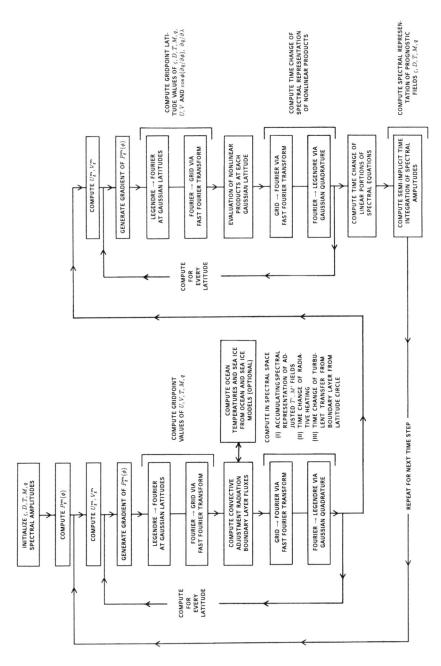

Fig. G.1 Flow diagram for the basic sequence of operations used in the Australian spectral model, with the addition of ocean and sea ice calculations. [Based on Bourke et al. (1977).]

Unit Abbreviations

Å	–	Ångstroms
°C	–	degrees Centigrade
°E	–	degrees east longitude
cm	–	centimeters
g	–	grams
J	–	joules
K	–	kelvins
kg	–	kilograms
km	–	kilometers
kPa	–	kiloPascals
m	–	meters
mb	–	millibars
MJ	–	megajoules
mm	–	millimeters
μm	–	micrometers
°N	–	degrees north latitude
Pa	–	Pascals
s	–	seconds
°S	–	degrees south longitude
sr	–	steradians
W	–	watts
°W	–	degrees west longitude
$'$	–	minutes ($60' = 1°$)
%	–	percent
‰	–	parts per thousand

Physical Constants in International System of Units (SI)

g = acceleration of gravity = 9.806 m s^{-2}

Ω = angular velocity of the earth = 7.292 × 10^{-5} s^{-1}

a = mean radius of earth = 6.37122 × 10^6 m

R = gas constant for dry air = 287.04 J kg^{-1} K^{-1}

c_p = specific heat capacity of dry air at constant pressure = 1.00464 × 10^3 J kg^{-1} K^{-1}

c_{pw} = specific heat of water = 4.19 × 10^3 J kg^{-1} K^{-1} for pure water at 15°C; 3.93 × 10^3 J kg^{-1} K^{-1} for sea water at salinity 30 ‰

c_w = volumetric heat capacity of water = 4.19 MJ m^{-3} K^{-1}

κ = R/c_p

σ = Stefan-Boltzmann constant = 5.67 × 10^{-8} W m^{-2} K^{-4}

L = latent heat of evaporation or condensation = 2.5104 × 10^6 J kg^{-1}

L_f = latent heat of fusion = 0.334×10^5 J kg^{-1}

L_s = latent heat of sublimation = 2.834×10^6 J kg^{-1}

ϵ = ratio of molecular weight of water vapor to dry air = 0.622

Q_s = heat of fusion of snow = 110 MJ m^{-3}

Q_i = heat of fusion of ice = 302 MJ m^{-3}

α_i = shortwave albedo of ice $\simeq 0.50$

ϵ_i = longwave emissivity of ice $\simeq 0.97$

Conversions

1 km	=	1000 m $= 10^5$ cm $= 10^6$ mm
1 m	=	10^6 micrometers (μm)
1 km	=	0.6214 statute mile
1 km	=	0.5396 nautical mile
1 mile	=	1.609 km
1 in	=	2.54 cm
1 kg	=	1000 g
1 kg	=	2.205 lb
1 Langley day^{-1}	=	0.485 W m^{-2}
1 min	=	60 s
1 hr	=	3600 s
1 day	=	86,400 s
1 dyn	=	10^{-5} N (newtons)
1 dyn cm^{-2}	=	10^{-1} N m$^{-2} = 10^{-1}$ Pa (Pascals)
1 J (joule)	=	10^7 ergs $= 0.24$ cal
1 cal	=	4.184 J
1 W (watt)	=	1 J s^{-1}
1 N (newton)	=	1 kg m s^{-2}
1 mb	=	100 N m^{-2}
1 m s^{-1}	=	2.24 mi hr$^{-1} = 1.94$ knots

Greek Alphabet

Name of Letter	Greek Alphabet		Name of Letter	Greek Alphabet	
Alpha	A	α	Nu	N	ν
Beta	B	β	Xi	Ξ	ξ
Gamma	Γ	γ	Omicron	O	o
Delta	Δ	δ	Pi	Π	π
Epsilon	E	ϵ	Rho	P	ρ
Zeta	Z	ζ	Sigma	Σ	σ
Eta	H	η	Tau	T	τ
Theta	Θ	θ	Upsilon	Υ	υ
Iota	I	ι	Phi	Φ	ϕ
Kappa	K	κ	Chi	X	χ
Lambda	Λ	λ	Psi	Ψ	ψ
Mu	M	μ	Omega	Ω	ω

Acronyms

AIDJEX:	Arctic Ice Dynamics Joint Experiment
ANMRC:	Australian Numerical Meteorology Research Centre
CCC:	Canadian Climate Centre
CCM:	Community Climate Model (NCAR)
CLIMAP:	Climate: Long-Range Investigation, Mapping and Prediction
CNRM:	Centre National de Recherches Météorologiques
COHMAP:	Cooperative Holocene Mapping Project
CRREL:	Cold Regions Research and Engineering Laboratory (U.S. Army Corps of Engineers)
DOE:	Department of Energy
ECMWF:	European Centre for Medium-Range Weather Forecasts
ENSO:	El Niño and the Southern Oscillation
ERBE:	Earth Radiation Budget Experiment

FGGE:	First GARP Global Experiment
FMS:	French Meteorological Service
GARP:	Global Atmospheric Research Program
GCM:	General Circulation Model
GFDL:	Geophysical Fluid Dynamics Laboratory (at Princeton; NOAA)
GISS:	Goddard Institute for Space Studies (NASA)
GLA:	Goddard Laboratory for Atmospheres
GLAS:	Goddard Laboratory for Atmospheric Sciences (later changed to Goddard Laboratory for Atmospheres, GLA)
GSFC:	Goddard Space Flight Center
IAP:	Institute of Atmospheric Physics
IAS:	Institute for Advanced Studies
JMA:	Japan Meteorological Agency
JOC:	Joint Scientific Committee
HUM:	Hamburg University
LANL:	Los Alamos National Laboratory
LMD/CNRS:	Laboratoire de Météorologie Dynamique du Centre National de Recherches Scientifiques
NAS:	National Academy of Sciences
NASA:	National Aeronautics and Space Administration
NCAR:	National Center for Atmospheric Research
NMC:	National Meteorological Center
NOAA:	National Oceanic and Atmospheric Administration
NRC:	National Research Council
NSF:	National Science Foundation
OSU:	Oregon State University
PNA:	Pacific-North American pattern
Reading/UK:	University of Reading, United Kingdom
RSMAS:	Rosenstiel School of Marine and Atmospheric Science, University of Miami
UCLA:	University of California, Los Angeles
UKMO:	United Kingdom Meteorological Office

USGS: United States Geological Survey
WMO: World Meteorological Organization
WOCE: World Ocean Circulation Experiment
WCRP: World Climate Research Programme

References

Abramowitz, M., and I.A. Stegun, 1964: *Handbook of Mathematical Functions with Formulas, Graphs, and Mathematical Tables,* National Bureau of Standards Appl. Math. Ser. 55, U.S. Govt. Printing Office, Washington, D.C., 1046 pp. (Reprinted by Dover Publications, Inc., New York, 1965.)

Ackley, S.F., 1981: A review of sea-ice weather relationships in the Southern Hemisphere. *Sea Level, Ice, and Climatic Change,* I. Allison, Ed., International Association of Hydrological Sciences, Guildford, Great Britain, 127–159.

Ackley, S.F., and T.E. Keliher, 1976: Antarctic sea ice dynamics and its possible climatic effects. *Aidjex Bulletin,* **33,** 53–76.

Aleksandrov, V.V., and G.L. Stenchikov, 1983: On the modelling of the climatic consequences of nuclear war. *Proc. Appl. Math.,* The Computing Center of the USSR Academy of Sciences, Moscow, 1–21.

Allison, I., 1981: Antarctic sea ice growth and oceanic heat flux. *Sea Level, Ice, and Climatic Change,* I. Allison, Ed., International Association of Hydrological Sciences, Guildford, Great Britain, 161–170.

Alpatev, A.M., 1954: *Moisture Exchange in Crops,* Gidrometeoizdat, Leningrad, 247 pp.

Alyea, F.M., 1972: Numerical simulation of an ice age paleoclimate, Atmospheric Sciences Paper No. 193, Dept. of Atmos. Sci., Colorado State University, Colo., 120 pp.

Anthes, R.A., 1977: A cumulus parameterization scheme utilizing a one-dimensional cloud model. *Mon. Wea. Rev.,* **105,** 270–286.

Anthes, R.A., 1984: An observational basis for cumulus parameterization. *Report of the Seminar on Progress in Numerical Modelling and the Understanding of Predictability as a Result of the Global Weather Experiment*, Sigtuna, Sweden, October 1984, *GARP Special Report No. 43*, pp. II-1 to II-24. (Available from Secretariat of the WMO, Case Postale No. 5, CH-1211 Geneva, Switzerland.)

Arakawa, A., 1966: Computational design for long-term numerical integrations of the equations of atmospheric motion. *J. Comput. Phys.*, 1, 119–143.

Arakawa, A., and V.R. Lamb, 1977: Computational design of the basic dynamical processes of the UCLA general circulation model. *Methods in Computational Physics*, Vol. 17, Academic Press, New York, 174–265.

Arakawa, A., and W.H. Schubert, 1974: Interaction of a cumulus cloud ensemble with the large-scale environment, Part I. *J. Atmos. Sci.*, 31, 674–701.

Arfken, G., 1966: *Mathematical Methods for Physicists*, Academic Press, New York, 654 pp.

Asselin, R.A., 1972: Frequency filter for time integrations. *Mon. Wea. Rev.*, 100, 487–490.

Augustsson, T., and V. Ramanathan, 1977: A radiative-convective model study of the CO_2 climate problem. *J. Atmos. Sci.*, 34, 448–451.

Bach, W., 1984: *Our Threatened Climate*, D. Reidel Publishing Co., Boston, 368 pp.

Baede, A.P.M., M. Jarraud and U. Cubasch, 1979: Adiabatic formulation and organization of ECMWF's spectral model, ECMWF Tech. Rept. No. 15, European Centre for Medium-Range Weather Forecasts, Reading, England, 40 pp.

Baer, F., 1972: An alternate scale representation of atmospheric energy spectra. *J. Atmos. Sci.*, 29, 649–664.

Baer, F., and G.W. Platzman, 1961: A procedure for numerical integration of the spectral vorticity equation. *J. Meteorol.*, 18, 393–401.

Barron, E.J., 1984: Ancient climates: Investigation with climate models. *Rep. Prog. Phys.*, 47, 1563–1599.

Barron, E.J., 1985: Climate models: Applications for the pre-Pleistocene. *Paleoclimate Analysis and Modeling*, A.D. Hecht, Ed., John Wiley & Sons, New York, 397–421.

Barron, E.J., and W.M. Washington, 1982: Atmospheric circulation during warm geologic periods: Is the equator-to-pole surface-temperature gradient the controlling factor. *Geology*, 10, 633–636.

Barron, E.J., and W.M. Washington, 1984: The role of geographic variables in explaining paleoclimates: Results from Cretaceous climate model sensitivity studies. *J. Geophys. Res.*, 89, 1267–1279.

Barry, R.G., 1983: Arctic Ocean ice and climate: perspectives on a century of polar research, *Annals of the Association of American Geographers*, 73, 485–501.

Barry, R.G., and R.J. Chorley, 1971: *Atmosphere, Weather and Climate*, Methuen & Co., London, 379 pp.

Battan, L.J., 1979: *Fundamentals of Meteorology*, Prentice-Hall, Inc., Englewood Cliffs, N.J., 321 pp.

Betts, A.K., 1986: A new convective adjustment scheme. Part I, Observational and theoretical basis. *Quart. J. Roy. Meteorol. Soc.*, **112**, 677–691.

Betts, A.K., and M.J. Miller, 1986: A new convective adjustment scheme. Part II, Single column tests using GATE wave, BOMEX, ATEX and Arctic air-mass data sets. *Quart. J. Roy. Meteorol. Soc.*, **112**, 693–709.

Bjerknes, J., 1966: A possible response of the atmospheric Hadley circulation to equatorial anomalies of ocean temperature. *Tellus*, **18**, 820–829.

Bjerknes, J., 1969: Atmospheric teleconnections from the equatorial Pacific. *Mon. Wea. Rev.*, **97**, 163–172.

Bjerknes, V., 1904: Das problem von der wettervorhersage, betrachtet vom standpunkt der mechanik und der physik. *Meteor. Z.*, **21**, 1–7.

Blackmon, M.L., 1985: Sensitivity of January climate response to the position of Pacific sea surface temperature anomalies. *Coupled Ocean-Atmosphere Models*, J.C.J. Nihoul, Ed., Elsevier Oceanography Series, 40, Elsevier, Amsterdam, 19–27.

Blackmon, M.L., J.E. Geisler and E.J. Pitcher, 1983: A general circulation model study of January climate anomaly patterns associated with interannual variation of equatorial Pacific sea surface temperatures. *J. Atmos. Sci.*, **40**, 1410–1425.

Boer, G.J., 1985: Modelling the atmospheric response to the 1982/83 El Niño. *Coupled Ocean-Atmosphere Models*, J.C.J. Nihoul, Ed., Elsevier Oceanography Series, 40, Elsevier, Amsterdam, 7–18.

Boer, G.J., N.A. MacFarlane, R. Laprise, J.D. Henderson and J.-P. Blanchet, 1984: The Canadian Climate Centre spectral atmospheric general circulation model. *Atmos.-Ocean*, **22**, 397–429.

Bourke, W., 1972: An efficient, one-level, primitive-equation spectral model. *Mon. Wea. Rev.*, **100**, 683–689.

Bourke, W., 1974: A multi-level spectral model. I. Formulation and hemispheric integrations. *Mon. Wea. Rev.*, **102**, 687–701.

Bourke, W., B. McAvaney, K. Puri and R. Thurling, 1977: Global modeling of atmospheric flow by spectral methods. *Methods in Computational Physics*, Vol. 17, Academic Press, New York, 267–324.

Brown, R.A., 1979: Planetary boundary layer modeling for AIDJEX. *Sea Ice Processes and Models*, R. S. Pritchard, Ed., University of Washington Press, Seattle, 387–401.

Bryan, K., 1963: A numerical investigation of a nonlinear model of a wind-driven ocean, *J. Atmos. Sci.*, **20**, 594–606.

Bryan, K., 1966: A scheme for numerical integration of the equation of motion on an irregular grid free of nonlinear instability. *Mon. Wea. Rev.*, **94**, 38–40.

Bryan, K., 1969a: A numerical method for the study of the circulation of the world ocean. *J. Comput. Phys.*, **4**, 347–376.

Bryan, K., 1969b: Climate and the ocean circulation: III. The ocean model. *Mon. Wea. Rev.*, **97**, 806–827.

Bryan, K., 1974: GFDL global oceanic model. *Modeling for the First GARP Global Experiment*, GARP Publications Series No. 14, June 1974, World Meteorological Organization, Geneva, 252–261.

Bryan, K., 1986: Poleward buoyancy transport in the ocean and mesoscale eddies, *J. Phys. Oceanogr.*, **16**, 927–933.

Bryan, K., and M.D. Cox, 1972: An approximate equation of state for numerical models of ocean circulation. *J. Phys. Oceanogr.*, **2**, 510–514.

Bryan, K., and L.J. Lewis, 1979: A water mass model of the world ocean. *J. Geophys. Res.*, **84**, 2503–2517.

Bryan, K., S. Manabe and R.L. Pacanowski, 1975: A global ocean-atmosphere climate model. Part II: The oceanic circulation. *J. Phys. Oceanogr.*, **5**, 30–46.

Bryan, K., F.G. Komro, S. Manabe and M.J. Spelman, 1982: Transient climate response to increasing atmospheric carbon dioxide. *Science*, **215**, 56–58.

Bryden, H.L., and M.M. Hall, 1980: Heat transport by currents across 25°N latitude in the Atlantic Ocean. *Science*, **207**, 884–885.

Budd, W.F., 1975: Antarctic sea-ice variations from satellite sensing in relation to climate. *J. Glaciology*, **15**, 417–427.

Budd, W.F., and I.N. Smith, 1982: Large-scale numerical modelling of the Antarctic ice sheet. *Annals of Glaciology*, **3**, 42-49.

Budyko, M.I., 1966: Polar ice and climate. *Proc. Symposium on the Arctic Heat Budget and Atmospheric Circulation*, J.O. Fletcher, Ed., Rand Corp. Memorandum RM-5233-NSF, Santa Monica, Calif., 3–22.

Budyko, M.I., 1974: *Climate and Life* (English edition edited by D.H. Miller), International Geophysical Series, Vol. 18, Academic Press, New York, 508 pp.

Burridge, D.M., and A.J. Gadd, 1977: The Meteorological Office Operational 10-level Numerical Weather Prediction Model (December 1975), Meteorological Office, Scientific Paper No. 34, Her Majesty's Stationery Office, London.

Campbell, W.J., 1964: *On the Steady-State Flow of Sea Ice*, Department of Atmospheric Sciences, University of Washington, Seattle, 167 pp.

Carsey, F.D., 1980: Microwave observation of the Weddell polynya. *Mon. Wea. Rev.*, **108**, 2032–2044.

Cavalieri, D.J., and C.L. Parkinson, 1981: Large-scale variations in observed Antarctic sea ice extent and associated atmospheric circulation. *Mon. Wea. Rev.*, **109**, 2323–2336.

Cess, R.D., and V. Ramanathan, 1972: Radiative transfer in the atmosphere of Mars and that of Venus above the cloud deck. *J. Quant. Spectrosc. Radiat. Transfer*, **12**, 933–945.

Charney, J.G., 1948: On the scale of atmospheric motions. *Geofys. Publik.*, **17**, 1–17.

Charney, J.G., 1971: Geostrophic turbulence. *J. Atmos. Sci.*, **28**, 1087–1095.

Charney, J.G., and G.R. Flierl, 1981: Oceanic analogues of large-scale atmospheric motions. *Evolution of Physical Oceanography*, The MIT Press, Cambridge, Mass., 504–548.

Charney, J.G., and N.A. Phillips, 1953: Numerical integration of the quasi-geostrophic equations for barotropic and simple baroclinic flows. *J. Meteorol.*, **10**, 71–99.

Charney, J.G., R. Fjørtoft and J. von Neumann, 1950: Numerical integration of the barotropic vorticity equation. *Tellus*, **2**, 237–254.

Chervin, R.M., 1981: On the comparison of observed and GCM-simulated climate ensembles. *J. Atmos. Sci.*, **38**, 885–901.

Chervin, R.M., 1986: Interannual variability and seasonal climate predictability. *J. Atmos. Sci.*, **43**, 233–251.

Colony, R., and A.S. Thorndike, 1984: An estimate of the mean field of Arctic sea ice motion. *J. Geophys. Res.*, **89**, 10,623–10,629.

Cooley, J.W., and J.W. Tukey, 1965: An algorithm for the machine calculation of complex Fourier series. *Math. Comput.*, **19**, 297–301.

Coon, M.D., 1980: A review of AIDJEX modeling. *Sea Ice Processes and Models*, R.S. Pritchard, Ed., University of Washington Press, Seattle, 12–27.

Coon, M.D., R. Colony, R.S. Pritchard and D.A. Rothrock, 1976: Calculations to test a pack ice model. *Numerical Methods in Geomechanics, II*, C.S. Desai, Ed., American Society of Civil Engineers, New York, 1210–1227.

Coon, M.D., G.A. Maykut, R.S. Pritchard, D.A. Rothrock and A.S. Thorndike, 1974: Modeling the pack ice as an elastic-plastic material. *Aidjex Bulletin*, **24**, 1–105.

Corby, G.A., A. Gilchrist and P.R. Rowntree, 1977: United Kingdom Meteorological Office Five-Level General Circulation Model. *Methods in Computational Physics*, Vol. 17, Academic Press, New York, 67–108.

Courant, R., K.O. Friedrichs and H. Lewy, 1928: Über die partiellen differenzengleichungen der mathematischen physik. *Math. Annalen*, **100**, 32–74.

Covey, C., S.H. Schneider and S.L. Thompson, 1984: Global atmospheric effects of massive smoke injections from a nuclear war: Results from general circulation model simulations. *Nature*, **308**, 21–25.

Covey, C., S.L. Thompson and S.H. Schneider, 1985: "Nuclear winter": A diagnosis of atmospheric general circulation model simulations. *J. Geophys. Res.*, **90**, 5615–5628.

Cox, M.D., 1975: A baroclinic numerical model of the world ocean: Preliminary results. *Numerical Models of Ocean Circulation*, National Academy of Sciences, Washington, D.C., 107–120.

Cox, M.D., 1984: A primitive equation, three-dimensional model of the ocean. GFDL Ocean Group, Tech. Rept. No. 1, 250 pp.

Cox, M.D., 1985: An eddy resolving numerical model of the ventilated thermocline. *J. Phys. Oceanogr.*, **15**, 1312–1324.

Cox, M.D., and K. Bryan, 1984: A numerical model of the ventilated thermocline. *J. Phys. Oceanogr.*, **14**, 674–687.

Crowley, T.J., 1983: The geologic record of climatic change. *Rev. Geophys. Space Phys.*, **21**, 828–877.

Crowley, T.J., D.A. Short, J.G. Mengel and G.R. North, 1986: Role of seasonality in the evolution of climate during the last 100 million years. *Science*, **231**, 579–584.

Crowley, W.P., 1968: A global numerical ocean model: Part I. *J. Comp. Phys.*, **3**, 111–147.

Crowley, W.P., 1970: A numerical model for viscous, free-surface, barotropic wind driven ocean circulations. *J. Comp. Phys.*, **5**, 139–168.

Crutzen, P.J., and J.W. Birks, 1982: The atmosphere after a nuclear war: Twilight at noon. *Ambio*, **11**, 114–125.

Cubasch, U., 1985: The mean response of the ECMWF global model to the composite El Niño anomaly in extended range prediction experiments. *Coupled Ocean-Atmosphere Models*, J.C.J. Nihoul, Ed., Elsevier Oceanography Series, 40, Elsevier, Amsterdam, 329–344.

Deardorff, J.W., 1966: The counter-gradient heat flux in the lower atmosphere and in the laboratory. *J. Atmos. Sci.*, **23**, 503–506.

Deardorff, J.W., 1972: Parameterization of the planetary boundary layer for use in general circulation models. *Mon. Wea. Rev.*, **100**, 93–106.

Demidovich, B.P., and I.A. Maron, 1976: *Computational Mathematics*, Mir Publishers, Moscow, 611 pp.

Déqué, M., and J.F. Royer, 1983: Présentation et validation d'un modéle de climate. *La Météorologie*, **32**, 77–88.

Dettman, J.W., 1969: *Mathematical Methods in Physics and Engineering*, McGraw-Hill Book Co., New York, 428 pp.

Dickinson, R.E., 1983: Land surfaces processes and climate-surface albedos and energy balance. *Adv. Geophys.*, **25**, 305–353.

Dickinson, R.E., 1984: Modeling evapotranspiration for three-dimensional global climate models. *Climate Processes and Climate Sensitivity*, J.E. Hansen and T. Takahashi, Eds., Maurice Ewing Series, Vol. 5, American Geophysical Union, Washington, D.C., 58–72.

Dickinson, R.E., 1985: The climate system and modeling of future climate. *An Assessment of the Role of Carbon Dioxide and of Other Radiatively Active Constituents in Climate Variations and Associated Impacts*, The International Meteorological Institute, Stockholm, 61 pp.

Dobson, G.M.B., 1968: *Exploring the Atmosphere*, Clarendon Press, Oxford, 209 pp.

Donner, L.J., H.-L. Kuo and E.J. Pitcher, 1982: The significance of thermodynamic forcing by cumulus convection in a general circulation model. *J. Atmos. Sci.*, **39**, 2159–2181.

Dunbar, M., and W. Wittman, 1963: Some features of ice movement in the Arctic basin. *Proc. Arctic Basin Symposium*, October 1962, Arctic Institute of North America, Hershey, Pa., 90–103.

Dutton, J.A., 1976: *The Ceaseless Wind: An Introduction to the Theory of Atmospheric Motion*, McGraw-Hill Book Co., New York, 579 pp.

Ekman, V.W., 1905: On the influence of the earth's rotation on ocean currents. *Arkiv. Matem., Astr. Fys.*, **2**, 11, 53 pp.

Eliasen, E., B. Machenhauer and E. Rasmussen, 1970: On a numerical method for integration of the hydrodynamical equations with a spectral representation of the horizontal fields, Rept. No. 2, Institute for Theoretical Meteorology, Copenhagen University, Copenhagen, 35 pp.

Eliassen, A., 1949: The quasi-static equations of motion with pressure as independent variable. *Geofys. Publik.*, **17**, No. 3, 44 pp.

Ellsaesser, H.W., 1966: Evaluation of spectral versus grid method of hemispheric numerical weather prediction. *J. Appl. Meteorol.*, **5**, 246–262.

Elsasser, W.M., and M.F. Culbertson, 1960: *Atmospheric Radiation Tables*, Meteorological Monograph No. 4, American Meteorological Society, Boston, Mass., 43 pp.

Esbensen, S.K., 1985: The response of the OSU two-level atmospheric general circulation model to a warm sea-surface temperature anomaly over the eastern equatorial Pacific Ocean. *Coupled Ocean-Atmosphere Models*, J.C.J. Nihoul, Ed., Elsevier Oceanography Series, 40, Elsevier, Amsterdam, 371–390.

Ewing, M., and W.L. Donn, 1956: A theory of ice ages. *Science*, **123**, 1061–1066.

Felzenbaum, A.I., 1958: The theory of the steady drift of ice and the calculation of the long period mean drift in the central part of the Arctic basin. *Problems of the North*, **2**, 5–44.

Fennessy, M.J., L. Marx and J. Shukla, 1985: GCM sensitivity to 1982–1983 equatorial Pacific sea-surface temperature anomalies. *Coupled Ocean-Atmosphere Models*, J.C.J. Nihoul, Ed., Elsevier Oceanography Series, 40, Elsevier, Amsterdam, 121–130.

Fleagle, R.G., and J.A. Businger, 1980: *An Introduction to Atmospheric Physics*, Academic Press, New York, 432 pp.

Fletcher, J.O., 1969: Ice Extent on the Southern Ocean and Its Relation to World Climate, Memorandum RM-5793-NSF, March, Rand Corp., Santa Monica, CA, 108 pp.

Fofonoff, N.P., 1962: Physical properties of sea water. *The Sea*, Vol. 1, Interscience, New York, 864 pp.

Fux-Rabinovich, M.S., 1974: Global 5-level general circulation model. *Modeling for the First GARP Global Experiment*, GARP Publications Series No. 14, World Meteorological Organization, Geneva, 101–112.

Gadd, A.J., 1978: A split explicit integration scheme for numerical weather prediction. *Quart J. Roy. Meteorol. Soc.*, **104**, 569–582.

Gadd, A.J., 1980: Two refinements of the split explicit integration scheme. *Quart. J. Roy. Meteorol. Soc.*, **106**, 215–220.

Gary, J.M., 1973: Estimate of truncation error in transformed coordinate, primitive equation atmospheric models. *J. Atmos. Sci.*, **30**, 223–233.

Gary, J.M., 1979: Nonlinear instability. *Numerical Methods Used in At-mospheric Models*, Vol. II, GARP Publications Series No. 17, World Meteorological Organization, Geneva, 476–499.

Gates, W.L., 1976a: Modeling the ice-age climate. *Science*, **191**, 1138–1144.

Gates, W.L., 1976b: The numerical simulation of ice-age climate with a global general circulation model. *J. Atmos. Sci.*, **33**, 1844–1873.

Gates, W.L., 1981: Paleoclimatic Modeling–A Review of Problems and Prospects for the Pre-Pleistocene, Climatic Research Institute Rept. No. 27, Oregon State University, Corvallis, Oregon, 44 pp.

Gates, W.L., and M.E. Schlesinger, 1977: Numerical simulation of the January and July global climate with a two-level atmospheric model. *J. Atmos. Sci.*, **34**, 36–76.

Gates, W.L., Y.-J. Han and M.E. Schlesinger, 1985: The global climate simulated by a coupled atmosphere-ocean general circulation model: Preliminary results. *Coupled Ocean-Atmosphere Models*, J.C.J. Ni-houl, Ed., Elsevier Oceanography Series, 40, Elsevier, Amsterdam, pp. 131–151.

Geisler, J.E., M.L. Blackmon, G.T. Bates, and S. Muñoz, 1985: Sensitivity of January climate response to the magnitude and position of a warm equatorial Pacific sea surface temperature anomaly. *J. Atmos. Sci.*, **42**, 1037–1049.

Gill, A.E., 1982: *Atmosphere-Ocean Dynamics*, Academic Press, New York, 662 pp.

Goody, R.M., 1952: A statistical model for water-vapour absorption. *Quart. J. Roy. Meteorol. Soc.*, **78**, 165–169.

Goody, R.M., 1964: *Atmospheric Radiation I: Theoretical Basis*, Oxford University Press (Clarendon Press), London and New York, 436 pp.

Gordienko, P., 1958: Arctic ice drift. *Proc. Conference on Arctic Sea Ice*, Publ. 598, National Academy of Sciences and National Research Council, Washington, D.C., 210–222.

Gordon, A.L., 1978: Deep Antarctic convection west of Maud Rise. *J. Phys. Oceanogr.*, **8**, 600–612.

Gordon, A.L., 1981: Seasonality of southern ocean sea ice. *J. Geophys. Res.*, **86**, 4193–4197.

Gordon, C.T., and W.F. Stern, 1982: A description of the GFDL Global Spectral Model. *Mon. Wea. Rev.*, **110**, 625–644.

Grammeltvedt, A., 1969: A survey of finite-difference schemes for the primitive equations for a barotropic fluid. *Mon. Wea. Rev.*, **97**, 384–405.

Grimmer, M., and D.B. Shaw, 1967: Energy–preserving integrations of the primitive equations on the sphere. *Quart. J. Roy. Meteorol. Soc.*, **93**, 337–349.

Hack, J.J., W.H. Schubert and P.L.S. Dias, 1984: A spectral cumulus parameterization for use in numerical models of the tropical atmosphere. *Mon. Wea. Rev.*, **112**, 704–716.

Haltiner, G.J., and R.T. Williams, 1980: *Numerical Prediction and Dynamic Meteorology*, John Wiley and Sons, New York, 477 pp.

Han, Y.-J., 1984a: A numerical world ocean general circulation model. Part I. Basic design and barotropic experiment. *Dyn. Atmos. Oceans*, **8**, 107–140.

Han, Y.-J., 1984b: A numerical world ocean general circulation model. Part II. A baroclinic experiment. *Dyn. Atmos. Oceans*, **8**, 141–172.

Han, Y.-J., M.E. Schlesinger and W.L. Gates, 1985: An analysis of the air-sea-ice interaction simulated by the OSU-coupled atmosphere-ocean general circulation model. *Coupled Ocean-Atmosphere Models*, J.C.J. Nihoul, Ed., Elsevier Oceanography Series, 40, Elsevier, Amsterdam, 167–182.

Hansen, J., D. Johnson, A. Lacis, S. Lebedeff, P. Lee, D. Rind and G. Russell, 1981: Climate impact of increasing atmospheric carbon dioxide. *Science*, **213**, 957–966.

Hansen, J., A. Lacis, D. Rind, G. Russell, P. Stone, I. Fung, R. Ruedy and J. Lerner, 1984: Climate sensitivity: Analysis of feedback mechanisms. *Climate Processes and Climate Sensitivity*, J.E. Hansen and T. Takahashi, Eds., Maurice Ewing Series, Vol. 5, American Geophysical Union, Washington, D.C., 130–163.

Hansen, J., G. Russell, D. Rind, P. Stone, A. Lacis, S. Lebedeff, R. Ruedy and L. Travis, 1983: Efficient three-dimensional global models for climate studies: Models I and II. *Mon. Wea. Rev.*, **111**, 609–662.

Hastenrath, S., 1980: Heat budget of tropical ocean and atmosphere. *J. Phys. Oceanogr.*, **10**, 159–170.

Haurwitz, B., 1940: The motion of atmospheric disturbances on the spherical earth. *J. Mar. Res.*, **3**, 254–267.

Haurwitz, B., 1941: *Dynamical Meteorology*, McGraw-Hill Book Co., New York, 365 pp.

Held, I.M., 1982: Climate models and the astronomical theory of the ice ages. *Icarus*, **50**, 449–461.

Hibler, W.D. III, 1979: A dynamic thermodynamic sea ice model. *J. Phys. Oceanogr.*, **9**, 815–846.

Hibler, W.D. III, and S.F. Ackley, 1983: Numerical simulation of the Weddell Sea pack ice. *J. Geophys. Res.*, **88**, 2873–2887.

Hibler, W.D. III, and K. Bryan, 1984: Ocean circulation: Its effects on seasonal sea-ice simulations. *Science*, **224**, 489–492.

Hibler, W.D. III, and J.E. Walsh, 1982: On modeling seasonal and interannual fluctuations of Arctic sea ice. *J. Phys. Oceanogr.*, **12**, 1514–1523.

Hoffman, G.-R., P.N. Swarztrauber and R.A. Sweet, 1984: Aspects of using multiprocessors for meteorological modeling. *Proc. Workshop on Using Multiprocessors in Meteorological Models*, 3–6 December, European Centre for Medium Range Weather Forecasts, Reading, U.K.

Holland, W.R., T. Keffer and P.B. Rhines, 1984: Dynamics of the oceanic general circulation: The potential vorticity field. *Nature*, **308**, 698–705.

Holton, J.R., 1979: *An Introduction to Dynamic Meteorology*, Academic Press, New York, 391 pp.

Hoskins, B.J., and A.J. Simmons, 1975: A multi-layer spectral model and the semi-implicit method. *Quart. J. Roy. Meteorol. Soc.*, **101**, 637–655.

Houghton, J. T., 1979: *The Physics of Atmospheres*, Cambridge University Press, London, 203 pp.

Hunkins, K., 1966: Ekman drift currents in the Arctic Ocean. *Deep Sea Res.*, **13**, 607–620.

Hunt, B.G., and N.C. Wells, 1979: An assessment of the possible future climatic impact of carbon dioxide increases based on a coupled one-dimensional atmospheric-oceanic model. *J. Geophys. Res.*, **84**, 787–791.

Julian, P.R., and R.M. Chervin, 1978: A study of the Southern Oscillation and Walker Circulation phenomenon. *Mon. Wea. Rev.*, **106**, 1433–1451.

Kalnay, E., R. Balgovind, W. Chao, D. Edelmann, J. Pfaendtner, L. Takacs, and K. Takano, 1983: Documentation of the GLAS Fourth Order General Circulation Model, Vol. 1: Model Documentation, Laboratory for Atmospheric Sciences, Global Modeling and Simulation Branch, Goddard Space Flight Center, Greenbelt, Maryland, NASA Technical Memorandum 86064.

Kalnay-Rivas, E., A. Bayliss and J. Storch, 1977: The 4th order GISS model of the global atmosphere. *Beitr. Phys. Atmos.*, **50**, 299–311.

Kanamitsu, M., K. Tada, T. Kudo, N. Sato and S. Isa, 1983: Description of the JMA Operational Spectral Model. *J. Meteorol. Soc. Japan*, **61**, 812–828.

Kasahara, A., 1974: Various vertical coordinate systems used for numerical weather prediction. *Mon. Wea. Rev.*, **102**, 504–522.

Kasahara, A., 1977: Computational aspects of numerical models for weather prediction and climate simulation. *Methods in Computational Physics*, Vol. 17, Academic Press, New York, 2–66.

Kasahara, A., and W.M. Washington, 1967: NCAR global general circulation model of the atmosphere. *Mon. Wea. Rev.*, **95**, 389–402.

Kasahara, A., and W.M. Washington, 1971: General circulation experiments with a six-layer NCAR model, including orography, cloudiness and surface temperature calculations. *J. Atmos. Sci.*, **28**, 657–701.

Keeling, C.D., R.B. Bacastow and T.P. Whorf, 1982: Measurements of concentration of carbon dioxide at Mauna Loa Observatory, Hawaii. *Carbon Dioxide Review*, W.C. Clark, Ed., Oxford University Press, 377–385.

Keeling, C.D., J.A. Adams, C.A. Ekdahl and P.R. Guenther, 1976a: Atmospheric carbon dioxide variations at the South Pole. *Tellus*, **28**, 552–564.

Keeling, C.D., R.B. Bacastow, A.E. Bainbridge, C.A. Ekdahl, P.R. Guenther, L.S. Waterman and J.F.S. Chin, 1976b: Atmospheric carbon dioxide variations at Mauna Loa Observatory, Hawaii. *Tellus*, **28**, 538–551.

Kellogg, W.W., 1975: Climatic feedback mechanisms involving the polar regions. *Climate of the Arctic*, G. Weller and S.A. Bowling, Eds., Geophysical Institute, University of Alaska, 111–116.

Keshavamurty, R.N., 1982: Response of the atmosphere to sea-surface temperature anomalies over the equatorial Pacific and the teleconnection of the Southern Oscillation. *J. Atmos. Sci.*, **39**, 1241–1259.

Kibel, I.A., 1940: Prilozhenie k meteorologii uravnenii mekhaniki baroklinnoi zhidkosti (Application of the baroclinic fluid mechanics equations to meteorology). *Izv. Akad. Nauk SSSR*, Ser. Geogr. Geofiz. No. 5.

Killworth, P.D., 1983: Deep convection in the world ocean. *Revs. Geophys. Space Phys.*, **21**, 1–26.

Knauss, J.A., 1978: *Introduction to Physical Oceanography*, Prentice-Hall, Inc., Englewood Cliffs, N.J., 338 pp.

Kreiss, H., and J. Oliger, 1973: Methods for the approximate solution of time dependent problems. WMO/ICSU Joint Organizing Committee, GARP Publications Series No. 10, World Meteorological Organization, Geneva, 107 pp.

Krichak, S.O., and M.S. Fux-Rabinovich, 1972: Baroclinic primitive hydrodynamic equation model for an extratropical part of the Northern Hemisphere with the inclusion of moisture exchange processes. *Meteorologia i Gidrologia*, No. 1, 26–36.

Krishnamurti, T.N., S.-L. Nam and R. Pasch, 1983: Cumulus parameterization and rainfall rates II. *Mon. Wea. Rev.*, **111**, 815–828.

Krishnamurti, T.N., V. Ramanathan, H.-L. Pan, R.J. Pasch and J. Molinari, 1980: Cumulus parameterization and rainfall rates I. *Mon. Wea. Rev.*, **108**, 465–472.

Kubota, S., M. Hirose, Y. Kikuchi and Y. Kurihara, 1961: Barotropic forecasting with the use of surface spherical harmonic representation. *Pap. Meteorol. Geophys.*, **12**, 199–215.

Kuo, H.L., 1974: Further studies of the parameterization of the influence of cumulus convection on large-scale flow. *J. Atmos. Sci.*, **31**, 1232–1240.

Kutzbach, J.E., 1981: Monsoon climate of the early Holocene: Climatic experiment using the Earth's orbital parameters for 9000 years ago. *Science*, **214**, 59–61.

Kutzbach, J.E., and P.J. Guetter, 1984a: Sensitivity of late-glacial and holocene climates to the combined effects of orbital parameter changes and lower boundary condition changes: 'Snapshot' simulations with a general circulation model for 18, 9 and 6 ka BP. *Annals of Glaciology*, **5**, 85–87.

Kutzbach, J.E., and P.J. Guetter, 1984b: The sensitivity of monsoon climates to orbital parameter changes for 9,000 years BP: Experiments with the NCAR general circulation model. *Milankovitch and Climate*, Part 2, A.L. Berger et al., Eds., D. Reidel Publishing Company, Dordrecht, 801–820.

Kutzbach, J.E., and B.L. Otto-Bliesner, 1982: The sensitivity of the African-Asian monsoonal climate to orbital parameter changes for 9000 yr BP in a low-resolution general circulation model. *J. Atmos. Sci.*, **39**, 1177–1188.

Kwizak, M., and A.J. Robert, 1971: A semi-implicit scheme for grid-point atmospheric models of the primitive equations. *Mon. Wea. Rev.*, **99**, 32–36.

Lacis, A.A., and J.E. Hansen, 1974: A parameterization for the absorption of solar radiation in the earth's atmosphere. *J. Atmos. Sci.*, **31**, 118–133.

Lacis, A., J. Hansen, P. Lee, T. Mitchell, S. Lebedeff, 1981: Greenhouse effect of trace gases, 1970–1980. *Geophys. Res. Lett.*, **8**, 1035–1038.

Langleben, M.P., 1972: The decay of an annual cover of sea ice. *J. Glaciology*, **11**, 337–344.

Lau, N.-C., and A.H. Oort, 1985: Response of a GFDL general circulation model to SST fluctuations observed in the tropical Pacific Ocean during the period 1962–1976. *Coupled Ocean-Atmosphere Models*, J.C.J. Nihoul, Ed., Elsevier Oceanography Series, 40, Elsevier, Amsterdam, 289–302.

Lau, N.-C., G.H. White and R.L. Jenne, 1981: Circulation Statistics for the Extratropical Northern Hemisphere Based on NMC Analyses, NCAR Tech. Note TN-171-STR, 138 pp.

Leith, C.E., 1965: Numerical simulation of the earth's atmosphere, *Methods in Computational Physics*, vol. 4, Applications in Hydrodynamics, Academic Press, New York, 1–28.

Levitus, S., 1982: Climatological Atlas of the World Ocean, NOAA Professional Paper No. 13, Dept. of Commerce, Rockville, Md. (U.S. GPO, Washington, D.C.), 173 pp.

Lilly, D.K., 1965: On the computational stability of numerical solutions of time-dependent non-linear geophysical fluid dynamics problems. *Mon. Wea. Rev.*, **93**, 11–26.

Ling, C.H., and C.L. Parkinson, 1986: Arctic sea ice extent and drift, modeled as a viscous fluid, *Ocean Sci. Engineering*, **11**, 71–98.

Ling, C.H., L.A. Rasmussen and W.J. Campbell, 1980: A continuum sea ice model for a global climate model. *Sea Ice Processes and Models*, R.S. Pritchard, Ed., University of Washington Press, Seattle, 187–196.

Liou, K.N., 1980: *An Introduction to Atmospheric Radiation*, Int. Geophys. Ser., Vol. 25, Academic Press, New York, 392 pp.

List, R.J., 1951: *Smithsonian Meteorological Tables*, 6th ed., Smithson. Misc. Collect. No. 114, Smithsonian Institution, Washington, D.C.

Lord, S.J., 1982: Interaction of a cumulus cloud ensemble with the large-scale environment. Part III: Semi-prognostic test of the Arakawa-Schubert parameterization. *J. Atmos. Sci.*, **39**, 88-103.

Machenhauer, B., 1979: The spectral method. *Numerical Methods Used in Atmospheric Models*, Vol. II, GARP Publications Series No. 17, World Meteorological Organization, Geneva, 121–275.

Machenhauer, B., and R. Daley, 1972: A baroclinic primitive equation model with a spectral representation in three dimensions. Rept. No. 4, Institute for Theoretical Meteorology, Copenhagen University, Copenhagen, 63 pp.

Malkmus, W., 1967: Random Lorentz band model with exponential-tailed S^{-1} line intensity distribution function. *J. Opt. Soc. Am.*, **57**, 323–329.

Malone, R.C., L.H. Auer, G.A. Glatzmaier, M.C. Wood and O.B. Toon, 1986: Nuclear winter: three-dimensional simulations including interactive transport, scavenging, and solar heating of smoke. *J. Geophys. Res.*, **91**, 1039–1053.

Manabe, S., 1969a: Climate and the ocean circulation: I. The atmospheric circulation and the hydrology of the earth's surface. *Mon. Wea. Rev.*, **97**, 739–774.

Manabe, S., 1969b: Climate and the ocean circulation: II. The atmospheric circulation and the effect of heat transfer by ocean currents. *Mon. Wea. Rev.*, **97**, 775–805.

Manabe, S., and A.J. Broccoli, 1985: The influence of continental ice sheets on the climate of an ice age. *J. Geophys. Res.*, **90**, 2167–2190.

Manabe, S., and K. Bryan, 1969: Climate calculations with a combined ocean-atmosphere model. *J. Atmos. Sci.*, **26**, 786–789.

Manabe, S., and K. Bryan, 1985: CO_2-induced change in a coupled ocean-atmosphere model and its paleoclimatic implications. *J. Geophys. Res.*, **90**, 11689–11707.

Manabe, S., and D.G. Hahn, 1977: Simulation of the tropical climate of an ice age. *J. Geophys. Res.*, **82**, 3889–3911.

Manabe, S., and D.G. Hahn, 1981: Simulation of atmospheric variability. *Mon. Wea. Rev.*, **109**, 2260–2286.

Manabe, S., and R.J. Stouffer, 1980: Sensitivity of a global climate model to an increase of CO_2 concentration in the atmosphere. *J. Geophys. Res.*, **85**, 5529–5554.

Manabe, S., and R.F. Strickler, 1964: On the thermal equilibrium of the atmosphere with a convective adjustment. *J. Atmos. Sci.*, **21**, 361–385.

Manabe, S., and R.T. Wetherald, 1967: Thermal equilibrium of the atmosphere with a given distribution of relative humidity. *J. Atmos. Sci.*, **24**, 241–259.

Manabe, S., and R.T. Wetherald, 1975: The effects of doubling the CO_2 concentration on the climate of a general circulation model. *J. Atmos. Sci.*, **32**, 3–15.

Manabe, S., and R.T. Wetherald, 1980: On the distribution of climate change resulting from an increase of CO_2-content of the atmosphere. *J. Atmos. Sci.*, **37**, 99–118.

Manabe, S., K. Bryan and M.J. Spelman, 1975: A global ocean-atmosphere climate model. Part 1: The atmospheric circulation. *J. Phys. Oceanogr.*, **5**, 3–29.

Manabe, S., K. Bryan and M.J. Spelman, 1979: A global ocean-atmosphere climate model with seasonal variation for future studies of climate sensitivity. *Dyn. Atmos. Oceans*, **3**, 393–426.

Manabe, S., J. Smagorinsky and R.F. Strickler, 1965: Simulated climatology of a general circulation model with a hydrologic cycle. *Mon. Wea. Rev.*, **93**, 769–798.

Marchuk, G.I., V.P. Dymnikov, V.N. Lykossov, V.B. Zalesny, V.Ya. Galin, 1984: *Mathematical Modelling of the General Circulation of the Atmosphere and Ocean*, Gidrometeoizdat, Leningrad, 320 pp.

Martinson, D.G., P.D. Killworth and A.L. Gordon, 1981: A convective model for the Weddell polynya. *J. Phys. Oceanogr.*, **11**, 466–488.

Mason, B.J., 1985: First Implementation Plan for the World Climate Research Programme, World Climate Research Programme (WCRP) Publications Series No. 5, WMO/TD–No. 80, World Meteorological Organization, Geneva, 123 pp.

Maykut, G.A., 1978: Energy exchange over young sea ice in the central Arctic. *J. Geophys. Res.*, **83**, 3646–3658.

Maykut, G.A., and N. Untersteiner, 1969: *Numerical Prediction of the Thermodynamic Response of Arctic Sea Ice to Environmental Changes*, Rand Corp. Memo. RM-6093-PR, Santa Monica, Calif., 173 pp.

Maykut, G.A., and N. Untersteiner, 1971: Some results from a time-dependent thermodynamic model of sea ice. *J. Geophys. Res.*, **76**, 1550–1575.

McAvaney, B.J., W. Bourke and K. Puri, 1978: A global spectral model for simulation of the general circulation. *J. Atmos. Sci.*, **35**, 1557–1583.

McPhee, M.G., 1975: Ice-ocean momentum transfer for the AIDJEX ice model. *Aidjex Bull.*, **29**, 93–111.

McPhee, M.G., 1986: The upper ocean. *The Geophysics of Sea Ice*, N. Untersteiner, Ed., Plenum Press, New York, 339–394.

McPhee, M.G., and J.D. Smith, 1976: Measurements of the turbulent boundary layer under pack ice. *J. Phys. Oceanogr.*, **6**, 696–711.

Meehl, G.A., 1987: The tropics and their role in the global climate system. *Geogr. J.*, **153**, 21–36.

Meehl, G.A., W.M. Washington and A.J. Semtner, Jr., 1982: Experiments with a global ocean model driven by observed atmospheric forcing. *J. Phys. Oceanogr.*, **12**, 301–312.

Merilees, P., 1976: Fundamentals of large-scale numerical weather prediction. *Weather Forecasting and Weather Forecasts: Models, Systems, and Users*, Vol. 1, Notes from a colloquium, summer 1976, NCAR/CQ-5+1976-ASP, coordinators, A. Murphy and D. Williamson, 2–138.

Mesinger, F., and A. Arakawa, 1976: Numerical methods used in atmospheric models. WMO/ICSU Joint Organizing Committee, GARP Publications Series No. 17, World Meteorological Organization, Geneva, 64 pp.

Mintz, Y., 1984: The sensitivity of numerically simulated climates to land-surface boundary conditions. *The Global Climate*, J.T. Houghton, Ed., Cambridge University Press, New York, 79–103.

Monin, A.S., 1972: *Weather Forecasting as a Problem in Physics*, The MIT Press, Cambridge, Mass., 199 pp.

Morrey, C.B., Jr., 1962: *University Calculus*, Addison-Wesley Publishing Co., Reading, Mass., p. 531.

Murray, F.W., 1967: On the computation of saturation vapor pressure. *J. Appl. Meteorol.*, **6**, 203–204.

Nansen, F., 1902: The oceanography of the north polar basin. *Norwegian North Pole Expedition 1893–1896, Scientific Results*, Vol. 3, Longmans, Green, Toronto, 357–386.

NAS, 1985: *The Effects on the Atmosphere of a Nuclear Exchange*, National Academy of Sciences Committee on the Atmospheric Effects of Nuclear Explosions, National Academy Press, Washington, D.C.

Neumann, G., and W.J. Pierson Jr., 1966: *Principles of Physical Oceanography*, Prentice-Hall, Englewood Cliffs, N.J., 545 pp.

Newell, R.E., J.W. Kidson, D.G. Vincent and G.J. Boer, 1972: *The General Circulation of the Tropical Atmosphere and Interactions with Extratropical Latitudes*, Vol. 1, The MIT Press, Cambridge, Mass., 258 pp.

North, G.R., J.G. Mengel and D.A. Short, 1983: Simple energy balance model resolving the seasons and the continents: Application to the astronomical theory of the ice ages. *J. Geophys. Res.*, **88**, 6576–6586.

NRC (National Research Council), 1975: *Understanding Climatic Change: A Program for Action*, U.S. Committee for the GARP, National Academy of Sciences, Washington, D.C., 239 pp.

Obukhov, A.M., 1946: Turbulence in an atmosphere with nonuniform temperature. *Tr. Akad. Nauk, SSSR Inst. Teoret. Geofi*, No. 1 (English translation in *Boundary Layer Meteorol.*, **2**, 7–29, 1971).

Oort, A.H., and T.H. Vonder Haar, 1976: On the observed annual cycle in the ocean-atmosphere heat balance over the Northern Hemisphere. *J. Phys. Oceanogr.*, **6**, 781–800.

Orszag, S.A., 1970: Transform method for calculation of vector-coupled sums: Application to the spectral form of the vorticity equation. *J. Atmos. Sci.*, **27**, 890–895.

Orszag, S.A., 1971: Numerical simulation of incompressible flows within simple boundaries: Accuracy. *J. Fluid Mech.*, **49**, 75–112.

Pacanowski, R.C., and S.H. Philander, 1981: Parameterization of vertical mixing in numerical models of tropical oceans. *J. Phys. Oceanogr.*, **11**, 1443–1451.

Palmer, T.N., 1985: Response of the UK Meteorological Office general circulation model to sea-surface temperature anomalies in the tropical Pacific Ocean. *Coupled Ocean-Atmosphere Models*, J.C.J. Nihoul, Ed., Elsevier Oceanography Series, 40, Elsevier, Amsterdam, 83–108.

Paltridge, G.W., and C.M.R. Platt, 1976: *Radiative Process in Meteorology and Climatology*, Elsevier, Amsterdam, 318 pp.

Parkinson, C.L., 1983: On the development and cause of the Weddell polynya in a sea ice simulation. *J. Phys. Oceanogr.*, **13**, 501–511.

Parkinson, C.L., 1985: Possible sea ice impacts on oceanic deep convection. *North Atlantic Deep Water Formation*, T. Bennett, W. Broecker, and J. Hansen, Eds., NASA Conference Publication 2367, National Aeronautics and Space Administration, Washington, D.C., 39–41.

Parkinson, C.L., and R.A. Bindschadler, 1984: Response of Antarctic sea ice to uniform atmospheric temperature increases. *Climate Processes and Climate Sensitivity*, J.E. Hansen and T. Takahashi,

Eds., Maurice Ewing Series, Vol. 5, American Geophysical Union, Washington, D.C., 254–264.

Parkinson, C.L., and D.J. Cavalieri, 1982: Interannual sea-ice variations and sea-ice/atmosphere interactions in the southern ocean, 1973–1975. *Annals of Glaciology*, **3**, 249–254.

Parkinson, C.L., and A.J. Gratz, 1983: On the seasonal sea ice cover of the Sea of Okhotsk. *J. Geophys. Res.*, **88**, 2793–2802.

Parkinson, C.L., and G.F. Herman, 1980: Sea ice simulations based on fields generated by the GLAS GCM. *Mon. Wea. Rev.*, **108**, 2080–2091.

Parkinson, C.L., and W.W. Kellogg, 1979: Arctic sea ice decay simulated for a CO_2-induced temperature rise. *Climatic Change*, **2**, 149–162.

Parkinson, C.L., and W.M. Washington, 1979: A large-scale numerical model of sea ice. *J. Geophys. Res.*, **84**, 311–337.

Parkinson, C.L., J.C. Comiso, H.J. Zwally, D.J. Cavalieri, P. Gloersen and W.J. Campbell, 1987: *Arctic Sea Ice, 1973–1976: Satellite Passive-Microwave Observations*, NASA SP-489, National Aeronautics and Space Administration, Washington, D.C., 296 pp.

Peixóto, J.P., and A.H. Oort, 1984: Physics of climate. *Rev. Mod. Phys.*, **56**, 365–429.

Peterson, G.M., T. Webb, III, J.E. Kutzbach, T. van Der Hammen, T.A. Wijmstra and F.A. Street, 1979: The continental record of environmental conditions at 18,000 yr BP: An initial evaluation. *Quat. Res.*, **12**, 47–82.

Phillips, N.A., 1956: The general circulation of the atmosphere: A numerical experiment. *Quart. J. Roy. Meteorol. Soc.*, **82**, 123–164.

Phillips, N.A., 1957: A coordinate system having some special advantages for numerical forecasting. *J. Meteorol.*, **14**, 184–185.

Phillips, N.A., 1959: An example of nonlinear computational instability. *The Atmosphere and Sea in Motion, Rossby Memorial Volume*, B. Bolin, Ed., Rockefeller Institute Press, New York, 501–504.

Phillips, N.A., 1966: The equations of motion for a shallow rotating atmosphere and the traditional approximation. *J. Atmos. Sci.*, **23**, 626.

Phillips, N.A., 1973: Principles of large-scale numerical weather prediction. *Dynamic Meteorology*, P. Morel, Ed., D. Reidel Publishing Co., Dordrecht, 1–96.

Phillips, N.A., 1974: *Application of Arakawa's energy-conserving layer model to operational numerical weather prediction*. National Meteorological Center Office: Note 104. National Weather Service, Washington, D.C., 40 pp.

Pitcher, E.J., R.C. Malone, V. Ramanathan, M.L. Blackmon, K. Puri and W. Bourke, 1983: January and July simulations with a spectral general circulation model. *J. Atmos. Sci.*, **40**, 580–604.

Pittock, A.B., L.A. Frakes, D. Jensen, J.A. Peterson and J.W. Zillman, 1978: *Climatic Change and Variability: A Southern Perspective*, Cambridge University Press, Cambridge, 455 pp.

Platzman, G.W., 1960: The spectral form of the vorticity equation. *J. Meteorol.*, **17**, 635–644.

Platzman, G.W., 1967: A retrospective view of Richardson's book on weather prediction. *Bull. Am. Meteorol. Soc.*, **48**, 514–550.

Pritchard, R.S., 1975: An elastic-plastic constitutive law for sea ice. *J. Appl. Mech.*, **42E**, 379–384.

Pritchard, R.S., 1976: An estimate of the strength of arctic pack ice. *Aidjex Bull.*, **34**, 94–113.

Pritchard, R.S., M.D. Coon and M.G. McPhee, 1977: Simulation of sea ice dynamics during AIDJEX. *J. Pressure Vessel Tech.*, **99**, 491–497.

Ramanathan, V., and J.A. Coakley, Jr., 1978: Climate modeling through radiative-convective models. *Rev. Geophys. Space Phys.*, **16**, 465–489.

Ramanathan, V., M.S. Lian and R.D. Cess, 1979: Increased atmospheric CO_2: Zonal and seasonal estimates of the effects on the radiation energy balance and surface temperature. *J. Geophys. Res.*, **84**, 4949–4958.

Ramanathan, V., R.J. Cicerone, H.B. Singh and J.T. Kiehl, 1985: Trace gas trends and their potential role in climate change. *J. Geophys. Res.*, **90**, 5547–5566.

Ramanathan, V., E.J. Pitcher, R.C. Malone and M.L. Blackmon, 1983: The response of a spectral general circulation model to refinements in radiative processes. *J. Atmos. Sci.*, **40**, 605–630.

Randall, D.A., J.A. Abeles and T.G. Corsetti, 1985: Seasonal simulations of the planetary boundary layer and boundary-layer stratocumulus clouds with a general circulation model. *J. Atmos. Sci.*, **42**, 641–676.

Rasmusson, E.M., and T.H. Carpenter, 1982: Variations in the tropical sea surface temperature and surface wind fields associated with the Southern Oscillation/El Niño. *Mon. Wea. Rev.*, **110**, 354–384.

Rasmusson, E.M., and J.M. Wallace, 1983: Meteorological aspects of the El Niño/Southern Oscillation. *Science*, **222**, 1195–1203.

Reed, R.J., and W.J. Campbell, 1960: *Theory and Observations of the Drift of Ice Station Alpha*, Dept. of Meteorology, University of Washington, Seattle, 255 pp.

Reid, G.C., and K.S. Gage, 1985: Interannual variations in the height of the tropical tropopause. *J. Geophys. Res.*, **90**, 5629–5635.

Richardson, L.F., 1922: *Weather Prediction by Numerical Process*, Cambridge University Press, reprinted Dover, New York, 1965, 236 pp.

Riehl, H., 1978: *Introduction to the Atmosphere*, 3rd ed., McGraw-Hill Book Co., New York, 410 pp.

Robert, A.J., 1966: The integration of a low order spectral form of the primitive meteorological equations. *Meteorol. Soc. Japan*, Ser. 2, **44**, 237–245.

Robert, A.J., 1969: The integration of a spectral model of the atmosphere by the implicit method. *Proc. WMO/IUGG Symposium on Numerical Weather Prediction*, Meteorological Society of Japan, Tokyo, VII-19–VII-24.

Rodgers, C.D., 1967: The use of emissivity in atmospheric radiation calculations. *Quart. J. Roy. Meteorol. Soc.*, **93**, 43–54.

Rodgers, C.D., 1968: Some extensions and applications of the new random model for molecular band transmission. *Quart. J. Roy. Meteorol. Soc.*, **94**, 99–102.

Rodgers, C.D., and C.D. Walshaw, 1966: The computation of infrared cooling rate in planetary atmospheres. *Quart. J. Roy. Meteorol. Soc.*, **92**, 67–92.

Romanova, E.N., 1954: The influence of forest belts on the vertical structure of the wind and on the turbulent exchange. *Study of the Central Geophysical Observatory*, No. 44(106), Glavnaia Geofizicheskaia Observatoriia, Trudy, Leningrad, 80–90.

Rossby, C.G., 1939: Relation between variations in the intensity of the zonal circulation of the atmosphere and the displacements of the semi-permanent centers of action. *J. Mar. Res.*, **2**, 38–55.

Rothrock, D.A., 1973: The steady drift of an incompressible Arctic ice cover. *Aidjex Bull.*, **21**, 49–77.

Rowntree, P.R., 1972: The influence of the tropical east Pacific Ocean temperatures on the atmosphere. *Quart. J. Roy. Meteorol. Soc.*, **98**, 290–321.

Sadourny, R., 1975: The dynamics of finite-difference models of the shallow-water equations. *J. Atmos. Sci.*, **32**, No. 4, 680–689.

Sadourny, R., and L.K. Laval, 1984: January and July performance of the LMD general circulation model. *New Perspectives in Climate Modelling*, A.L. Berger and C. Nicolis, Eds., Developments in Atmospheric Science, 16, Elsevier, Amsterdam, 173–197.

Sarkisyan, A.S., 1966: *Osnovy teorii i raschet okeanicheskyky techeny (Fundamentals of the Theory and Calculation of Ocean Currents)*, Gidrometeoizdat, Moscow.

Sarkisyan, A.S., S.G. Demyshev, G.K. Korotaev and V.A. Moiseenko, 1985: Numerical experiments on a four-dimensional analysis of polymode and 'sections' programmes—oceanographic data. *Coupled Ocean-Atmosphere Models*, J.C.J. Nihoul, Ed., Elsevier Oceanography Series, 40, Elsevier, Amsterdam, 659–674.

Sasamori, T., 1968: The radiative cooling calculation for application to general circulation experiments. *J. Appl. Meteorol.*, **7**, 721–729.

Schlesinger, M.E., and W.L. Gates, 1980: The January and July performance of the OSU two-level atmospheric general circulation model. *J. Atmos. Sci.*, **37**, 1914–1943.

Schlesinger, M.E., and J.F.B. Mitchell, 1985: Model projections of equilibrium climatic response to increased carbon dioxide. *The Potential Climatic Effects of Increasing Carbon Dioxide*, M.C. MacCracken and F.M. Luther, Eds., DOE/ER-0237, Dec. 1985, Washington, D.C., 81–147.

Schlesinger, M.E., W.L. Gates and Y.-J. Han, 1985: The role of the ocean in CO_2-induced climate warming: Preliminary results from the OSU coupled atmosphere-ocean general circulation model. *Coupled Ocean-Atmosphere Models*, J.C.J. Nihoul, Ed., Elsevier Oceanography Series, 40, Elsevier, Amsterdam, 447–478.

Schmidt, V., 1985: *Planet Earth and the New Geoscience*, Kendall/Hunt Publishing Co., Dubuque, Iowa, 554 pp.

Schutz, C., and W.L. Gates, 1971: Global climatic data for surface, 800 mb, 400 mb: January. Advanced Research Projects Agency, Rept. R-915-ARPA, Rand Corp., Santa Monica, 173 pp. [NTIS AD-736204].

Schutz, C., and W.L. Gates, 1972a: Supplemental global climatic data for surface, 800 mb, 400 mb: January. Advanced Research Projects Agency, Rept. R-915/1-ARPA, Rand Corp., Santa Monica, 41 pp. [NTIS AD-744633].

Schutz, C., and W.L. Gates, 1972b: Global climatic data for surface, 800 mb, 400 mb: July. Advanced Research Projects Agency, Rept. R-1029-ARPA, Rand Corp., Santa Monica, 180 pp. [NTIS AD-760283].

Sela, J.G., 1980: Spectral modeling at the National Meteorological Center. *Mon. Wea. Rev.*, **108**, 1279–1292.

Sellers, W.D., 1965: *Physical Climatology*, University of Chicago Press, Chicago, 272 pp.

Sellers, W.D., 1974: A reassessment of the effect of CO_2 variations on a simple global climate model. *J. Appl. Meteorol.*, **13**, 831–833.

Semtner, A.J., Jr., 1974: An oceanic general circulation model with bottom topography. Numerical Simulation of Weather and Climate, Tech. Rept. No. 9, University of California, Los Angeles, 99 pp.

Semtner, A.J., Jr., 1976: A model for the thermodynamic growth of sea ice in numerical investigations of climate. *J. Phys. Oceanogr.*, **6**, 379–389.

Semtner, A.J., Jr., 1984a: Modeling the ocean in climate studies. *Annals of Glaciology*, **5**, 133–140.

Semtner, A.J., Jr., 1984b: On modelling the seasonal cycle of sea ice in studies of climatic change. *Climatic Change*, **6**, 27–37.

Semtner, A.J., Jr., 1986: Finite-difference formulation of a world ocean model. *Proc. NATO Institute on Advanced Physical Oceanographic Numerical Modelling*, D. Reidel Publishing Co., Dordrecht, 187–231.

Semtner, A.J., Jr., 1987: A numerical study of sea ice and ocean circulation in the Arctic. *J. Phys. Oceanogr.*, in press.

Semtner, A.J., Jr., and Y. Mintz, 1977: Numerical simulation of the Gulf Stream and mid-ocean eddies. *J. Phys. Oceanogr.*, **7**, 208-230.

Shukla, J., and J.M. Wallace, 1983: Numerical simulation of the atmospheric response to equatorial Pacific sea surface temperature anomalies. *J. Atmos. Sci.*, **40**, 1613–1630.

Shuleikin, V. V., 1938: The drift of ice fields. *Comptes Rendus (Doklady) de l'Acad. des Sci. de l'USSR*, **19**, 589–594.

Shuman, F.G., 1962: Numerical experiments with the primitive equations. *Proc. International Weather Prediction in Tokyo*, 7–13 Nov. 1960, Meteorological Society of Japan, Tokyo, March 1962, 85–107.

Silberman, I.S., 1954: Planetary waves in the atmosphere. *J. Meteorol.*, **11**, 27–34.

Simmons, A., 1983: The introduction of a new forecast model and revised orographic representation. ECMWF Newsletter No. 20, European Centre for Medium-Range Forecasts, Reading, U.K., 1–6.

Smagorinsky, J., 1963: General circulation experiments with the primitive equations. 1. The basic experiment. *Mon. Wea. Rev.*, **91**, 98–164.

Smagorinsky, J., 1969: Numerical simulation of the global atmosphere. *The Global Circulation of the Atmosphere*, G.A. Corby, Ed., Royal Meteorological Society, 24–41.

Smagorinsky, J., S. Manabe and J.L. Holloway, 1965: Numerical results from a nine-level general circulation model of the atmosphere. *Mon. Wea. Rev.*, **93**, 727–768.

Somerville, R.C.J., and L.A. Remer, 1984: Cloud optical thickness feedbacks in the CO_2 climate problem. *J. Geophys. Res.*, **89**, 9668–9672.

Sommerfeld, A., 1949: *Partial Differential Equations in Physics*, Academic Press, New York, 335 pp.

Staley, D.O., and G.M. Jurica, 1970: Flux emissivity tables for water vapor, carbon dioxide and ozone. *J. Appl. Meteorol.*, **9**, 365–372.

Starr, V.P., 1946: A quasi-Lagrangian system of hydrodynamical equations. *J. Meteorol.*, **2**, 227–237.

Stephens, G.L. 1984: The parameterization of radiation for numerical weather prediction and climate models. *Mon. Wea. Rev.*, **112**, 826–867.

Stephens, G.L., G.G. Campbell and T.H. Vonder Haar, 1981: Earth radiation budgets. *J. Geophys. Res.*, **86**, 9739–9760.

Stone, H.M., and S. Manabe, 1968: Comparison among various numerical models designed for computing infrared cooling. *Mon. Wea. Rev.*, **96**, 735–741.

Storch, H. von, 1984: An accidental result: The mean 1983 January 500 mb height field significantly different from its 1967–81 predecessors. *Beitr. Phys. Atmos.*, **57**, 440–444.

Storch, H. von, and E. Roeckner 1983: Methods for the verification of general circulation models applied to the Hamburg University GCM. Part I: Test of individual climate states. *Mon. Wea. Rev.*, **111**, 1965–1976.

Storch, H. von, E. Roeckner and U. Cubasch, 1985: Intercomparison of extended-range January simulations with general circulation models: Statistical assessment of ensemble properties. *Beitr. Phys. Atmosph.*, **58**, 477–497.

Suarez, M.J., 1985: A GCM study of the atmospheric response to tropical SST anomalies. *Coupled Ocean-Atmosphere Models*, J.C.J. Nihoul, Ed., Elsevier Oceanography Series, 40, Elsevier, Amsterdam, 749–764.

Suarez, M.J., A. Arakawa and D.A. Randall, 1983: The parameterization of the planetary boundary layer in the UCLA general circulation model: Formulation and results. *Mon. Wea. Rev.*, **111**, 2224–2243.

Sud, V.C., and M.J. Fennessy, 1982: An observational-data based evapotranspiration function for general circulation models. *Atmos.-Ocean*, **20**, 301–316.

Sundqvist, H., 1979: Vertical coordinates and related discretization. *Numerical Methods Used in Atmospheric Models*, GARP Publications Series No. 17, World Meteorological Organization, Geneva, 1–50.

Sverdrup, H.U., 1928: The wind-drift of the ice on the North Siberian Shelf. *Norwegian North Polar Expedition with the 'Maud', 1918–1925, Scientific Results*, Vol. 4, No. 1, Griegs Boktrykkeri, Bergen, Norway, 46 pp.

Swarztrauber, P.N., 1984: FFT algorithms for vector computers. *Parallel Computing*, **14**, 45–63.

Thompson, P.D., 1961: *Numerical Weather Analysis and Prediction*, Macmillan, New York, 170 pp.

Thompson, P.D., 1978: The mathematics of meteorology. *Mathematics Today, Twelve Informal Essays*, L.A. Steen, Ed., Springer-Verlag, New York, 125–152.

Thompson, S.L., V.V. Aleksandrov, G.L. Stenchikov, S.H. Schneider, C. Covey and R.M. Chervin, 1984: Global climatic consequences of nuclear war: Simulations with three-dimensional models. *Ambio*, **13**, 236–243.

Tiedtke, M., 1984: The effect of penetrative cumulus convection on the large-scale flow in a general circulation model. *Beitr. Phys. Atmos.*, **57**, 216–239.

Tolstikov, Ye. I., Ed., 1966: *Atlas of Antarctica*, Vol. 1, U.S.S.R. Academy of Sciences, Moscow, 225 plates.

Tourre, Y., M. Déqué and J.F. Royer, 1985: Atmospheric response of a general circulation model forced by a sea-surface temperature distribution analogous to the winter 1982–83 El Niño. *Coupled Ocean-Atmosphere Models*, J.C.J. Nihoul, Ed., Elsevier Oceanography Series, 40, Elsevier, Amsterdam, 479–490.

Trenberth, K.E., 1979: Mean annual poleward energy transports by the oceans in the Southern Hemisphere. *Dyn. Atmos. Oceans*, **4**, 57–64.

Turco, R.P., O.B. Toon, T.P. Ackerman, J.B. Pollack and C. Sagan, 1983: Nuclear winter: Global consequences of multiple nuclear explosions. *Science*, **222**, 1283–1292.

UNESCO, 1981: Tenth report of the joint panel on oceanographic tables and standards. UNESCO Technical Papers in Marine Sci. No. 36, UNESCO, Paris.

Untersteiner, N., 1964: Calculations of temperature regime and heat budget of sea ice in the central Arctic. *J. Geophys. Res.*, **69**, 4755–4766.

Untersteiner, N., 1975: Sea ice and ice sheets and their role in climatic variations. *The Physical Basis of Climate and Climate Modelling*, GARP Publications Series No. 16, 206–224.

van Loon, H., 1986: The characteristics of sea level pressure and surface temperature during the development of a Warm Event in the Southern Oscillation. A talk given at the Namias Symposium on Short Period Climatic Variations, Scripps Institution of Oceanography, October 22, 1985, Manuscript NCAR/0305-86-2, 15 pp.

Walker, G.T., 1924: Correlation in seasonal variations of weather. I. A further study of world weather. *Mem. Indian Meteorol. Dep.*, **24**, 275–332.

Wallace, J.M., and D.S. Gutzler, 1981: Teleconnections in the geopotential height field during the Northern Hemisphere winter. *Mon. Wea. Rev.*, **109**, 784–812.

Walsh, J.E., and C.M. Johnson, 1979: Interannual atmospheric variability and associated fluctuations in Arctic sea ice extent. *J. Geophys. Res.*, **84**, 6915–6928.

Walsh, J.E., and J.E. Sater, 1981: Monthly and seasonal variability in the ocean-ice-atmosphere systems of the North Pacific and the North Atlantic. *J. Geophys. Res.*, **86**, 7425–7445.

Warren, B.A., and C. Wunsch, Eds., 1981: *Evolution of Physical Oceanography*, The MIT Press, Cambridge, Mass., 623 pp.

Washington, W.M., 1968: Computer simulation of the earth's atmosphere. *Science J.*, 36-41.

Washington, W.M., and A. Kasahara, 1970: A January simulation experiment with the two-layer version of the NCAR global circulation model. *Mon. Wea. Rev.*, **95,** 559–580.

Washington, W.M., and G.A. Meehl, 1983: General circulation model experiments on the climatic effects due to a doubling and quadrupling of carbon dioxide concentration. *J. Geophys. Res.*, **88**, 6600–6610.

Washington, W.M., and G.A. Meehl, 1984: Seasonal cycle experiment on the climate sensitivity due to a doubling of CO_2 with an atmospheric general circulation model coupled to a simple mixed-layer ocean model. *J. Geophys. Res.*, **89**, 9475–9503.

Washington, W.M., and L. VerPlank, 1986: A description of coupled general circulation models of the atmosphere and oceans used for CO_2 studies. NCAR Tech. Note, NCAR/TN-271+EDD, Boulder, Colo., 28 pp.

Washington, W.M., and D.L. Williamson, 1977: A description of the NCAR global circulation models. *Methods in Computational Physics*, Vol. 17, Academic Press, New York, 111–172.

Washington, W.M., A.J. Semtner, Jr., C.L. Parkinson and L. Morrison, 1976: On the development of a seasonal change sea-ice model. *J. Phys. Oceanogr.*, **6**, 679–685.

Washington, W.M., A.J. Semtner, Jr., G.A. Meehl, D.J. Knight and T.A. Mayer, 1980: A general circulation experiment with a coupled atmosphere, ocean, and sea ice model. *J. Phys. Oceanogr.*, **10**, 1887–1908.

Webster, F., 1984: *An Ocean Climate Research Strategy*, National Academy Press, Washington, D.C., 66 pp.

Weeks, W.F., 1976: Sea ice conditions in the Arctic. *Aidjex Bull.*, **34**, 173–205.

Williams, J., R.G. Barry and W.M. Washington, 1974: Simulation of the atmospheric circulation using the NCAR global circulation model with ice age boundary conditions. *J. Appl. Meteorol.*, **133**, 305–317.

Williamson, D.L., 1983: Description of the NCAR Community Climate Model (CCMØB), NCAR Tech. Note, NCAR/TN-210+STR, Boulder, Colo., 88 pp.

Williamson, D.L., and P.N. Swarztrauber, 1984: A numerical weather prediction model—Computational aspects on the CRAY-1. *Proc. IEEE*, Vol. 72, No. 1, 56–67.

Wittmann, W.I., and J.J. Schule, Jr., 1966: Comments on the mass budget of Arctic pack ice. *Proc. Symposium on the Arctic Heat Budget and Atmospheric Circulation*, J.O. Fletcher, Ed., Rand Corp., Santa Monica, 215–246.

WMO, 1972: Parameterization of sub-grid scale processes. GARP Publications Series No. 8, World Meteorological Organization, Geneva, 101 pp.

Zeng Q.-C., and Ji Z.-Z., 1981: Problems on nonlinear computational instability, *Acta Mechanica Sinica*, **3**, 70–77.

Zeng Q.-C., and Zhang X.-H., 1982: Perfectly energy-conservative time-space finite difference schemes and the consistent split method to solve the dynamical equations of compressible fluid, *Scientia Sinica*, **XXV, 8**, 866–880.

Zeng Q.-C., Yuan C.-G. and Zhang X.-H., 1984: A numerical coupled atmosphere-ocean model for the simulation of long-term variation. The 16th International Liége colloquium on ocean hydrodynamics, 16.

Zubov, N. N., 1944: *Arctic Ice*, Moscow (translated by U.S. Navy Electronics Laboratory, 1963), 491 pp.

Zwally, H.J., J.C. Comiso, C.L. Parkinson, W.J. Campbell, F.D. Carsey and P. Gloersen, 1983: *Antarctic Sea Ice, 1973–1976: Satellite Passive-Microwave Observations*, NASA SP-459, National Aeronautics and Space Administration, Washington, D.C., 206 pp.

Index